Computer Graphics for Surveying

Computer Graphics for Surveying

P.H. MILNE
Department of Civil Engineering
University of Strathclyde, UK

LONDON NEW YORK
E. & F.N. SPON

First published in 1987 by
E. & F.N. Spon Ltd
11 New Fetter Lane, London EC4P 4EE
Published in the USA by
E. & F.N. Spon
29 West 35th Street, New York NY 10001

Printed in Great Britain by
St Edmundsbury Press Ltd, Bury St Edmunds, Suffolk

ISBN 0 419 14080 8 (hardback)
ISBN 0 419 14070 0 (paperback)

British Library Cataloguing in Publication Data

Milne, P.H.
Computer graphics for surveying.
1. Surveying——Data processing 2. Computer graphics
I. Title
526.9'028'566 TA549

ISBN 0-419-14080-8
ISBN 0-419-14070-0 Pbk

Library of Congress Cataloging-in-Publication Data

Milne, P.H.
Computer graphics for surveying.

Bibliography: p.
Includes index.
1. Surveying—Data processing. 2. Computer graphics.
I. Title.
TA549.M53742 1987 526.9'028'566 87-9752
ISBN 0-419-14080-8
ISBN 0-419-14070-0 (pbk.)

Contents

Preface

This book is a sequel to *BASIC Programs for Land Surveying* (published in 1984) which dealt with numerous facets of computation in land surveying. The introduction and development of low-cost microcomputer peripherals, colour graphics monitors, plotters, and digitizing tablets has created a market for computer graphics software. Not all aspects of land surveying are conducive to interactive graphic displays, but the three most popular areas are dealt with in the core of this book: plan areas and earthworks, surface modelling and highway alignment design.

The format for this book has followed that of *BASIC Programs for Land Surveying* where, after a general discussion of the program, the reader is presented with a breakdown of the program into subroutines. All numeric variables and string variables are given for information before the listing of the program. Example computer printouts are presented, together with computer graphic screen dumps to dot-matrix printers and plotters. I am grateful to my editor, Mr Michael Dunn, for his helpful contact and advice, especially in connection with the graphics presentations and computer listings.

Since the publication of the previous book, where the programs were written in Applesoft BASIC for the Apple II series of microcomputers, there has been a steady drift towards IBM PCs or compatibles for computing in the business world. As a result the programs in this book are written for use under BASICA on an IBM PC or GW-BASIC on an IBM PC compatible, and the programs have already been run successfully on a variety of machines: Amstrad PC-1512, Data General One, Epson-PC, IBM PC, Olivetti M21 and M24. Similarly, various plotters have been used: Apple 410, Benson 1002, Epson HI-80 and Hewlett-Packard HP-7475A.

As some readers may wish to convert these programs to either the Apple II series or the BBC B, a conversion table of graphic commands is included. All printout routines follow a standard format with no ESCape codes so they should run on any IBM PC compatible printer. Users of my previous programs on plan areas, earthworks or road curves, will find the new computer graphics routines will dovetail into their previous programs and therefore line numbers have been condensed to fit within this framework.

It has been my good fortune to receive very generous assistance from fellow research workers and surveying equipment manufacturers who have either given demonstrations or supplied equipment for evaluation. I would like to thank all those who kindly provided information or illustrations to complement the text (all equipment or photographs provided are duly acknowledged in the captions).

Over the years it has been my privilege to attend meetings of the Royal Institution of Chartered Surveyors, the Institution of Civil Engineering Surveyors and the Institution of Civil Engineers whose dissemination of knowledge has been of great benefit. I am indebted to the Institution of Civil Engineering Surveyors for their support, allowing me to attend the 18th FIG (Fédération Internationale Geomètres) Congress in Toronto in June 1986 as the Chairman of their Automation (Computer) Committee. I would also like, through the Institution of Civil Engineers, to acknowledge the receipt of a Public Works and Municipal Services Congress and Exhibition Council award for research into expert systems for earthworks.

I am extremely grateful to Professor D.I.H. Barr of the Department of Civil Engineering at the University of Strathclyde for his continued encouragement in publishing the results of my research work. I would also like to express my thanks to all the undergraduate and post-graduate students who have assisted directly or indirectly in developing or debugging these programs. Gratitude is also due to the technicians in the Department for their help and assistance, especially Mr Dick Lamb and Mr 'Mac' MacMillan in the laboratory, and Mr Sam Rennie on site during students' survey exercises, etc.

I also wish to thank Mrs Sheena Nelson of the Mechanical, Civil and Chemical Engineering Drawing Office at the University of Strathclyde for assistance in the preparation of the illustrations for the book, and especially Mrs Eleanor Caldwell for the hand-drawn figures and Mrs Isabel Mungall for the photographs.

Finally I would like to thank my family for their patience, support and understanding, especially my younger son, Gordon. I would also like to thank my older son, Robert, for his assistance with the preparation of the colour graphics presentations for the programs, particularly the surface modelling program in Chapter 3 where the zoom and rotate options are due to his direct efforts. This book could not have been completed without the wholehearted support of my wife, Helen, whom I thank most sincerely for her helpful suggestions and perseverance in typing the complete manuscript.

Department of Civil Engineering,
University of Strathclyde,
John Anderson Building,
Glasgow G4 0NG,
Scotland.

P.H. Milne
January 1987

PROGRAM DISCS AVAILABLE

This book contains many lengthy program listings. To save typing them into your computer, the programs are available on 5¼″ discs for the IBM PC and compatibles. For details of prices, etc., and availability of other micro-computer software/hardware formats, readers should contact the author.

Glossary of abbreviations and technical terms used with microcomputers

Access time The length of time between a request for information and its availability.

Algorithm A set of procedures for solving a problem. Derived from the name of a 9th-century Arab mathematician called Al Khwaruznu.

Alphanumeric A descriptive term for a set of characters consisting of alphabetic, numeric and special characters.

Application program A set of computer instructions written for a specific user application.

Array An arrangement of elements (numbers, characters, etc.) in rows and columns.

ASCII Acronym for American Standard Code for Information Interchange. The ASCII code is used extensively in data communications.

Backup It is normal practice to keep a backup copy of all discs to cover in the event of damage or failure.

BASIC Beginner's All-purpose Symbolic Instruction Code. A high-level computer language, with most commands in recognizable English.

BASICA IBM BASIC for the IBM PC.

Baud The rate of data transmission, often designed so that one baud equals one binary bit of data.

Command A code or series of codes which when processed by the CPU causes one of the peripherals to perform some action.

Computer Aided Design (CAD) Special software to generate graphics shapes with special terminals to display or print them.

Configure Design and set up a system containing elements of hardware and/or software.

Central Processing Unit (CPU) This component, right at the heart of the microprocessor, performs the tasks of data manipulation and the organization of peripheral components.

Cursor 1. An electronically generated symbol (usually a line or a square) appearing on the screen of a video display terminal to indicate where the next character will appear. 2. A device which can be moved on a digitizing tablet to record position.

Data Information stored in numerical or text format, used as transients in programs, for calculations or information storage.

Database A pool of shared data, supported by utilities for editing, sorting, entering new data and so on.

Data processing The manipulation of data by the execution of a sequence of instructions. Synonymous with information processing.

Digital Elevation Model A digital representation of a surface. See also Digital Ground Model.

Digital Ground Model A digital representation of relief (ground surface). Similar to a Digital Elevation Model, but often enhanced by the addition of planimetric information.

Digital Terrain Model Synonymous with Digital Ground Model.

Digitizing The process of converting analogue maps and other sources to a computer readable form.

Digitizing tablet A device consisting of a flat surface that can detect the position of a cursor.

Disc drive A unit which contains a reading and writing head for loading data on to a disc, or reading data from a disc. The drive also contains the motor for rotating the discs. Hard discs of much greater storage capacity are usually housed in sealed units, whereas flexible discs are easily swapped.

DOS (Disc Operating System) A series of routines which need to be loaded into the CPU to enable it to initialize, save to and read from the disc, plus numerous other associated refinements.

Dump Transfer amounts of data straight to a peripheral, like a printer or disc.

Execute To carry out an operation in a program, or 'run' a program.

Field A specified part of a record containing a unit of data, such as the date of digitizing. The unit of data may be a data element or a data item.

File An organized collection of related records. The records on a file may be related by a specific purpose, format, or data source, and the records may or may not be arranged in sequence. A file may be made up of records, fields, words, bytes, characters or bits.

Format The specified arrangement of data. For example, the layout of a printed document, the arrangement of the parts of a computer instruction, the arrangement of data in a record.

Geographic Information System (GIS) A system for capturing, storing, checking, integrating, analysing and displaying data that is spatially referenced to the Earth. This is normally considered to involve a spatially referenced computer database and appropriate applications software.

Grid The planimetric frame of reference. For example, the National Grid.

GW-BASIC A version of Microsoft BASIC for use on an IBM PC compatible.

Hardcopy A printed paper copy of a program or its graphic results produced by a printer or plotter connected to the microcomputer.

Hardware Generic term for all manufactured computer equipment, i.e. the physical parts as contrasted with the programs (software).

HP Hewlett-Packard.

HPGL Hewlett-Packard Graphics Language, a set of two-letter mnemonic commands by means of which a computer controls a plotter or other output device.

Interface A device for linking one component with another, such as a printer, plotter or digitizing tablet and a microcomputer, to permit transfer of data.

Load To transfer data from an external data storage device, such as a disc, into a microcomputer.

Loop A sequence of instructions repeated until the loop is terminated.

Mainframe A very large computer, capable of handling many jobs at any one time and many terminals.

Memory Any device used to store data or instructions for the computer. Memory devices are compared in terms of storage capacity, access time and cost.

Menu A list of options presented to the operator during execution of a program.

Microcomputer A small computer employing a microprocessor as the central processing unit (CPU).

Microprocessor Electronic circuitry on a single integrated chip that can perform data manipulation.

Monitor A cathode ray tube (CRT) display screen for text and/or graphics, often called a VDU or visual display unit.

MS-DOS *M*icro*S*oft *D*isk *O*perating *S*ystem.

Off-line Data collected online during a real-time program is often rerun off-line for post-survey analysis.

Online Data collected during an interactive real-time program is often stored for subsequent processing, printing and plotting, i.e. no longer in a real-time mode.

Operating system A program which organizes the internal operation of the computer.

Parallel communications The standard character and ASCII code transmission method where bits are sent on eight lines at a time in parallel, normally used for printer communications.

PC-DOS Disc operating system for IBM PC.

Peripheral An external accessory connected to the microcomputer and used for input/output or storage of data, e.g. visual display units (VDUs), keyboards, printers, plotters, digitizing tablets, floppy and hard discs.

Pixel The smallest addressable picture element or point on a VDU, generally given as a number of horizontal and vertical points, e.g. 320 × 200 for SCREEN 1, medium resolution.

Plotter Hardcopy device with a resolution much superior to that of a graphics screen with straight diagonal lines.

Program A list of computer instructions connected in a logical format directing the computer to perform specific operations.

Real-time program A real-time program is one which, during its execution, is subjected to real-time constraints, i.e. input of data (such as distance or time generated by some external activity). The program interacts with the data by means of online devices.

Record A group of data items in a file treated as a unit.

RS-232C Serial communications interface for plotters etc.

Serial communications The standard character and ASCII code transmission method where bits are sent, one at a time, in sequence, normally used for plotters and digitizing tablets.

Software Generic term for computer programs and digitized information which is used to issue instructions to the computer hardware and peripherals for specific applications.

Spatial data Data which includes a reference to a two or three dimensional position in space as one of its attributes.

Terminal An input/output device, normally consisting of a visual display unit and connected to a central mainframe computer for use in time-sharing mode.

Transfer format The format used to transfer data between computer systems.

Visual display unit (VDU) see **Monitor**

1

Introduction to computer graphics

1.1 INTRODUCTION

This chapter is an introduction to computer graphics with specific reference to the graphics capabilities of the IBM PC and its compatibles. It explains the advantages of interactive computer graphics displays for today's personal and professional needs in building, civil engineering, construction and surveying.

Also provided is an overview of the workings of the computer graphics hardware and software of the IBM PC and its compatibles, using an interactive high-level language like BASICA or GW-BASIC to create and control the graphics output.

Each of the BASIC computer programs in this book can be run, either independently by the first time user, or in conjunction with programs from the previous book, *BASIC Programs for Land Surveying*. For example, in Chapter 3, with respect to topographic mapping, data collected by tacheometric surveys (dealt with in the previous book) can be saved to disc for access by the contour plotting programs presented in this book.

1.2 COMPUTER GRAPHICS

The term *computer graphics* is used to describe the presentation of two- or three-dimensional data on screen or plotter. There are many advantages in using computer graphics. For example, in the generation of an engineering drawing, the computer, having been appropriately programmed, will execute the instructions repeatedly, without fatigue or complaint. The computer can also use the same three-dimensional (3-D) data to produce either a contour drawing or an isometric projection of the ground surface, i.e. surface modelling. The drawing can then be scaled on the screen with different perspectives and different colours, until we are satisfied with the result before obtaining a hardcopy of the output on a printer or plotter.

Computer modelling of complex 3-D surfaces is difficult because of the need to present a 3-D isometric projection on a 2-D display. It therefore requires both sophisticated computer programs (software) and special computer graphics displays (hardware). It is only in recent years that these facilities have been available to the microcomputer user, as a result of the increase in computer memory capacity and the reduction in end user costs.

1.3 HARDWARE AND SOFTWARE

To be able to display any of the computer graphics programs discussed in this book on screen, the user will require a Color/Graphics Adapter. It is not essential, however, to have a colour monitor as the user is presented with a choice of colour or monochrome displays. If the user has only a Monochrome Adapter for text, then a plotter will be required to obtain a hardcopy of the graphics output.

All the programs presented allow the user to obtain a text printout of the data and results. It is assumed that a parallel printer interface is available, using the address **LPT1:**. If the user has a Color/Graphics Adapter and a dot-matrix printer, then it is possible to dump the graphics screen output to the printer for record purposes. In this case the program 'graphics' has to be loaded from PC-DOS or MS-DOS before running BASICA or GW-BASIC.

Whereas the IBM PC graphics display commands will run on any IBM PC or compatible under BASICA or GW-BASIC, there is unfortunately no standard graphics plotter language. However, many manufacturers have adopted the Hewlett-Packard Graphics Language (HPGL) for their plotters, and HPGL is used in the programs in this book. The changes required to make these programs run on two other popular plotters, the Apple 410 or the Epson HI-80 Colour Plotter are also given.

In general, initial data input to the programs is via the keyboard, with data files being saved to disc for access by other programs. An alternative form of input is to use a digitizer when two dimensional (2-D) data, i.e. X and Y coordinates, can be entered. The third dimension, the Z coordinate, has to be entered from the keyboard (Fig. 1.1).

Mention has been made of text and graphics modes on the display screen. If a Color/Graphics Adapter is fitted to the computer, the user has a choice of four modes as shown in Table 1.1.

1.4 COMPUTER LANGUAGE

To be able to use computer graphic commands, it is essential to use IBM Advanced BASIC (commonly known as BASICA) on an IBM PC, or GW-BASIC on an IBM PC compatible. Both of these BASIC programs are supplied by Microsoft Inc. and they are very similar; programs written under either BASICA or GW-BASIC should run on IBM PC compatible computers. The programs developed in this book have, for example, been run under BASICA on an IBM PC and on a variety of compatibles

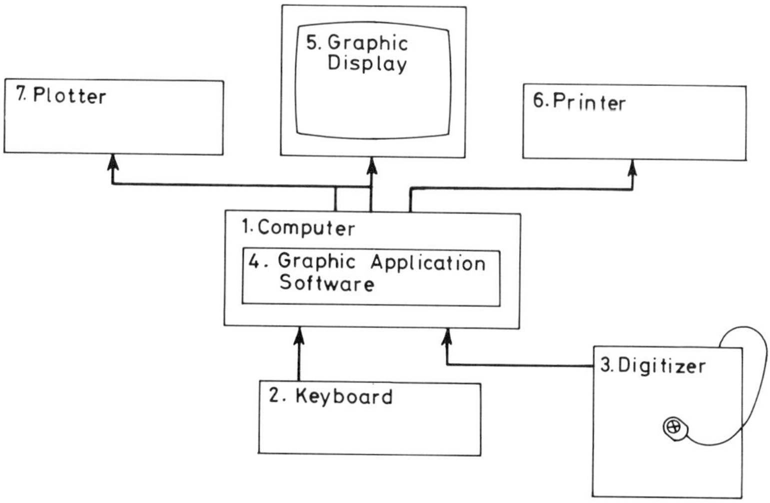

Fig. 1.1 Typical computer graphics system. The graphic application software is the heart of the system accepting and processing input from the keyboard or digitizer with output generated on a graphic display (colour/monochrome), printer and plotter.

Table 1.1 Supported modes of the Color/Graphics Adapter

Mode	*Text/ graphics*	*Hori-zontal reso-lution*	*Vertical reso-lution*	*Charac-ters per row*	*Number of rows of charac-ters*	*Number of pixels*	*Colour*
1. Low-resolution text	Text	40	25	40	25		16 foreground 8 background 16 border
2. High-resolution text	Text	80	25	80	25		16 foreground 8 background 16 border
3. Medium-resolution graphics	Graphics	320	200	40	25	64 000	3 foreground 16 background
4. High-resolution graphics	Graphics	640	200	80	25	128 000	Black & white

(Amstrad PC-1512, Data General One, Epson-PC, Olivetti M21 and M24) using GW-BASIC.

The language BASIC itself can be used to plot points and draw lines, but it is very crude and time consuming. Microsoft have therefore incorporated

Table 1.2 Graphics commands

CIRCLE	Draws circles, ellipses and arcs
CLS	Clears the screen
COLOR	Sets the colour of plotted points and the background colour
DRAW	Draws using a graphics control language
GET	Stores a block of screen in an array
LCOPY	Prints a screen dump
LINE	Draws lines given the end points
PAINT	Fills an area of the screen with colour
PMAP	Converts coordinates
POINT	Returns point settings
PRESET	Resets points to the background colour
PSET	Sets points on the screen
PUT	Restores blocks saved with **GET**
SCREEN	Selects graphics mode screens
VIEW	Sets a viewport on the screen
WINDOW	Defines the logical size of the screen window

several graphics commands into both BASICA and GW-BASIC as listed in Table 1.2. The **SCREEN** command sets the mode, whether low-, medium- or high-resolution, in text or graphics. Modes 1 to 4, listed in Table 1.1, are selected with the following commands:

Mode	*Commands*
1. low-resolution, text	**SCREEN Ø: WIDTH 4Ø**
2. high-resolution, text	**SCREEN Ø: WIDTH 8Ø**
3. medium-resolution, graphics	**SCREEN 1**
4. high-resolution, graphics	**SCREEN 2**

In both modes 3 and 4 it is possible to mix text and graphics on the screen, the size of the text characters being defined by the screen resolution, i.e. 40 columns in mode 3 and 80 columns in mode 4.

All programs should be saved to disc before running by choosing an 8-letter file name, such as "**PC-AREAS**".

1.5 SCREEN GRAPHICS

As detailed in Table 1.1 it is possible to plot on the screen using two different graphics modes. The user has to choose between medium-resolution (colour) or high-resolution (monochrome) graphics. The medium-resolution mode (**SCREEN 1**) allows a resolution of 320 points across the screen and 200 points down the screen, whilst the high-resolution (**SCREEN 2**) has 640 points across and 200 points down the screen.

Note that reference is made to measuring across and down the screen. This is because the zero (0,0) is in the top left-hand corner of the screen. The

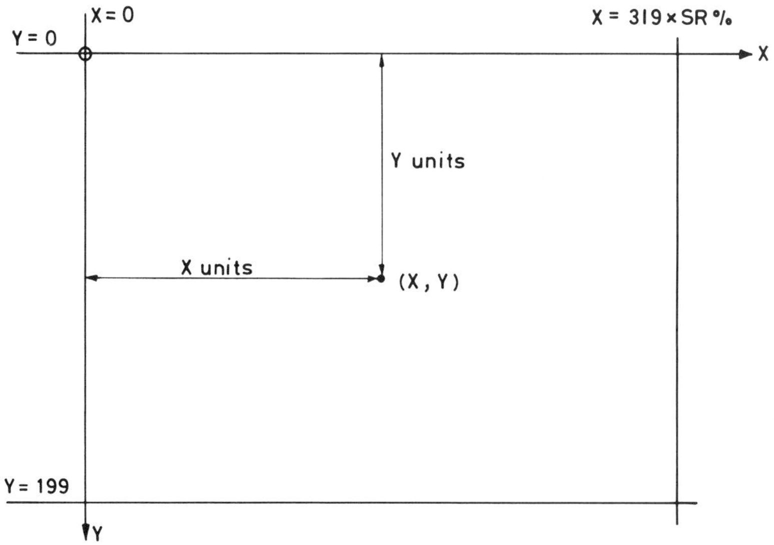

Fig. 1.2 Default graphics screen coordinates with zero (0, 0) in top left-hand corner of screen.

maximum Y-value is thus 199 points but the X-value will depend on whether Screen 1 or Screen 2 graphics are being used. In medium-resolution, using 4 colours from Table 1.1, the maximum X-value is 319 and if in high-resolution (monochrome), the maximum X-value is 639. To simplify the graphics plotting in the programs in this book, the user is given a choice of selecting at the outset either medium-resolution (**SR%=1**) or high-resolution (**SR%=2**), then by using the variable **SR%** in the plotting commands, adjustment is made automatically. The graphics screen in Fig. 1.2 would be set up using the command

WINDOW SCREEN (Ø,Ø) – (319*SR%,199)

However, in general plotting it is usual to have the zero in the bottom left-hand corner, and this is the way all plotters are normally set up. Accordingly, in the programs in this book, the zero of the graphics screen is taken to be in the bottom left-hand corner of the screen (see Fig. 1.3) by using the command

WINDOW (Ø,Ø) – (319*SR%,199)

As the above commands define the whole area of the screen, it is often desirable to shrink the graphics screen to allow text to be displayed either at the top and right-hand side of the screen (Fig. 1.4) or at the top and bottom of the screen (Fig. 1.5). Examples of both types of screen displays are to be

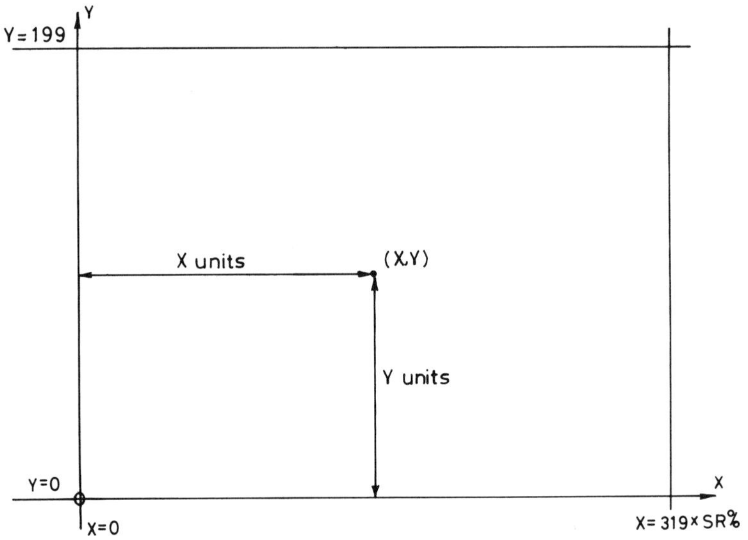

Fig. 1.3 Normal plotting screen coordinates with zero (0, 0) in lower left-hand corner of screen to be compatible with digitizers and plotters.

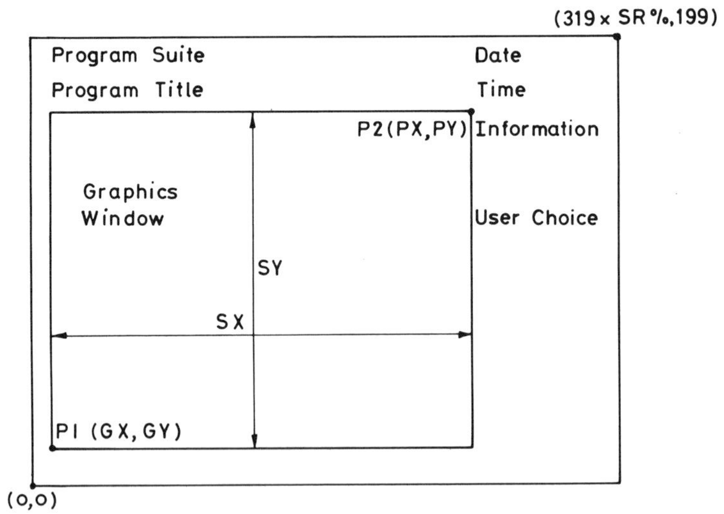

Fig. 1.4 Combined graphics window and text screen with text at top and right-hand side of graphics display.

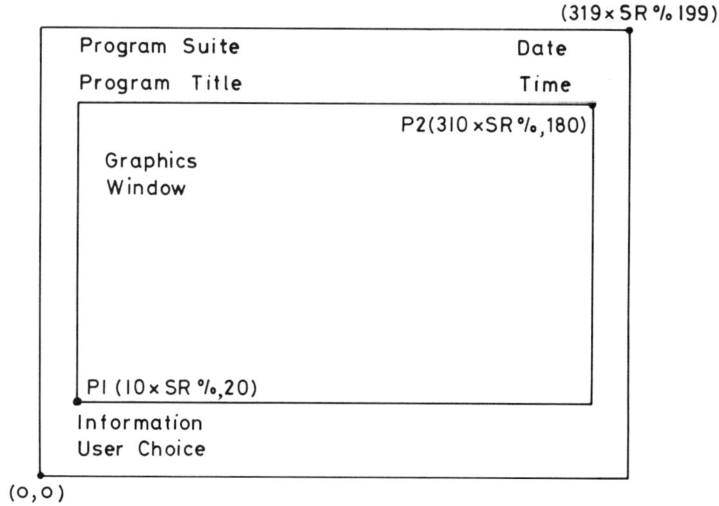

Fig. 1.5 Combined graphics window and text screen with text at top and bottom of graphics display – useful for wider plots and multiple user choices.

found in this book. In the first case it is desirable to leave two text lines clear at the top of the screen. To enable these programs to be transferred to other microcomputers (Table 1.3) with different resolutions, the view screen coordinates are referred to using variables, i.e. left-hand corner P_1 (X-min, Y-min) as **(GX,GY)** and right-hand corner as P_2 (X-max, Y-max) as **(PX,PY)** where the size of the view screen is given by **SX** and **SY**. The variables are then defined at the start of the program, for example

GX=5*SR% : GY=5 : SX=240*SR% : SY=180

This plotting area then covers three quarters of the horizontal width of the

Table 1.3 Comparison of computer graphics commands

IBM PC	*Apple II series*	*BBC B*
CLS	**HOME**	**CLS**
COLOR	**COLOR**	**COLOUR**
LCOPY	—	***GDUMP** (requires printer ROM)
LINE	**HPLOT**	**DRAW**, **PLOT**
PAINT	—	**PLOT** [80–87]
PSET	**HTAB, VTAB**	**MOVE**, **PLOT**
SCREEN	**HGR**, **HGR2**	**VDU** [codes]
VIEW	—	**VDU** [codes]
WINDOW	—	**VDU** [codes]

screen with graphics allowing 10 text characters in Screen 1 and 20 in Screen 2, with two lines at the top for text, e.g. title headings, date, time, etc. (Fig. 1.4). It is suggested that title headings, etc., are stored in string variables for easy recall. One advantage of the IBM PC and its compatibles is that they have a built-in calendar/clock, and if **DATE$** is used as a variable the date is returned in the form MM–DD–YYYY (i.e. in the US date format). Similarly, if **TIME$** is used as a variable it returns the time from the built-in clock in the form HH:MM:SS, thus allowing files and computer print-outs to be date and time stamped for future reference.

If, however, a wider plotting width is required then the second choice (Fig. 1.5) should be used and a small border left at either edge of the screen, for example

VIEW SCREEN (10*SR%,20) – (310*SR%,180)

with two lines for text at the top for the headings, etc., and two lines at the foot for instructions and user choices.

To delineate the graphics plotting area, the user has a choice of drawing a box (**B**) round the area either using a **LINE** command, such as

LINE (GX,GY) – (PX,PY), RGB%, B

where **RGB%** is a variable for the required colour; or alternatively using a **VIEW SCREEN** command, such as

VIEW SCREEN (10*SR%,20) – (310*SR%,180),, RGB%

The two commas before the selected colour for the box, **RGB%**, indicates that the area is to be left blank; if a number is inserted here, the box will be filled with the designated colour.

If a colour monitor is being used, the choice of background and text colour is most important. These programs have been written using the command

COLOR 8,0

to give yellow text on a grey background which is restful to the eye. The **8** selects the grey colour and the **0** selects the palette of colours shown in Table 1.4.

Table 1.4

	Mode	
Colour number	*Screen 1*	*Screen 2*
1	Green	White
2	Red	Black
3	Brown/yellow	White

Thus if red is used in Screen 1, its number must be changed when a monochrome monitor is used, as black would not be seen on monochrome; hence the storing of the colour in the variable **RGB%** which allows it to be changed depending on the screen resolution. It should also be pointed out that program lines with the command **COLOR** must be avoided if using a monochrome monitor with Screen 2 as they will generate an **Illegal function** error.

All the programs in this book clear the screen using the **CLS** command, and as no editing takes place during the running of the programs, the special function key display which normally appears at the foot of the screen is removed with the command **KEY OFF**.

If the user has a dot-matrix graphics printer it is possible to obtain a screen dump of the graphics screen display. This requires the MS-DOS routine 'graphics' to be loaded prior to loading BASICA or GW-BASIC. Then, to obtain a screen dump, the user can either use the command **LCOPY** in the program, or press the SHIFT and Print Screen Prt sc keys simultaneously. Alternatively, if a graphics printer is not available, the graphics screen can be saved to disc with the following commands:

```
DEF SEG = &HB8ØØ
PIC$ = F$ + ".PIC"
BSAVE PIC$,Ø, &H4ØØØ
```

where **F$** is the filename for the display.

To recall a saved picture screen, the following routine should be used:

```
PRINT "File to display:";
INPUT F$
PIC$ = F$ + ".PIC"
CLS
DEF SEG = &HB8ØØ
BLOAD PIC$,Ø
```

However, care should be taken before using this recall routine to check which screen resolution (i.e. Screen 1 or Screen 2) was used for the original to ensure compatibility between the saved and recalled display screens.

1.6 PLOTTER GRAPHICS

Because of the relatively low resolution of the screen graphics of the IBM PC and its compatibles, diagonal lines show characteristic stair-stepping effects. To produce good hardcopies of screen graphics it is often desirable to use a graphics plotter. There are numerous models on the market, some of which, like the Apple 410 and Epson HI-80, use a natural plotting language, whilst others like Hewlett-Packard use HPGL (Hewlett-Packard Graphics

Language). This is an industry standard and other manufacturers, such as Benson and Hitachi, often offer the user the choice of HPGL emulation. For those unfamiliar with HPGL, it is a set of two-letter mnemonic commands by which a computer controls a plotter or other output device. The syntax of HPGL commands has a standard format, with the mnemonic followed by parameters necessary to specify it. The mnemonic and parameters are separated by delimiters (,) and the whole command is completed with a terminator character (;).

As mentioned previously, the zero (0, 0) of all plotters is in the lower left-hand corner. Distances are therefore measured from the lower and left-hand edges of the plotter. The resolution of a plotter is, however, greatly superior to that of a graphics screen. For example, the horizontal and vertical point resolution of an A4 plotter can be 10 120 by 6 800 and of an A3 plotter, 15 370 by 10 602 points respectively. Each plotter unit therefore represents 0.025 mm, which gives 40 plotter units per millimeter, or approximately 1016 plotter units per inch.

When a plotter is turned on, it is normally initialized to specify the parameters of P_1 (X-min,Y-min) and P_2 (X-max,Y-max). For example, if using a popular A3 plotter like the Hewlett-Packard HP-7475A (Fig. 1.6) the initialization command would be

```
"IP 170,602,15370,10602;"
```

giving a total of 15 200 plotter units in the X direction and 10 000 plotter units in the Y direction. This plotting area can then be scaled into user units to match the graphics screen plotting coordinates used by P_1 and P_2 in Fig. 1.4, i.e.

```
"SC";GX;",";PX;",";GY;",";PY;";"
```

The line plotting routine is, however, more complex than that for the screen, discussed previously. For example, the command used to generate a box was

```
LINE (GX,GY) – (PX,PY), RGB%,B
```

To plot the same rectangle on an HP-7475A plotter would require the following commands

```
"PU;PA";GX;",";GY;";"
"PD;PA" ;PX;",";GY;";"
"PA";PX;",";PY;";"
"PA";GX;",";PY;";"
"PA";GX;",";GY;";PU;"
```

where **"PU"** stands for Pen Up, **"PA"** for Pen Absolute and **"PD"** for Pen Down.

The commands for the Apple 410 colour plotter are very similar where the

Fig. 1.6 Popular Hewlett-Packard HP-7475A graphics plotter, with choice of six pens and A3 or A4 paper sizes.

"PU;PA" command is changed to **"MA"** (Move Absolute) and the **"PD;PA"** command is **"DA"** (Draw Absolute). The programs in this book assume the plotter uses an RS-232C interface with address **"COM1:"**, i.e. the primary asynchronous communications interface. It is up to the user to enter the correct plotter communication setting at the appropriate line number. It is suggested that to save repetitive keyboard work the plotter communication setting is stored at line 50 000, and that the plotter initialization commands are stored in routines at lines 51 000, etc. If the following computer listing was saved in ASCII mode as **"PLOTTERS",A** then it would be possible to **MERGE** it into each program when required. Thus these routines which are called by the programs in this book are not repeated in each computer listing.

1.6.1 PLOTTERS – initialization routines for plotters

```
49995 REM ***** COM1: Setting for Plotters *****
50000 OPEN "COM1:9600,E,7,1,CS2000,DS15000" FOR OUTPUT AS #2
50010 RETURN
50495 REM ***** COM2: Setting for Digitisers *****
50500 OPEN "COM2:9600,O,8,1,CS2000,DS15000" FOR INPUT AS #5
50510 RETURN
50995 REM ***** HPGL Plotter (HP-7475A) *****
51000 PRINT#2,"PS3;"
51010 PRINT#2,"IP170,602,15370,10602;"
51020 PRINT#2,"SC-10,180,-5,120;"
51030 PRINT#2,"SP1;"
51040 PRINT#2,"PU;PA";GX;",";GY;";"
51050 PRINT#2,"PD;PA";PX;",";GY;";"
51060 PRINT#2,"PA";PX;",";PY;";"
51070 PRINT#2,"PA";GX;",";PY;";"
51080 PRINT#2,"PA";GX;",";GY;";"
51090 RETURN
52995 REM ***** Apple 410 Plotter *****
53000 PRINT#2,"SP3;"
53020 PRINT#2,"WD-5,-5,192,125;"
53030 PRINT#2,"PS1;"
53040 PRINT#2,"MA";GX;",";GY;";"
53050 PRINT#2,"DA";PX;",";GY;";"
53060 PRINT#2,"DA";PX;",";PY;";"
53070 PRINT#2,"DA";GX;",";PY;";"
53080 PRINT#2,"DA";GX;",";GY;";"
53090 RETURN
53995 REM ***** Epson HI-80 Plotter *****
54000 PRINT#2,"IW0,0,2670,1920"
54010 PRINT#2,"MA0,0"
54020 PRINT#2,"OR"
54030 PRINT#2,"SP1":PRINT#2,"LT0"
54040 PRINT#2,"MA";GX;",";GY*MX
54050 PRINT#2,"DA";PX*MX;",";GY*MX
54060 PRINT#2,"DA";PX*MX;",";PY*MX
54070 PRINT#2,"DA";GX;",";PY*MX
54080 PRINT#2,"DA";GX;",";GY*MX
54090 RETURN
```

1.7 DIGITIZER GRAPHICS

Where data has already been collected and plotted by hand, the user has a choice of re-entering the data via the keyboard or using a digitizing tablet (Fig. 1.7). There are several makes of tablet available, from A4 up to A0 sizes depending on the drawing size being used. Digitizing tablets, like the plotters discussed earlier, have the zero (0, 0) in the lower left-hand corner. If results are to be plotted on the graphics screen it is necessary to change the graphics display to give the zero in the lower left-hand corner, as shown in Fig. 1.3, using the appropriate commands. It is then necessary to set up the digitizer by entering the scaled coordinates for P_1 (X-min,Y-min) and P_2 (X-max,Y-max), Fig. 1.4, from the keyboard and then digitizing the points on the drawing. Once this has been done, all digitized points will be in scaled coordinates and can be plotted on screen, or a plotter if enough input/output interfaces are available, and also stored on disc. Most manufacturers offer a

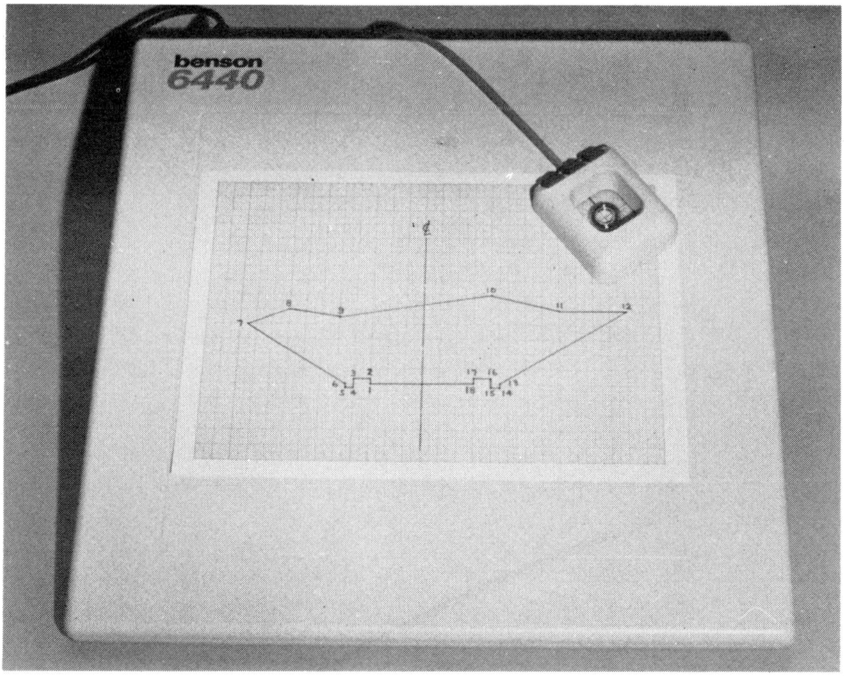

Fig. 1.7 Digitizing tablet with cursor for entering cross-sectional data. (Benson 6440 Tablet kindly supplied for evaluation by Benson Electronics Ltd.)

choice of a pen or a cursor with cross-hairs, the latter being more accurate for digitizing drawings. The programs in this book assume a digitizer uses an RS-232C interface with address **"COM2:"**, i.e. the secondary asynchronous communications interface. It is up to the user to enter the correct digitizer communication settings at the appropriate line number.

1.8 PERFORMANCE CHARACTERISTICS

Microcomputer users are often confused by the terminology to describe computers, either 8-bit, 16-bit or 32-bit. This terminology refers to the number of data lines connected to the system's microprocessor, either 8, 16 or 32. One would therefore expect 16-bit and 32-bit computers to run much faster than an 8-bit machine, and this in general is true. However, some so-called 16-bit machines are really 'pseudo' 16-bit micros as the data is handled in two batches of 8-bits. These computers, like the IBM PC and identical compatibles, have a 4.77 MHz clock rate, whereas more modern microcomputers like the Amstrad PC-1512 and Olivetti M24 are 'true' 16-bit machines running at 8 MHz. Recently introduced are the IBM 6150 and the

Compaq 386, both 32-bit machines running at 16.0 MHz. Thus if extensive use is made of computation during a program, for example the iteration of data points for plotting contours, then the time taken will vary depending upon the clock speed of the microcomputer being used.

As screen graphics are very much faster than plotter graphics, it is suggested that users, when given a choice, check the output on the screen before using the plotter. Some plotters can be run at a baud rate of 9600 to speed up the plotting, but some require a slower rate, e.g. 1200 baud, and thus large A3 contour drawings can take a considerable length of time. Some of the faster microcomputers, running at 8 MHz are too fast for some plotters and give a **"Device Timeout error"**. If this happens, a delay routine has to be incorporated into the program. The following routine can be stored as a **"DELAY"** routine for incorporation into programs. The choice of a high program line number is to keep it clear of other program lines. The value of **TSEC** is set before calling the subroutine, and will depend on the particular combination of microcomputer and plotter being used.

1.8.1 DELAY – routine to use with slow plotters

```
60000 REM  Delay Time Routine for HPGL Plotters
60010 REM  For TSEC Seconds
60020 GOSUB 60100
60030 VALUET=VALUE+TSEC
60040 GOSUB 60100
60050 IF VALUE < VALUET THEN 60040
60060 RETURN
60100 VALUE$=TIME$
60110 VALUEH=VAL(LEFT$(VALUE$,2))
60120 VALUEM=VAL(MID$(VALUE$,4,2))
60130 VALUES=VAL(RIGHT$(VALUE$,2))
60140 VALUE=VALUEH*3600+VALUEM*60+VALUES
60150 RETURN
```

1.9 GRAPHICS INSTALLATION

To simplify the selection of graphics screens from within the programs in this book, a graphics installation program **"GINSTALL"** is included to allow the user to store on disc the preferred options for colour or monochrome and whether medium or high-resolution displays. Once the user has selected the correct configuration a data file **"PCSCREEN"** is saved to disc for recall by subsequent graphics programs. If the user has only a monochrome text adapter no graphics displays will be possible, but hardcopies will still be available on a plotter.

1.9.1 Graphics installation – subroutine index

	Line numbers	*Function*
(a)	10–90	Initialization and control
(b)	100–380	Selection of screen display
(c)	500–740	Configuration display
(d)	1000–1100	Storage of screen details on disc
(e)	9000–9840	PC-Graphics installation screen

1.9.2 Graphics installation – numeric variables

CHG%	= Variable to identify changes
DELAY	= Delay routine variable
I	= Variable used in loop
J	= Variable used in loop
SCD%	= Screen display selection
SCI%	= Variable to install screen display
SCR%	= Variable to select screen resolution
TITLE	= Title routine variable

1.9.3 Graphics installation – string variables

CON$	= Configuration selection
F$	= **"PCSCREEN"**, installation data file
SCD$	= **"Monochrome"** or **"Colour"**
SCR$	= **"TEXT"**, **"Medium-resolution"** or **"High resolution"**

1.9.4 Graphics installation – BASIC program

```
10  REM  <GINSTALL>  Program for IBM PC and Compatibles
15  REM  Version PC-1.0, 01-01-87 *****
20  REM  (C)  Copyright  1987  P.H. Milne
25  REM  All Rights Reserved
30  REM  Title screen at program start
50 CLEAR
60 SCREEN 0
70 WIDTH 40
80 KEY OFF
90 CHG%=0:SCI%=0:SCD%=0:SCR%=0:TITLE=0
100 CLS
110 GOSUB 9000
120 PRINT
130 PRINT:PRINT TAB(3);"Select SCREEN";TAB(20);"<1> TEXT only"
140 PRINT:PRINT TAB(20);"<2> GRAPHICS ";
150 INPUT SCI%
160 IF SCI%<1 OR SCI%>2 THEN 100
170 IF SCI%=1 THEN SCR%=0
```

```
180 IF SCI%=2 THEN 300
190 IF CHG%=1 THEN 500
200 PRINT:PRINT:PRINT TAB(3);"Select DISPLAY";TAB(20);"<1> Monochrome"
210 PRINT:PRINT TAB(20);"<2> Colour ";
220 INPUT SCD%
230 IF SCD%<1 OR SCD%>2 THEN 200
240 GOTO 500
300 PRINT:PRINT:PRINT TAB(3);"Select DISPLAY";TAB(20);"<1> Colour -"
310 PRINT TAB(20);"    low resolution"
320 PRINT:PRINT;TAB(20);"<2> Monochrome -"
330 PRINT TAB(20);"    high resolution"
340 PRINT:PRINT;TAB(20);"<3> Monochrome -"
350 PRINT TAB(20);"    low resolution"
360 PRINT:PRINT TAB(20);:INPUT SCR%
370 IF SCR%<1 OR SCR%>3 THEN 300
380 IF SCR%=1 THEN SCD%=2 ELSE SCD%=1
500 CLS
510 GOSUB 9000
520 PRINT:PRINT TAB(3);"Installation Details :-"
530 IF SCR%=0 THEN SCR$="TEXT"
540 IF SCR%=1 THEN SCR$="Low resolution"
550 IF SCR%=2 THEN SCR$="High resolution"
560 IF SCR%=3 THEN SCR$="Low resolution"
570 IF SCD%=1 THEN SCD$="Monochrome"
580 IF SCD%=2 THEN SCD$="Colour"
600 PRINT:PRINT TAB(3);"Screen MODE";TAB(20);"- ";SCR$
610 PRINT:PRINT TAB(3);"Screen DISPLAY";TAB(20);"- ";SCD$
620 PRINT:PRINT:PRINT TAB(3);"Configuration correct (y/n) ";
630 INPUT CON$
640 IF CON$="y" THEN 1000
650 IF CON$<>"n" THEN 500
660 PRINT:PRINT:PRINT TAB(3);"Change <1> MODE, <2> DISPLAY,"
670 PRINT:PRINT TAB(3);"       <3> Both, <4> Neither ";
680 INPUT CHG%
690 IF CHG%<1 OR CHG%>4 THEN 660
700 ON CHG% GOTO 100,710,100,500
710 IF SCR%=0 THEN 200
720 CLS
730 GOSUB 9000
740 GOTO 300
1000 CLS:GOSUB 9000
1010 PRINT:PRINT:PRINT " Installation data being stored on disc"
1020 IF SCR$="Low resolution" THEN SCR%=1
1030 F$="PCSCREEN"
1040 OPEN "O",#3,F$
1050 PRINT#3,SCR%
1060 PRINT#3,SCD%
1070 CLOSE#3
1100 PRINT:PRINT
1110 END
9000 T$="PC-Graphics"
9010 LOCATE 1,2:PRINT T$
9020 LOCATE 1,31:PRINT DATE$
9030 LOCATE 2,2:PRINT "Installation"
9040 LOCATE 2,31:PRINT TIME$
9050 PRINT " ======================================"
9060 PRINT
9070 IF TITLE=1 THEN 9090
9080 GOTO 9700
9090 RETURN
9695 REM ***** PC Title Page *****
9700 LOCATE 6,1
9710 FOR I=6 TO 20
9720 PRINT SPACE$(4);
9730 FOR J=1 TO 2
```

```
9740 PRINT "survey software ";
9750 NEXT J
9760 PRINT
9770 NEXT I
9780 LOCATE 23,5:PRINT "Copyright (C) 1987 by P.H.Milne"
9800 TITLE=1
9810 FOR DELAY=1 TO 1500
9820 NEXT DELAY
9830 CLS
9840 GOTO 9000
```

2

Plan areas and earthworks

2.1 INTRODUCTION

The calculation of plan and cross-sectional areas either by planimeter or by plotting on gridded paper and counting squares has largely been superseded by computational methods. The use of *coordinate theory* which has now been adopted as the accepted method of calculating areas uses the coordinates of points lying on the boundary of the area or cross-section. Previous methods of computation often utilized either the *trapezoidal rule* or *Simpson's rule* where the boundary was curved. This can now be simplified by taking points on the boundary which, when connected by straight lines, represent the shape of the area to be measured. The coordinates of these points can either be entered from the keyboard or, if the boundary is very irregular, more conveniently by using a digitizer.

The basic programs in this chapter were originally developed in Chapter 4 of *BASIC Programs for Land Surveying* and have now been adapted to use a digitized input and both a graphics and a plotter display. The graphics display may be obtained first for dumping to a printer or plotter before returning to the main program again.

Two programs are presented in this chapter, PC-Areas, for calculating areas by coordinate theory, and PC-Earthworks, which includes cross-sections, a mass-haul diagram and a longitudinal section.

2.2 PC-AREAS

2.2.1 PC-Areas – subroutine index

	Line numbers	*Function*
(a)	10–90	Initialization and control
(b)	100–280	Program selection
(c)	300–390	Data input selection method
(d)	500–680	Entry of area location, etc.
(e)	1000–1490	Area from coordinates program

	Line numbers	Function
(f)	1500–1880	Digitizer input
(g)	4000–4380	Graphics display and screen dump
(h)	6000–6080	Routine to print area locations etc.
(i)	6250–6420	Routine to print area solution
(j)	7000–7160	Plotter initialization and selection
(k)	7200–7490	Plotter routine for HPGL
(l)	7500–7790	Plotter routine for Apple 410
(m)	7800–8090	Plotter routine for Epson HI-80
(n)	9000–9070	Routine to clear screen
(o)	9100–9130	Routine to display error message
(p)	9500–9620	Termination of program

(a) Initialization and control Line numbers 10–90

All numeric variables and required string variables are initialized with **REM**arks to identify the program, etc.

(b) Program selection Line numbers 100–280

At the start of each program selection the screen is cleared using **GOSUB 9ØØØ** to identify on screen the program suite and title.

(c) Data input selection method Line numbers 300–390

This routine is optional and depends upon whether the user has a digitizer available to enter the coordinates of the points, or whether all the data input will be via the keyboard.

(d) Entry of area location Line numbers 500–680

This routine is presented for record purposes to annotate the printout with the title of the area, the operator's name and date. This title is subsequently displayed on screen during the graphics displays, and also on the hardcopy output (printer or plotter) for identification.

(e) Area from coordinates program Line numbers 1000–1490

After the number (N) of points in the figure whose area is to be measured have been entered, the required numeric arrays are **DIM**ensioned in line 1100. If the data is to be entered via a digitizer, the program then branches to line 1500.

Alternatively the data can be entered from the keyboard. The point identification numbers and their X-coordinate (or easting) and Y-coordinate (or northing) are then entered in clockwise rotation (Fig. 2.1). The total area A for the N stations is then given by

$$A = [(N_1E_2 + N_2E_3 + N_3E_4 + \ldots + N_{N-1}E_N + N_NE_1) \\ - (E_1N_2 + E_2N_3 + E_3N_4 + \ldots + E_{N-1}N_N + E_NN_1)]/2 \quad (2.1)$$

where the subscripts 1,2, . . . N refer to the point numbers. The total area is then calculated by the program and the points coordinates and the total area displayed in lines 1300–1420, with an option to obtain a printout of the data. The user is then given an option to obtain a plot of the data either on the graphics screen (line 4000) or on a plotter (line 7000).

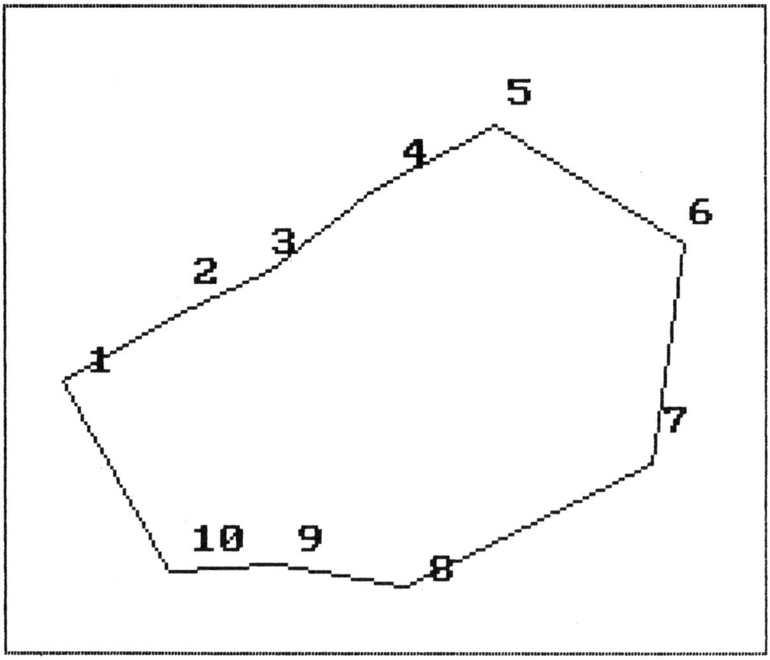

Fig. 2.1 Graphics screen dump of Stepps Traverse using Epson dot-matrix printer.

(f) Digitizer input Line numbers 1500–1880

The initial lines of this routine are used to set up the digitizer by entering the coordinates of P_1(X-min,Y-min) and P_2(X-max,Y-max) as shown in Fig. 1.4. The lines from 1500–1663 are also accessed by the graphics display and plotter routines to enter the coordinates of P_1 and P_2.

As most digitizers use an RS-232C interface, the communication setting in line 50500 has assumed the secondary asynchronous port is used, i.e. **"COM2:"**. The communication settings in line 50500 will possibly need to be altered by the user as it is not feasible to set these using variables from the keyboard. Both the BASIC manual and the digitizer manual should be consulted to ensure the correct 'handshaking' arrangement. The setting shown is the default setting for the Benson 6440 digitizer shown in Fig. 1.7. On completion of the digitizing, the program returns to the computation and display procedure discussed in (e) above.

(g) Graphics display and screen dump Line numbers 4000–4380

This display screen chosen for the presentation of data is shown in Fig. 1.4 with text at the top and to the right-hand side of the graphics window. The initial part of this routine (lines 4010–4090) sets up the graphics window with the zero in the bottom left-hand corner. Each of the points is then plotted using the **PSET** command in line 4140 and a line drawn to the next point in the figure (line 4150). After all the points have been plotted, a line is drawn joining the last and first points (line 4180) to complete the figure.

The second part of this routine (lines 4200–4240) then labels each of the points with its text identification number. This is a complex routine to calculate the position of each point on the screen and then its row and column location, remembering that the X, Y coordinates are calculated from the lower left-hand corner.

Finally the user is given an opportunity to obtain a screen dump of the graphics screen. It should be noted that this option is only available if the user, prior to running BASICA or GW-BASIC, has remembered to load the 'graphics' routine from the MS-DOS disc supplied with the computer!

(h) Routine to print area location, etc. Line numbers 6000–6080

This routine is provided for record purposes to annotate the printout. As this and the following routine (i) do not use escape codes, they can be printed on any IBM PC compatible printer, either dot-matrix or daisywheel, using a parallel interface, location **"LPT1:"**, which is the default location for printers connected to the IBM PC and its compatibles.

(i) Routine to print area solution Line numbers 6250–6420

This routine prints out the boundary points and their coordinates (either fed in from the keyboard or digitizer) together with the area.

(j) Plotter initialization and selection Line numbers 7000–7160

If the coordinates of P_1 and P_2 were not entered earlier when using a digitizer, they are now collected from routine (f) above. The plotter details are set up in a similar fashion to the screen display, and the area of the plot scaled to fit the plotter, depending on the values of P_1 and P_2 and the A3 or A4 size. In this program three plotter routines are provided for the Apple 410, Epson HI-80 and the Hewlett-Packard HP-7475A plotter.

To save space and repetition, the plotter communications setting and initialization routines have been placed in **GOSUB** routines and will therefore not be repeated in future programs. Once the plotter is selected, the user is reminded to place paper on the plotter and to switch it on. The communication setting in line 50000 (see Section 1.6) assumes the plotter is connected to the primary asynchronous port **"COM1:"** at 9600 baud. The remaining communication settings will possibly need to be altered by the user to suit the specific plotter being used.

(k) Plotter routine for HPGL Line numbers 7200–7490

This routine is suitable for Hewlett-Packard plotters using HPGL and should run with the modern versions, HP-7470A, HP-7475A, and their equivalents, such as Benson 1002 and Hitachi 672. Some of these plotters have different locations for P_1 and P_2 in plotter units and line 51010 may need to be altered to suit a specific plotter. The default setting given is that of the HP-7475A.

After scaling the plotter into user units in line 51020, pen 1 (black) is selected to draw a box round the plotting area delineated by the points **(GX,GY)** and **(PX,PY)**. Pen 4 (blue) is then chosen to draw a line joining the points on the boundary of the area to be measured. Finally the points are

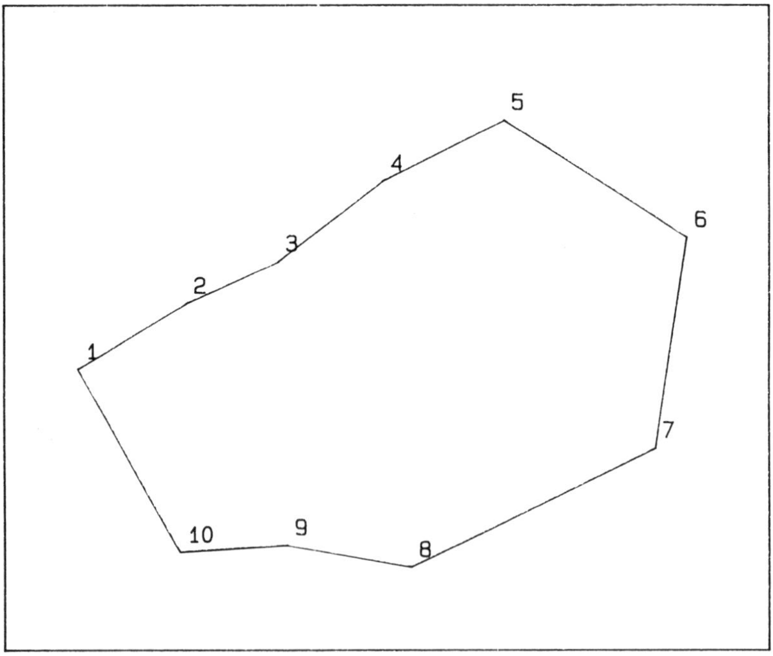

Fig. 2.2 Detailed plot of Stepps Traverse using plan area software for comparison with Fig. 2.1.

numbered and the plot labelled with the name of the area and the date of the calculation using pen 2 (red) (Fig. 2.2).

(l) Plotter routine for Apple 410 Line numbers 7500–7790

This routine for the Apple 410 colour plotter follows the same format as (j) above to allow users to see the difference in the commands where the same variables are used for the location of P_1 and P_2. A comparison of the plotter commands is listed in Table 2.1 for Hewlett-Packard, Apple 410 and Epson HI-80 colour plotters.

(m) Plotter routine for Epson HI-80 Line numbers 7800–8090

This routine is identical to (k) and (l) above. However, it is not possible to scale the plotter units into user units; hence a scaling factor **MX** has to be calculated initially to convert the user units to plotter units, which explains the multiplication factor **MX** being used for plotting. The software commands are listed in Table 2.1.

(n) Routine to clear screen Line numbers 9000–9070

This routine is used to clear the screen and provide two lines of text at the head of each screen giving the title of the software suite and the selected program together with the date and time of the calculation. If the machine being used does not have an on-board battery to maintain the time, it is up to

Table 2.1 Comparison of plotter commands

	Software command		
Description	*HPGL*	*Apple 410*	*Epson HI-80*
Select A3 paper size	**"PS3;"**	**"SP3;"**	–
Scale plotting area	**"SC...;"**	**"WD...;"**	**"IW..."**
Select pen 1	**"SP1;"**	**"PS1;"**	**"SP1"**
Move absolute (i.e. do not draw)	**"PU;PAX1,Y1;"**	**"MAX1,Y1;"**	**"MAX1,Y1"**
Draw absolute (draw to new point)	**"PD;PAX1,Y1;"**	**"DAX1,Y1;"**	**"DAX1,Y1"**
Choose letter size	**"SIØ.2,Ø.3;"**	**"LS2;"**	**"SI25,2Ø"**
Label with text	**"LBtext"**	**"PLtext"**	**"LAtext"**
Label terminator	**CHR$(3)**	**CHR$(13)** **CHR$(1Ø)** **CHR$(3)**	

the user to enter the time and date correctly at the outset using PC-DOS or MS-DOS. It should be noted that this on-board calendar only works in the American calendar format of month–day–year, as discussed in Section 1.5.

(o) Routine to display error message Line numbers 9100–9130

If a wrong keyboard response has been made in the program this error message is displayed to prevent the program crashing, and the user is requested to re-enter the data.

(p) Termination of program Line numbers 9500–9620

This program can be terminated either from the initial screen header, or on completion of the calculations. After an end of program message the user is returned to command level.

2.2.2 PC-Areas – numeric variables

A	= Area
BYTE(3)	= Array for digitiser output
CCOL	= Column number for text display on graphics screen
CROW	= Row number for text display on graphics screen
DISP%	= Choice of display – screen or plotter
EP1	= Easting of P_1 on digitizer
EP2	= Easting of P_2 on digitizer
E(N–1)	= Eastings array of stations
GX	= X-coordinate at lower left-hand corner of plot
GY	= Y-coordinate at lower left-hand corner of plot
HP%	= Integer variable to identify plotter
I	= Integer loop variable
J	= Integer loop variable
K	= Integer loop variable
MG	= Multiplication factor for plotted display
MGD	= Multiplication factor for general resolution
MH	= Multiplication factor for horizontal resolution
MHD	= Multiplication factor for horizontal on digitizer
MV	= Multiplication factor for vertical resolution
MVD	= Multiplication factor for vertical on digitizer
MX	= Multiplication factor for Epson HI-80 plotter
N	= Number of stations/offsets
NP1	= Northing of P_1 on digitizer
NP2	= Northing of P_2 on digitizer
N(N–1)	= Northings array of stations
PP	= Screen display variable
PX	= X-coordinate at upper right-hand corner of plot
PY	= Y-coordinate at upper right-hand corner of plot
PLOT%	= Integer variable for plotter selection

RGB%	= Integer variable to select screen colour
SX	= Horizontal display width in user units
SY	= Vertical display height in user units
S(N–1)	= Number array of stations
SCR%	= Integer variable to select screen or plotter
SEL%	= Integer variable to select keyboard or digitizer
SR%	= Integer variable for screen resolution
T	= Selection of program from menu
T1,T2	= Temporary storage variables for calcuating area
XD	= Variable used to store *X*-location of plotted points
XP1	= *X*-location of digitized point P_1
XP2	= *X*-location of digitized point P_2
XP(N–1)	= Array for *X*-locations of plotted points
YD	= Variable used to store *Y*-location of plotted points
YP1	= *Y*-location of digitized point P_1
YP2	= *Y*-location of digitized point P_2
YP(N–1)	= Array for *Y*-locations of plotted points

2.2.3 PC-Areas – string variables

A$	Storage for location of area
B$	Storage for operator's name
C$	Storage for date of calculation
DATE$	Variable used to return the date of the computer's calendar/clock in US format MM-DD-YYYY
EO$	= **"SORRY, DATA ERROR... PLEASE RE-ENTER"**
GD$	Entry for graphics dump on printer
P$	= **"DO YOU WISH PRINTOUT OF ABOVE DATA (Y/N)"**
Q$	= Question response (Y/N)
T$	= **"PC-Graphics"**
TIME$	Variable used to return the time of the built-in clock in HH:MM:SS format
X$	= **"ENTER X(E) COORD.OF POINT"**
Y$	= **"ENTER Y(N) COORD.OF POINT"**

2.2.4 PC-Areas – BASIC program

```
10  REM  <PC-AREAS> Program for IBM PC and Compatibles
12  REM  BASIC Programs for Land Surveying, Chapter 4
13  REM  + Computer Graphics for Surveying, Chapter 2
14  REM  (C)  Copyright  1984-87  P.H.Milne
16  REM  All rights reserved
20  REM  Version PC-1.0, 01-01-87
22  REM  Plot on <1> Apple 410, <2> Epson HI-80, or <3> HPGL Plotter
25 CLEAR:KEY OFF
30 DEFDBL A,E,N,T,X,Y
35 E = 0:I = 0:J = 0:K = 0:N = 0:P1 = 0:T = 0:C = 0
40 A = 0:T1 = 0:T2 = 0:SR%=1:DISP%=0
```

```
50 PCF$="PCSCREEN"
52 OPEN "I",#3,PCF$
54 INPUT#3,SCR%
56 INPUT#3,SCD%
58 CLOSE#3
60 ON (SCR%+1) GOTO 62,64,66
62 SCREEN 0:WIDTH 40:GOTO 80
64 SCREEN 1:GOTO 68
66 SCREEN 1:GOTO 80
68 IF SCD%=1 THEN 80
70 COLOR 8,0
80 E0$ = "SORRY, DATA ERROR ... PLEASE RE-ENTER."
85 P$ = "DO YOU WISH PRINTOUT OF ABOVE DATA (Y/N)"
90 X$ = " ENTER X(E) COORD. OF POINT ":Y$ = " ENTER Y(N) COORD. OF PO
INT "
95  REM ***************************************************
100  GOSUB 9000
190  PRINT
200  PRINT " SELECT PROGRAM FROM MENU :-"
210  PRINT
220  PRINT "     <1> AREAS FROM CO-ORDINATES"
230  PRINT
250  PRINT "     <2> QUIT                ";
260  INPUT T
270  IF T < 1 OR T > 2 THEN 190
280  IF T = 2 THEN 9500
310  PRINT
320  PRINT " Select DATA Input method :-"
330  PRINT
340  PRINT "     <1> Keyboard"
350  PRINT:PRINT "     <2> Digitiser"
360  PRINT:PRINT "     <3> QUIT      ";
370  INPUT SEL%
380  IF SEL%< 1 OR SEL%> 3 THEN 300
390  IF SEL%= 3 THEN 9500
500  GOSUB 9000
540  PRINT
550  PRINT "ENTER LOCATION OF WORK, ETC. WHEN       REQUESTED, AND PR
ESS <Enter>"
560  PRINT:PRINT "ENTER LOCATION OF AREA :-"
570  INPUT A$
580  PRINT "ENTER OPERATOR'S NAME :-"
590  INPUT B$
600  PRINT "ENTER DATE OF CALCULATION AS DD/MM/YY :-"
610  INPUT C$
640  PRINT
650  PRINT "DO YOU WISH PRINTOUT OF LOCATION ETC.,  (Y/N) ";
660  INPUT Q$
670  IF Q$ = "Y" THEN 6000
680  IF Q$ <  > "N" THEN 650
995  REM ****************************************************
1000  GOSUB 9000
1010  PRINT "<1> AREA FROM CO-ORDINATES PROGRAM"
1020  PRINT "****************************************"
1030  PRINT:PRINT "ENTER NUMBER OF STATIONS IN FIGURE ";
1040  INPUT N
1050  IF N =  > 3 THEN 1100
1070  GOSUB 9100
1080  PRINT "THREE OR MORE POINTS ARE REQUIRED !"
1090  GOTO 1020
1100  DIM E(N-1),N(N-1),S(N-1),XP(N-1),YP(N-1),BYTE(3)
1105 IF SEL%=2 THEN 1500
1110  GOSUB 9000
1120  PRINT
1130  PRINT "ENTER STN. NUMBERS IN CLOCKWISE ROTATION"
```

```
1140  PRINT "****************************************"
1150  FOR I = 0 TO N - 1
1160  PRINT "ENTER STATION NUMBER ";: INPUT S(I)
1170  PRINT:PRINT X$;S(I);: INPUT E(I)
1180  PRINT Y$;S(I);: INPUT N(I)
1190  PRINT
1200  NEXT I
1205  REM  NOW CALCULATE AREA FROM VARIABLES T1 AND T2
1210  FOR J = 0 TO N - 2
1220 T1 = T1 + (N(J)) * (E(J + 1))
1230 T2 = T2 + (E(J)) * (N(J + 1))
1240  NEXT J
1250 T1 = T1 + (N(N - 1)) * (E(0))
1260 T2 = T2 + (E(N - 1)) * (N(0))
1270 A = (T1 - T2) / 2
1280 A =  INT (A * 1000 + .5) / 1000
1300  GOSUB 9000
1310  PRINT "<1> AREA FROM CO-ORDINATES :-"
1320  PRINT "****************************************"
1330  PRINT "STN.NO."; TAB( 10);"EASTING"; TAB( 25);"NORTHING"
1340  PRINT "========================================"
1350  FOR I = 0 TO N - 1
1355 E(I)=INT(E(I)*1000+.5)/1000:N(I)=INT(N(I)*1000+.5)/1000
1360  PRINT S(I); TAB( 10);E(I); TAB( 25);N(I)
1370  NEXT I
1380  PRINT "========================================"
1390  PRINT
1400  PRINT "TOTAL AREA = ";A;" SQ. UNITS"
1410  PRINT
1420  PRINT "****************************************"
1430  PRINT:PRINT P$;: INPUT Q$:IF Q$="Y" THEN 6250
1440  IF Q$ <  > "N" THEN 1430
1450 GOSUB 9000
1460 PRINT:PRINT " Do you wish plot of data on :-"
1470 PRINT:PRINT TAB(10);"<1> Graphics Screen":PRINT:PRINT TAB(10);"<
2> Plotter"
1480 PRINT:PRINT TAB(10);"<3> Neither (Quit) ";:INPUT DISP%:IF DISP%<
1 OR DISP%>3 THEN 1450
1490 ON DISP% GOTO 4000,7000,9500
1495 REM  ***** Digitiser Input *****
1500  GOSUB 9000
1510  PRINT "<1> AREA FROM CO-ORDINATES PROGRAM"
1520  PRINT "****************************************"
1530 PRINT:IF SEL%=1 THEN 1560
1540 GOSUB 50500
1550 PRINT " Place Drawing on Digitiser Tablet"
1560 PRINT:PRINT " Enter Co-ordinates for P1 (Xmin,Ymin)":PRINT " fro
m keyboard"
1570 PRINT:PRINT X$;" P1 ";:INPUT EP1
1580 PRINT Y$;" P1 ";:INPUT NP1
1585 IF SEL%=1 THEN 1640
1590 PRINT:PRINT " Click cursor button on P1"
1600 FOR K=1 TO 3
1610 INPUT#5,BYTE(K)
1620 NEXT K
1630 XP1=BYTE(1):YP1=BYTE(2):CLOSE#5
1640 PRINT:PRINT " Enter Co-ordinates for P2 (Xmin,Ymin)":PRINT " fro
m keyboard"
1650 PRINT:PRINT X$;" P2 ";:INPUT EP2
1660 PRINT Y$;" P2 ";:INPUT NP2
1662 IF SEL%=1 AND DISP%=1 THEN 4010
1663 IF SEL%=1 AND DISP%=2 THEN 7010
1665 GOSUB 50500
1670 PRINT:PRINT " Click cursor button on P2"
1680 FOR K=1 TO 3
```

```
1690 INPUT#5,BYTE(K)
1700 NEXT K
1710 XP2=BYTE(1):YP2=BYTE(2):CLOSE#5
1720 MHD=(EP2-EP1)/(XP2-XP1)
1730 MVD=(NP2-NP1)/(YP2-YP1)
1740 IF MHD<=MVD THEN MGD=MHD ELSE MGD=MVD
1750 GOSUB 9000
1760  PRINT:PRINT "ENTER STN. NUMBERS IN CLOCKWISE ROTATION"
1770  FOR I = 0 TO N - 1
1780 PRINT:PRINT " Enter STN. No. ";:INPUT S(I)
1785 GOSUB 50500
1790 PRINT " Click cursor button on ";S(I)
1800 FOR K=1 TO 3
1810 INPUT#5,BYTE(K)
1820 NEXT K
1830 XP(I)=BYTE(1):YP(I)=BYTE(2):CLOSE#5
1840 E(I)=(XP(I)-XP1)*MGD+EP1
1850 N(I)=(YP(I)-YP1)*MGD+NP1
1860 NEXT I
1870 GOTO 1210
3995 REM ***** Screen Plot of Data *****
4000 IF EP2=0 AND NP2=0 THEN 1500
4010 GX=5*SR%:GY=5:SX=240*SR%:SY=180:PP=0:XD=0:YD=0
4020 MH=SX/(EP2-EP1):MV=SY/(NP2-NP1)
4030 IF MH<=MV THEN MG=MH ELSE MG=MV
4040 GOSUB 9000:WINDOW (0,0) - (320*SR%,200)
4050 IF SR%=1 THEN RGB%=2 ELSE RGB%=3
4060 PX=MG*(EP2-EP1):PY=MG*(NP2-NP1)
4070 LINE (GX,GY)-(PX,PY),RGB%,B
4080 IF SR%=1 THEN 4100
4090 LINE (GX-1,GY)-(PX-1,PY),RGB%,B
4100 FOR I=0 TO N-1
4110 XP(I)=(E(I)-EP1)*MG:YP(I)=(N(I)-NP1)*MG
4120 IF XD=0 THEN XD=XP(I)*SR%
4130 IF YD=0 THEN YD=YP(I)
4140 PSET(XD,YD)
4150 LINE -(XP(I)*SR%,YP(I))
4160 XD=XP(I)*SR%:YD=YP(I)
4170 NEXT I
4180 LINE (XD,YD)-(XP(0)*SR%,YP(0))
4190 IF PP=1 THEN 4300
4200 FOR J=0 TO N-1
4210 CCOL=(XP(J)*SR%+GX)/8:CROW=(YP(J)+GY)/8
4215 IF CCOL<2*SR% THEN CCOL=2*SR%
4220 LOCATE (25-CROW),CCOL
4230 PRINT S(J)
4240 NEXT J
4250 PP=1:GOTO 4070
4300 LOCATE 4,32*SR%:PRINT "SCREEN"
4310 LOCATE 5,32*SR%:PRINT "DUMP(y/n)"
4320 LOCATE 6,32*SR%:PRINT "          "
4330 BEEP
4340 LOCATE 6,32*SR%:INPUT GD$
4350 IF GD$="n" THEN 1450
4360 IF GD$<>"y" THEN 4300
4370 LCOPY
4380 GOTO 1450
5985  REM  **************************************************
5990  REM  PRINTOUT OF LOCATION, ETC
6000 LPRINT "LOCATION OF AREA :- ";A$
6010 LPRINT "*******************************************"
6020 LPRINT "OPERATOR'S NAME :-      ";B$
6030 LPRINT "DATE OF CALCULATION :- ";C$
6060 LPRINT "*******************************************"
6070 LPRINT
```

```
6Ø8Ø  GOTO 1ØØØ
6243  REM  ***************************************************
6245  REM  PRINTOUT OF COORDINATES
625Ø LPRINT "<1> AREA FROM CO-ORDINATES :-"
626Ø P7=1:REM PRINTER CODE
629Ø LPRINT "*****************************************"
63ØØ LPRINT "STN.NO.";TAB(1Ø);"EASTING";TAB(25);"NORTHING"
631Ø LPRINT "========================================="
632Ø  FOR J = Ø TO N - 1
633Ø LPRINT S(J);TAB(1Ø);E(J);TAB(25);N(J)
634Ø  NEXT J
635Ø LPRINT "========================================="
637Ø LPRINT "TOTAL AREA = ";A;" SQ. UNITS"
639Ø LPRINT "*****************************************"
64ØØ LPRINT
64Ø5  REM  RETURN TO SCREEN AT END OF PROGRAM
642Ø  GOTO 145Ø
6995 REM ***** Plotter Output *****
7ØØØ IF EP2=Ø AND NP2=Ø THEN 15ØØ
7Ø1Ø GX=Ø:GY=Ø:MG=Ø:MH=Ø:MV=Ø:SX=192:SY=12Ø:XD=Ø:YD=Ø:MX=267Ø/SX
7Ø2Ø MH=SX/(EP2-EP1):MV=SY/(NP2-NP1)
7Ø3Ø IF MH <= MV THEN MG=MH ELSE MG=MV
7Ø4Ø PX=MG*(EP2-EP1):PY=MG*(NP2-NP1)
7Ø5Ø GOSUB 9ØØØ
7Ø6Ø PRINT:PRINT " Select Plotter -";TAB(18);"<1> Apple 41Ø":PRINT:PR
INT TAB(18);"<2> Epson HI-8Ø"
7Ø7Ø PRINT:PRINT TAB(18);"<3> Hewlett-Packard":PRINT:PRINT TAB(18);"<
4> QUIT          ";:INPUT PLOT%
7Ø8Ø IF PLOT% <1 OR PLOT% >4 THEN 7Ø5Ø
7Ø9Ø IF PLOT%=4 THEN 145Ø
71ØØ IF PLOT%=1 THEN HP%=3
711Ø IF PLOT%=2 THEN HP%=4
712Ø IF PLOT%=3 THEN HP%=1
713Ø PRINT:PRINT " Place paper on Plotter and switch on"
714Ø PRINT:PRINT " To Continue Press <Enter> ":WHILE R$="":R$=INKEY$:
WEND
715Ø GOSUB 5ØØØØ
716Ø ON HP% GOTO 72ØØ,72ØØ,75ØØ,78ØØ
7195 REM ***** HPGL Plotter *****
72ØØ GOSUB 51ØØØ
729Ø PRINT#2,"SP4;"
73ØØ FOR I=Ø TO N-1
731Ø XP(I)=(E(I)-EP1)*MG:YP(I)=(N(I)-NP1)*MG
732Ø IF XD=Ø THEN XD=XP(I)
733Ø IF YD=Ø THEN YD=YP(I)
734Ø PRINT#2,"PU;PA";XD;",";YD;";"
735Ø PRINT#2,"PD;PA";XP(I);",";YP(I);";"
736Ø XD=XP(I):YD=YP(I)
737Ø NEXT I
738Ø PRINT#2,"PU;PA";XD;",";YD;";"
739Ø PRINT#2,"PD;PA";XP(Ø);",";YP(Ø);";"
74ØØ PRINT#2,"SP2;"
741Ø FOR J=Ø TO N-1
742Ø PRINT#2,"PU;PA";(XP(J)-1);",";(YP(J)+1);";"
743Ø PRINT#2,"SIØ.18,Ø.22;":PRINT#2,"LB";S(J):PRINT#2,CHR$(3)
744Ø NEXT J
745Ø PRINT#2,"SIØ.2,Ø.3;"
746Ø PRINT#2,"PU;PA5,-3;"
747Ø PRINT#2,"LBPC-Areas : ";A$;" : Date : ";DATE$
748Ø PRINT#2,CHR$(3)
749Ø CLOSE#2:GOTO 145Ø
7495 REM ***** Apple 41Ø Plotter *****
75ØØ GOSUB 53ØØØ
759Ø PRINT#2,"PS4;"
76ØØ FOR I=Ø TO N-1
```

```
7610 XP(I)=(E(I)-EP1)*MG:YP(I)=(N(I)-NP1)*MG
7620 IF XD=0 THEN XD=XP(I)
7630 IF YD=0 THEN YD=YP(I)
7640 PRINT#2,"MA";XD;",";YD;";"
7650 PRINT#2,"DA";XP(I);",";YP(I);";"
7660 XD=XP(I):YD=YP(I)
7670 NEXT I
7680 PRINT#2,"MA";XD;",";YD;";"
7690 PRINT#2,"DA";XP(0);",";YP(0);";"
7700 PRINT#2,"PS2;":PRINT#2,"LS2;"
7710 FOR J=0 TO N-1
7720 PRINT#2,"MA";(XP(J)-1);",";(YP(J)+1);";"
7730 PRINT#2,"PL";S(J):PRINT#2,CHR$(13) CHR$(10) CHR$(3)
7740 NEXT J
7750 PRINT#2,"LS2;"
7760 PRINT#2,"MA5,-3;"
7770 PRINT#2,"PLPC-Areas : ";A$;" : Date : ";DATE$
7780 PRINT#2,CHR$(13) CHR$(10) CHR$(3)
7790 CLOSE#2:GOTO 1450
7795 REM ***** Epson HI-80 Plotter *****
7800 GOSUB 54000
7890 PRINT#2,"SP4"
7900 FOR I=0 TO N-1
7910 XP(I)=(E(I)-EP1)*MG*MX:YP(I)=(N(I)-NP1)*MG*MX
7920 IF XD=0 THEN XD=XP(I)
7930 IF YD=0 THEN YD=YP(I)
7940 PRINT#2,"MA";XD;",";YD
7950 PRINT#2,"DA";XP(I);",";YP(I)
7960 XD=XP(I):YD=YP(I)
7970 NEXT I
7980 PRINT#2,"MA";XD;",";YD
7990 PRINT#2,"DA";XP(0);",";YP(0)
8000 PRINT#2,"SP2":PRINT#2,"SI25,20"
8010 FOR J=0 TO N-1
8020 PRINT#2,"MA";(XP(J)-1*MX);",";(YP(J)+1*MX)
8030 PRINT#2,"LA";S(J)
8040 NEXT J
8050 PRINT#2,"SI40,30"
8060 PRINT#2,"MA";5*MX;",";(-3)*MX
8070 PRINT#2,"LAPC-Areas : ";A$;" : Date : ";DATE$
8090 CLOSE#2:GOTO 1450
8985  REM  ******************************************
8990  REM  GOSUB ROUTINE TO CLEAR SCREEN
9000  CLS:T$="PC-Graphics"
9010 LOCATE 1,2:PRINT T$
9020 LOCATE 1,30:PRINT DATE$
9030 LOCATE 2,2:PRINT "PC-Areas  : ";A$
9040 LOCATE 2,31:PRINT TIME$
9045 IF DISP%=1 THEN 9060
9050 PRINT "========================================"
9060 PRINT
9070  RETURN
9093  REM  ***************************************************
9095  REM  GOSUB ROUTINE FOR ERRORS
9100 BEEP
9110  PRINT : PRINT E0$
9120  PRINT
9130  RETURN
9493  REM  ***************************************************
9495  REM  END OF PROGRAM
9500  GOSUB 9000
9510  PRINT
9520  PRINT "   END PC-AREAS PROGRAM"
9530  PRINT
9540  PRINT "*****************************************"
```

```
9542  FOR I = 1 TO 500
9544  NEXT I
9550  REM  PROGRAM PREPARED BY
9560  REM  DR. P.H. MILNE
9610  REM  ***************************
9620  END
```

2.2.5 PC-Areas – computer printout

Coordinate details for Figs 2.1 and 2.2

```
LOCATION OF AREA :- Stepps
*****************************************
OPERATOR'S NAME :-       P.H. Milne
DATE OF CALCULATION :- 01/01/87
*****************************************

<1> AREA FROM CO-ORDINATES :-
*****************************************
STN.NO.  EASTING          NORTHING
=========================================
 1         264817.996       667743.484
 2         264989.603       667834.143
 3         265134.316       667893.038
 4         265296.458       668005.428
 5         265484.968       668090.544
 6         265773.481       667926.879
 7         265723.009       667632.285
 8         265340.257       667465.937
 9         265148.211       667497.305
 10        264979.418       667487.465
=========================================
TOTAL AREA =  378355.4  SQ. UNITS
*****************************************
```

2.3 PC-EARTHWORKS

In the computation of volumes for earthworks in building or civil engineering projects, it is essential to be able to calculate the areas of cross-sections. These might lie, for example, at set chainage intervals along the centre line of a new road.

Cross-sections may be regular or irregular in shape depending on the information available. A regular cross-section would have constant ground slopes whereas an irregular cross-section would have an irregular ground profile. Cross-sections with known ground slopes can be easily calculated using standard formulae, as presented in Chapter 4 of the previous book, *BASIC Programs for Land Surveying*. Irregular cross-sections, especially if they have a large number of points, are best calculated using coordinate theory as discussed previously with regard to plan areas. There are therefore considerable advantages in being able to use a digitizer, such as that shown in Fig. 1.7, to enter irregular cross-sections.

Six types of cross-sections are dealt with in this program:

⟨1⟩ One-level (cut or fill)
⟨2⟩ Two-level (cut or fill)
⟨3⟩ Two-level (cut + fill)
⟨4⟩ Three-level (cut or fill)
⟨5⟩ Irregular (cut or fill)
⟨6⟩ Irregular (cut + fill)

After entering the cross-sectional area data the user can plot the cross-section on screen or plotter. On completion of all the cross-sections, the user is given a choice of volume computations with a final opportunity to plot the mass-haul diagram on screen or plotter. This program is similar to that presented in the previous book, *BASIC Programs for Land Surveying*, with the addition of graphics display routines in lines 12 000–26 930.

2.3.1 PC-Earthworks – subroutine index

	Line numbers	*Function*
(a)	10–90	Initialization and control
(b)	100–110	Screen header display
(c)	300–400	Program selection from menu
(d)	500–680	Entry of earthworks location, etc.
(e)	700–790	Entry of earthworks data
(f)	800–990	Cross-section selection from menu
(g)	1000–1190	⟨A1⟩ One-level section (cut or fill)
(h)	1200–1440	⟨A2⟩ Two-level section (cut or fill)
(i)	1450–1690	⟨A3⟩ Two-level section (cut + fill)
(j)	1700–1990	⟨A4⟩ Three-level section (cut or fill)
(k)	2000–2350	⟨A5⟩ Irregular section (cut or fill)
(l)	2500–2630	⟨A6⟩ Irregular section (cut + fill)
(m)	2650–2690	Equivalent area (curvature correction) for ⟨A2⟩
(n)	2700–2780	Equivalent area (curvature correction) for ⟨A3⟩
(o)	2800–2980	Entry of cross-section data for earthworks
(p)	3000–3080	Volume calculation selection from menu
(q)	3200–3380	⟨V1⟩ Mean area rule
(r)	3400–3660	⟨V2⟩ End areas rule
(s)	3700–3890	Extrapolation of cut/fill lengths
(t)	4000–4280	⟨V3⟩ Prismoidal rule
(u)	5000–5180	Entry of volume data for mass-haul
(v)	5200–5680	Mass-haul aggregate volume computations
(w)	6000–6080	Routine to print earthworks location, etc.
(x)	6100–6390	Data printout for cross-sections ⟨A1–A4⟩
(y)	6400–6590	Data printout for cross-sections ⟨A5 and A6⟩
(z)	6600–6720	Printout summary of cross-sectional areas

	Line numbers	Function
(aa)	6800–6880	Data printout of volumes for ⟨V1⟩
(bb)	7000–7150	Data printout of volumes for ⟨V2⟩
(cc)	7200–7350	Data printout of volumes for ⟨V3⟩
(dd)	7400–7550	Printout summary of volumes
(ee)	7600–7790	Mass-haul aggregate volume printout
(ff)	9000–9010	Routine to clear screen
(gg)	9100–9130	Routine to display error message
(hh)	9500–9620	Termination of program
(ii)	12 000–12 300	Digitizer input
(jj)	20 000–20 090	Selection of CS display screen/plotter
(kk)	20 100–20 190	Set up for CS graphics display
(ll)	20 200–20 290	Set up array to display ⟨A1⟩
(mm)	20 300–20 490	CS graphics display and screen dump
(nn)	20 500–21 150	Set up array to display ⟨A2⟩–⟨A4⟩
(oo)	21 200–21 290	Set up array to display ⟨A5⟩ and ⟨A6⟩
(pp)	22 000–22 200	Plotter selection and initialization
(qq)	22 250–22 490	Plotter routine for HPGL
(rr)	22 500–22 740	Plotter routine for Apple 410
(ss)	22 750–22 990	Plotter routine for Epson HI-80
(tt)	23 000–23 450	Routines to label plot output
(uu)	24 000–24 280	Mass-haul display initialization
(vv)	24 500–24 900	MH graphics display and screen dump
(ww)	25 000–25 200	Plotter selection
(xx)	25 500–25 950	Plotter routine for HPGL
(yy)	26 000–26 450	Plotter routine for Apple 410
(zz)	26 500–26 930	Plotter routine for Epson HI-80
(aaa)	27 000–27 280	Longitudinal section display initialization
(bbb)	27 500–27 990	LS graphics display and screen dump
(ccc)	28 000–28 200	Plotter selection
(ddd)	28 500–28 950	Plotter routine for HPGL
(eee)	29 000–29 450	Plotter routine for Apple 410
(fff)	29 500–29 950	Plotter routine for Epson HI-80

(a) Initialization and control Line numbers 10–90

All numeric variables and required string variables are initialized and dimensioned. Note the maximum array for cross-sections and cross-sectional points is 31. If a larger number is required the arrays should be redimensioned.

(b) Screen header display Line numbers 100–110

At the start of each program selection the screen is cleared using **GOSUB 9ØØØ** to identify on screen the program suite and title.

(c) Program selection from menu Line numbers 300–400

A choice of three entry points to the PC-Earthworks program is given depending on the data provided:

⟨1⟩ Areas of cross-section
⟨2⟩ Volumes from cross-sections
⟨3⟩ Mass-haul from volumes

During the running of both the areas and mass-haul selections, the user is given the opportunity to obtain plots on a graphics printer or plotter for record purposes.

(d) Entry of earthworks location, etc. Line numbers 500–680

This routine is provided for record purposes to annotate the printout with a title of location, operator's name and date of calculation.

(e) Entry of earthworks data Line numbers 700–790

Each of the programs chosen in (c) above uses this routine to enter the number of cross-sections, the chainage of the first cross-section, and the distance between the cross-sections, before branching to the selected program.

(f) Cross-section selection from menu Line numbers 800–990

A choice of six types of cross-sections is given:

⟨1⟩ One-level (cut or fill)
⟨2⟩ Two-level (cut or fill)
⟨3⟩ Two-level (cut + fill)
⟨4⟩ Three-level (cut or fill)
⟨5⟩ Irregular (cut or fill)
⟨6⟩ Irregular (cut + fill)

After entry of each cross-section, the program returns to the menu to select the next type of cross-section before offering a printout summary in line 960.

(g) ⟨A1⟩ One-level section (cut or fill) Line numbers 1000–1190

When the ground surface is level, ⟨A1⟩ can be used for either cut or fill, where H is the formation depth/height. If the formation width is B and the side slope is 1:M, then W is the width of the excavation of embankment toe on either side of the centre line. The program prompts with the cross-section number and chainage and requests whether cut or fill, and asks for B, H and M. The side width W and the area A are then calculated and displayed. In this and all subsequent cross-sections the user is given an option to obtain a printout of the results for record purposes. The final area, either cut or fill, is stored in an array for later volume computations, before plotting the resulting cross-section on screen (Fig. 2.3) or plotter.

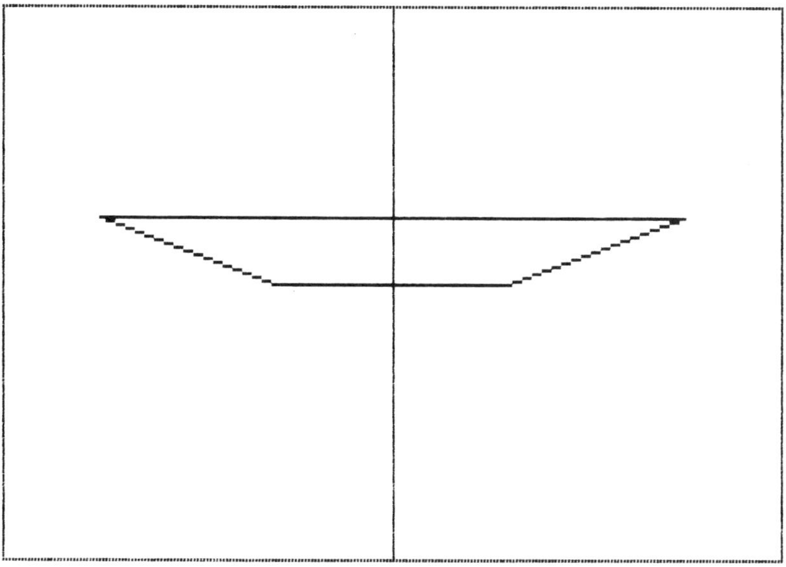

Fig. 2.3 Graphics screen dump of cross-section type ⟨A1⟩.

(h) ⟨A2⟩ Two-level section (cut or fill) Line numbers 1200–1440

Where the ground surface has a constant slope, transverse gradient or crossfall relative to the centre line, determined by two levels taken at right-angles to the centre line, then the side-widths will be unequal. To ensure the correct graphical display and plot, the user has to enter whether the crossfall is positive or negative (P/N) from left to right (L–R). After evaluating the side-widths, the area A is computed and an option given to allow curvature corrections to be applied if necessary, as discussed later in (m), before plotting the resulting cross-section on the screen (Fig. 2.4) or plotter.

(i) ⟨A3⟩ Two-level section (cut + fill) Line numbers 1450–1690

The crossfall K and its slope from left to right is determined in the same way as ⟨A2⟩ above. Depending upon the relationship between the formation level of construction and the original ground surface, there will often be both cut and fill in each cross-section. As the cut and fill side slopes are entered separately they can be varied depending on the ground conditions. The critical information required for this area calculation is whether there is cut or fill at the centre line to ensure the correct calculation of the side-widths. The formation depth is H, M is the fill side slope, and M_2 is the cut side slope. After computing the side widths and the cut and fill areas, these are stored in arrays for future volume computations. As in ⟨A2⟩ a curvature correction is

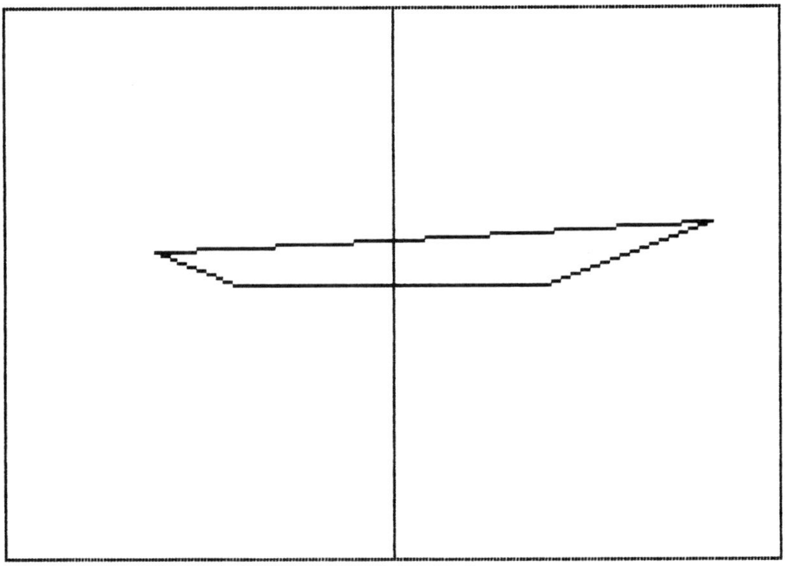

Fig. 2.4 Graphics screen dump of cross-section type ⟨A2⟩.

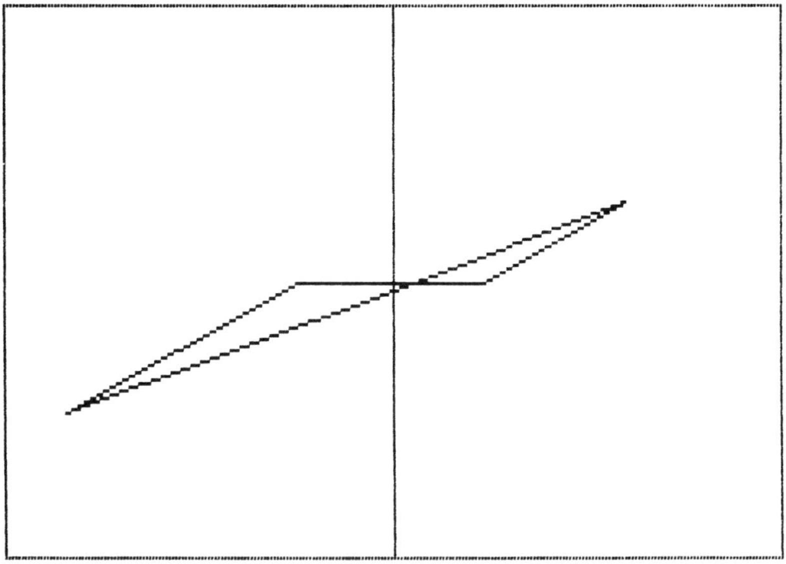

Fig. 2.5 Graphics screen dump of cross-section type ⟨A3⟩.

offered as an option, before plotting the resulting cross-section on the screen (Fig. 2.5) or plotter.

(j) ⟨A4⟩ Three-level section (cut or fill) Line numbers 1700–1990

In this cross-section there is a change of slope or crossfall at the centre line. The user is therefore asked to enter whether the crossfall is positive (P/N) from the centre line to ensure the correct graphical orientation of the cross-section. This section is so-called from the three levels required to calculate the two ground slopes and can be used for either cut or fill. Where the ground surface is very irregular, either ⟨A5⟩ or ⟨A6⟩ should be used. The side-widths are calculated in a similar fashion to ⟨A2⟩ and ⟨A3⟩ discussed previously. After calculating the area of cut or fill, the results are stored in an array and the user is able to obtain a printout of the results and a plot of the resulting cross-section on the screen (Fig. 2.6) or plotter.

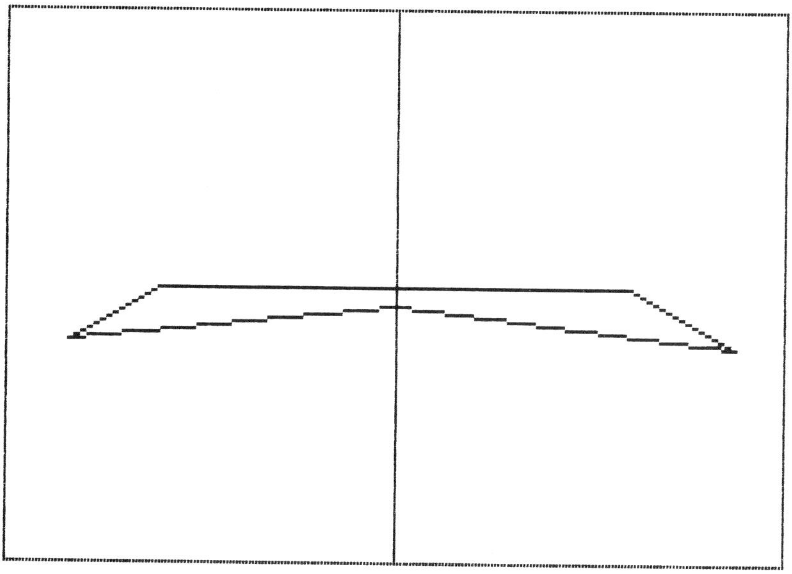

Fig. 2.6 Graphics screen dump of cross-section type ⟨A4⟩.

(k) ⟨A5⟩ Irregular section (cut or fill) Line numbers 2000–2350

In both this and the next section ⟨A6⟩ the user is given the opportunity to enter the data, either from the keyboard or a digitizer (Fig. 1.7). The scale of the drawing on the digitizer does not matter as it can be scaled using the digitizer from the known formation width of the cross-section. The calculation of the area is computed using an adaption of the method used for calculating a plan area from coordinates, as used in Section 2.2 where the points are entered in a clockwise direction. The cross-sectional area is calculated either by entering the offset from the centre line, x, and the difference in height, y, from the formation level using the keyboard or by

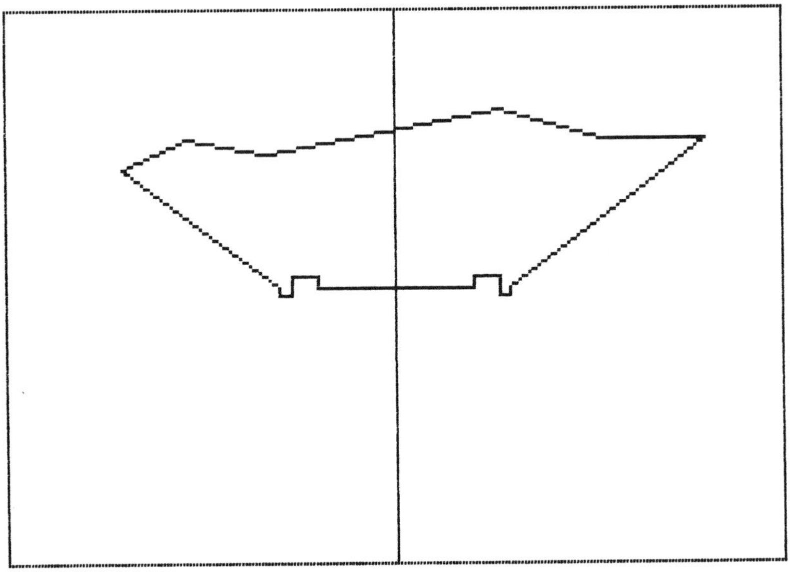

Fig. 2.7 Graphics screen dump of cross-section type ⟨A5⟩, see input information on digitizer in Fig. 1.7.

using the digitizer. Points to the left of the centre line have negative offsets, and points below the formation level are negative. After calculation, storage and display (Fig. 2.7), a computer printout of the points and their offsets etc. can be obtained for record purposes.

(l) ⟨A6⟩ Irregular section (cut + fill) Line numbers 2500–2630

This routine is similar to (k) above with a choice of entering the data from the keyboard or using a digitizer. The points are entered in a clockwise direction round both the cut and fill areas before display (Fig. 2.8).

(m) Equivalent area (curvature correction) for ⟨A2⟩ Line numbers 2650–2690

In computing volumes from cross-sections, it is normally assumed that the cross-sections are taken on a straight road. However, where there is a horizontal curve between two straights, the cross-sections will no longer be parallel to one another and errors can result in volume calculations, unless a curvature correction is applied. To ensure that the calculated volumes are as representative of the ground situation as possible, the equivalent area of the cross-section is calculated using *Pappus' theorem*. This theorem states that a volume swept out by a plane constant area revolving about a fixed axis is given by the product of the cross-sectional area and the distance moved by the centre of gravity of the section from the centre line.

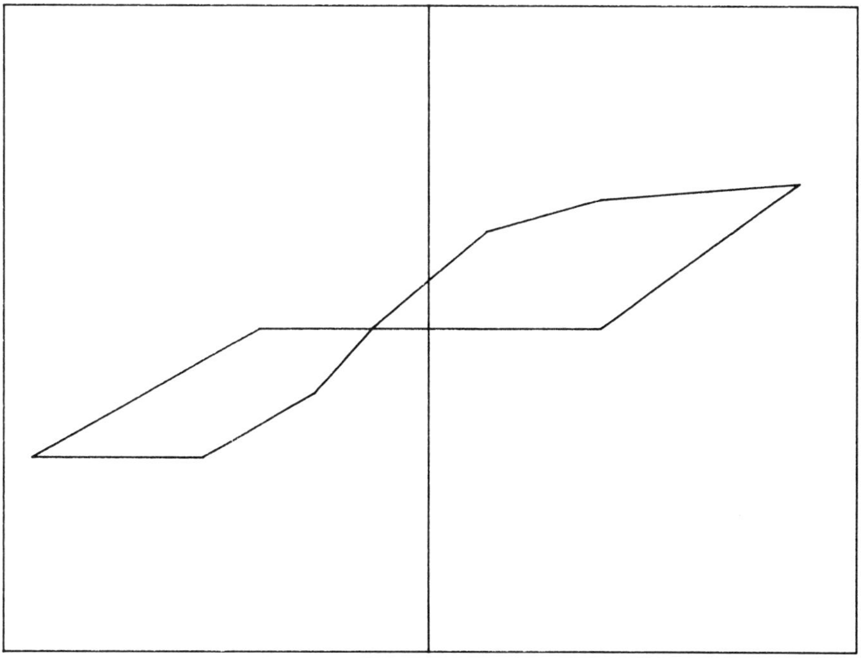

Fig. 2.8 Detailed plot of cross-section type ⟨A6⟩ with both cut and fill.

In a symmetrical cross-section, i.e. one-level ⟨A1⟩, the centroid of the area will lie on the centre line. However for an unsymmetrical cross-section, e.g. two-level ⟨A2⟩, the centroid of the area will be on the side of the higher ground level. The distance of the centroid from the centre line is called the eccentricity (E_C) and is calculated in line 2650. The cut and fill areas are then adjusted depending on the radius of curvature and stored in an array for future volume computations.

(n) Equivalent area (curvature correction) for ⟨A3⟩

Line numbers 2700–2780

In this two-level cross-section of the ⟨A3⟩ type there is both cut and fill so both of these areas need to be treated separately in calculating the centroids of their areas. There are now two eccentricities E_1 and E_2 which are calculated in lines 2700–2710.

(o) Entry of cross-section data for earthworks

Line numbers 2800–2980

If no cross-sectional areas have been computed, then they can be entered in this section prior to selecting the method of volume computation in the next section (p). If previous cross-sectional areas have been computed, this section is skipped from line 2830.

(p) Volume calculation selection from menu Line numbers 3000–3080

All volume calculations return to this menu, which allows a selection of volume calculations, or a final mass-haul aggregate volume calculation as shown:

⟨1⟩ Mean area rule
⟨2⟩ End areas rule
⟨3⟩ Prismoidal rule
⟨4⟩ Mass-haul calculations
⟨5⟩ Quit

If both end-area volumes and prismoidal rule volumes have been computed, the mass-haul aggregate volumes can be determined for both before quitting the program.

(q) ⟨V1⟩ Mean area rule Line numbers 3200–3380

This is a simple formula for calculating volume by multiplying the mean of the cross-sectional areas by the distance between the end sections. However, it is not very accurate as it assumes that the cross-sections are of similar shape and area. Since this method of volume calculation only gives the total volume it is not suitable for mass-haul aggregate volume computations.

(r) ⟨V2⟩ End areas rule Line numbers Line numbers 3400–3660

This method of volume computation assumes that any section taken midway between the two end areas is the mean of the end areas. Provided there are no large variations in cross-sectional shape between successive cross-sections, this method is more accurate than (q). After calculation of both cut and fill volumes, the results are stored in an array for display and then printout, prior to the mass-haul aggregate volume calculations.

(s) Extrapolation of cut/fill lengths Line numbers 3700–3890

In the computation of volumes by the end areas ⟨V2⟩ or prismoidal ⟨V3⟩ methods, should a zero area be encountered in either cut or fill areas, then the program branches to this routine. Either the specific area highlighted, cross-section CS(I), has just reduced to zero, or it could have reached zero somewhere between the previous/next cross-section and the present one. If zero is to be used, the program returns to the volume computations.

However, if the cut or fill has been steadily increasing/reducing in previous cross-sections, it is possible to extrapolate the new chainage of the zero area, but this will only work if C_2 is more than twice the area of C_1 as shown in line 3770.

(t) ⟨V3⟩ Prismoidal rule Line numbers 4000–4280

This method of volume computation is more accurate then ⟨V1⟩ and ⟨V2⟩ as the volume of earth between two successive cross-sections is considered as a *prismoid*, which is a solid made up of two end faces parallel to one another. It should be noted that the faces of the prismoid must be formed by straight continuous lines running from one end face to the other. Thus unless the ground profile is regular both transversely and longitudinally, it is likely that errors will be introduced in assuming that the figure is prismoidal over its entire length. However, since these errors are invariably small, the volume obtained using the prismoidal rule can be considered a good approximation of the volume to be excavated or filled.

The prismoidal formula assumes that the cross-sectional area M of the section mid-way between the end faces is also known. If A_1 and A_2 are the end face areas, distance L apart, then the volume, V, is given by:

$$V_{1-2} = (A_1 + 4M + A_2)L/6 \tag{2.2}$$

It is therefore normal practice in earthworks to use the prismoidal rule over three cross-sections A_1, A_2, A_3, again distance L apart, to give:

$$V_{1-3} = (A_1 + 4A_2 + A_3)L/3 \tag{2.3}$$

as used in lines 4070 and 4100 for cut and fill respectively.

Extending Equation 2.3 to the case of N cross-sections, where N must be *odd*, the total volume V_T is given by:

$$V_T = [\text{sum of end areas } + 4 \times \text{sum of even areas} + 2 \times \text{sum of odd areas}]L/3 \tag{2.4}$$

There is therefore a check in line 4050 of the program to see if N entered at the outset for the number of cross-sections is odd. If not, then the program calculates up to the previous cross-section, and the prismoidal cut and/or fill is stored in arrays for subsequent mass-haul aggregate volume computations. All the results are displayed with an opportunity to obtain a printout for record purposes.

(u) Entry of volume data for mass-haul Line numbers 5000–5180

The program can be entered at this point from the original menu if the volumes are already known. If previous cross-sections or volumes have been entered, this routine is skipped from line 5030.

(v) Mass-haul aggregate volume computations Line numbers 5200–5680

At the start of this section, there is a choice of using the previously computed end-area or prismoidal volumes. Both can be used to obtain mass-haul aggregate volumes if required. The aggregate volumes are computed on the premise that cut volumes are positive and fill volumes are negative. The computed volumes are tabulated against chainage, to allow a simple drawing of a mass-haul diagram where the positive volumes are plotted above the base line, as discussed in sections (uu)–(zz).

(w) Routine to print earthworks location, etc. Line numbers 6000–6080

This routine is provided for record purposes to annotate the printout, assuming a standard IBM PC type printer. It may require alteration for different microcomputers and printers.

(x) Data printout for cross-sections ⟨A1–A4⟩ Line numbers 6100–6390

This standard printout is provided to keep a record of entered data: formation width, depth/height, side-slopes, crossfall, etc., together with the computed areas of cut and/or fill. If a curvature correction has been made, the radius *R* is also printed.

(y) Data printout for cross-sections ⟨A5–A6⟩ Line numbers 6400–6590

A separate printout is provided for the irregular cross-sections giving the actual offsets and heights used in the area computations. If a digitizer has been used to enter the data, this provides a check on the digitized points.

(z) Printout summary of cross-sectional areas Line numbers 6600–6720

This routine is provided at the end of the area calculations prior to the volume computations.

(aa) Data printout of volumes for ⟨V1⟩ Line numbers 6800–6880

This is a simple printout of the total cut and fill computed for record purposes.

(bb) Data printout of volumes for ⟨V2⟩ Line numbers 7000–7150

This is the volume printout for the end-area computations between each cross-section. The total cut and fill volumes are also printed.

(cc) Data printout of volumes for ⟨V3⟩ Line numbers 7200–7350

The prismoidal volumes computed between odd cross-sections are presented, together with the total cut and fill volumes.

(dd) Printout summary of volumes Line numbers 7400–7550

This routine is provided to record volumes entered directly into the mass-haul aggregate volume computation section.

(ee) Mass-haul aggregate volume printout Line numbers 7600–7790

This final print routine can be used by either end-area or prismoidal volumes for printing out the computed totals and the aggregate volume for plotting the mass-haul diagram.

(ff) Routine to clear screen Line numbers 9000–9070

This routine is used to clear the screen and provide two lines of text at the head of each screen as described in routine (n) of the PC-Areas program in Section 2.2.

(gg) Routine to display error message Line numbers 9100–9130

If a wrong keyboard response has been made, the error is trapped and this error message displayed and re-entry requested.

(hh) Termination of program Line numbers 9500–9620

This program can be terminated either from the initial screen header (line 370) or after the area or volume computations are complete (line 3060). An end-of-program message is displayed to advise the operator that computations are complete.

(ii) Digitizer input Line numbers 12000–12300

The initial lines of this routine are used to set up the scale of the digitizer by entering the left-hand side (LHS) and right-hand side (RHS) of the formation width by clicking on the points with the cursor button. The other points are then entered in a clockwise direction.

(jj) Selection of CS display screen/plotter Line numbers 20000–20090

This routine offers the user an opportunity to plot the relevant cross-section either on the graphics screen or on a plotter. This routine can be used to plot cross-sections at different scales, for example, before returning to the main computations.

(kk) Set up for CS graphics display Line numbers 20100–20190

This routine sets up a graphics window on the screen like Fig. 1.4 and asks the user to select the maximum offset to be used to scale the window. At the end of the routine the program branches to the specific cross-section to be plotted ⟨A1⟩–⟨A6⟩.

(ll) Set up array to display ⟨A1⟩ Line numbers 20200–20290

This and the other cross-sections ⟨A1⟩–⟨A4⟩ can be plotted from six points, so an array is set up for X and Y to contain these points for plotting on either the graphics screen or the plotter.

(mm) CS graphics display and screen dump Line numbers 20300–20490

The display screen chosen for the presentation of the cross-section is shown in Fig. 1.4 with text at the top and to the right-hand side of the graphics window. The formation width is centred on the screen and the cross-section is plotted relative to this, that is cross-sections in cut will be plotted in the top half of the screen and fill sections will be plotted in the lower half of the screen. At the completion of the plot the cross-section number and chainage are given at the right-hand side with an opportunity to obtain a screen dump on a dot-matrix graphics printer (Figs. 2.3–2.8).

(nn) Set up arrays to display ⟨A2⟩–⟨A4⟩ Line numbers 20500–21150

The routines for each of these cross-sections are similar to (ll) above, where care has to be taken to check if in cut or fill and whether the crossfall has a positive or negative slope.

(oo) Set up array to display ⟨A5⟩–⟨A6⟩ Line numbers 21200–21290

The numbers of points in these sections will depend on the irregularity of the ground surface. The cross-sections are plotted in a clockwise direction in the order of entering the points.

(pp) Plotter selection and initialization Line numbers 22000–22200

A choice of three plotters is given in this menu. Like the screen graphics discussed earlier, the user has to enter the maximum offset from the centre line to select the plotter scale. After initialization of the plotter variables, the program branches to the relevant plotter routines.

(qq) Plotter routine for HPGL Line numbers 22250–22490

This routine is suitable for Hewlett-Packard plotters and their equivalents using HPGL as discussed in routine (k) of Section 2.2. After scaling the plotter into user units, a frame is drawn round the plotting area by pen 1 (black). Pen 3 (green) is chosen to draw the centre line of the cross-section and pen 4 (blue) chosen to draw the cross-section (Fig. 2.8).

(rr) Plotter routine for Apple 410 Line numbers 22500–22740

This routine is similar to (qq) where the plotter commands are altered as shown in Table 2.1.

(ss) Plotter routine for Epson HI-80 Line numbers 22750–22990

The same applies to this routine as (rr).

(tt) Routines to label plot output Line numbers 23000–23450

On completion of plotting the cross-section, the plot is labelled with pen 2 (red) in the lower left-hand corner with the location of the earthworks, the cross-section number, its chainage and the date of calculation or plot. The HPGL routine starts at line 23100, with those for the Apple 410 and Epson HI-80 routines at lines 23300 and 23400 respectively.

(uu) Mass-haul display initialization Line numbers 24000–24280

The user is first given a choice of displaying the mass-haul diagrams on either the graphics screen or a plotter. The next part of this routine, lines

24100–24260, then finds the maximum (**MHMAX**) and minimum (**MHMIN**) values to be plotted to select the vertical scale for the display before branching to either the graphics screen or plotter.

(vv) MH graphics display and screen dump Line numbers 24500–24900

The mass-haul diagram uses the same display screen used for the cross-sections discussed in (mm) and shown in Fig. 1.4. After setting up the graphics window, the zero of the vertical scale of the mass-haul diagram is plotted, followed by a vertical line at every odd cross-section before plotting the diagram. On completion, the user can obtain a screen dump on a dot-matrix graphics printer (Fig. 2.9).

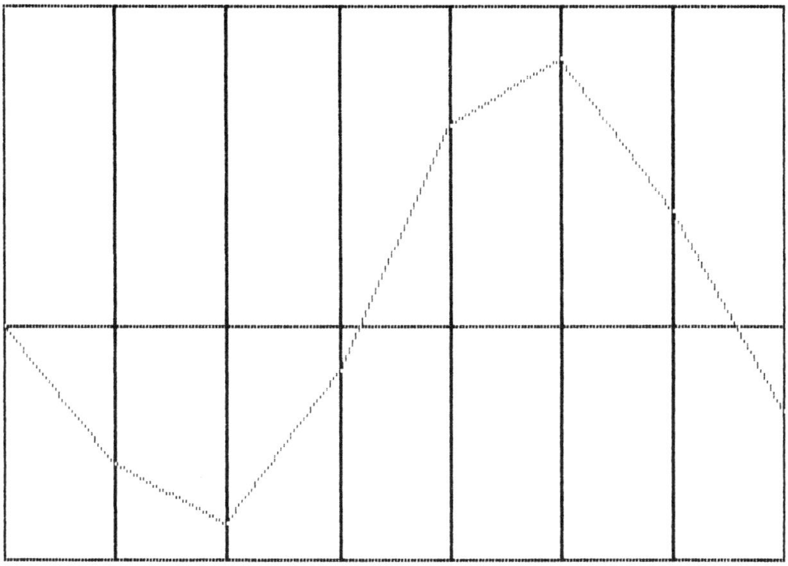

Fig. 2.9 Graphics screen dump of mass-haul diagram for earthworks.

(ww) Plotter selection Line numbers 25000–25200

This routine gives the user a choice of plotter, including a reminder to switch it on and load with paper. The plotter variables are then initialized and the zero of the vertical scale calculated, before branching to the selected plotter routine.

(xx) Plotter routine for HPGL Line numbers 25500–25950

This routine draws a frame as discussed in (qq) using pen 1 (black). The zero line of the earthwork quantity scale is then plotted with pen 3 (green) and vertical lines drawn at every odd cross-section using pen 4 (blue). Pen 7 (red) is then chosen to label the cross-sections with the chainage, and finally the plot is labelled in the lower left-hand corner with the location of the mass-haul diagram and the date of calculation or plot (Fig. 2.10).

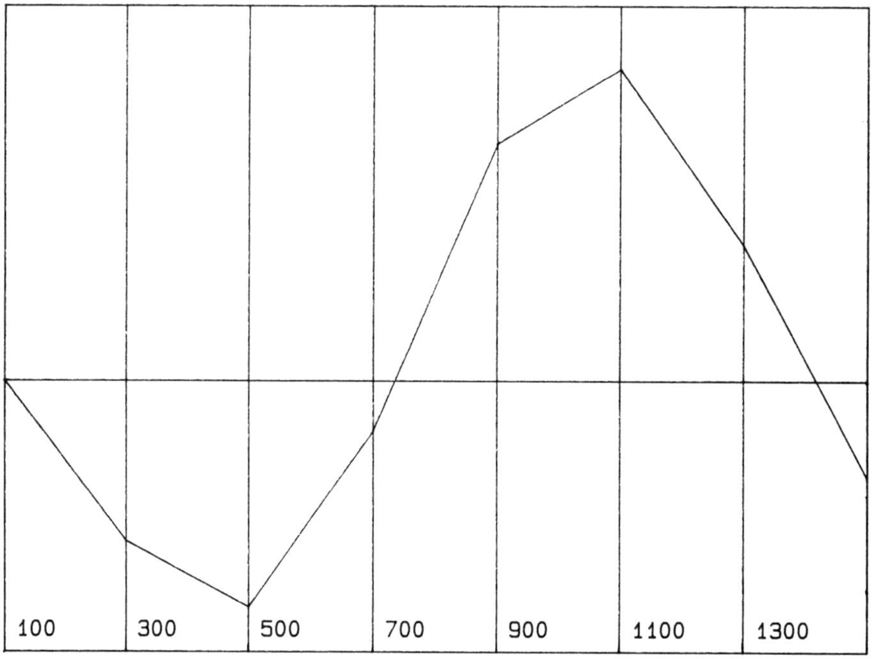

Fig. 2.10 Detailed plot of mass-haul diagram for earthworks for comparison with Fig. 2.9.

(yy) Plotter routine for Apple 410 Line numbers 26000–26450

This routine is similar to (xx) with the plotter commands changed as shown in Table 2.1.

(zz) Plotter routine for Epson HI-80 Line numbers 26500–26930

The same applies to this routine as to (yy).

(aaa) Longitudinal section display initialization Line numbers 27000–27280

This routine is similar to (uu) where the maximum and minimum ground levels are **LSMAX** and **LSMIN** respectively.

(bbb) LS graphics display and screen dump Line numbers 27500–27990

This routine is similar to (vv) with vertical lines at the odd cross-sections. If colour graphics are used, the formation level is plotted in red and the ground level in green. A screen dump is available on a dot-matrix graphics printer (Fig. 2.11).

(ccc) Plotter selection Line numbers 28000–28200

This routine is similar to (ww).

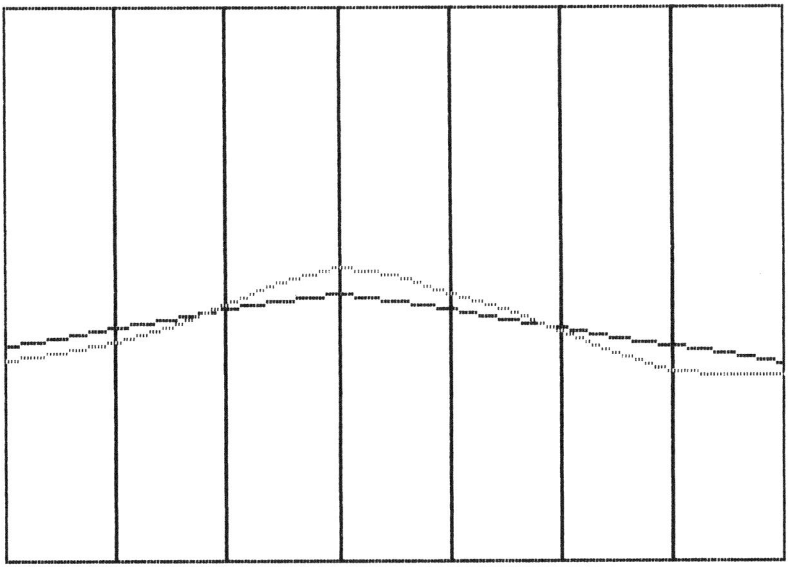

Fig. 2.11 Graphics screen dump of longitudinal section along centre line for earthworks shown in Fig. 2.9.

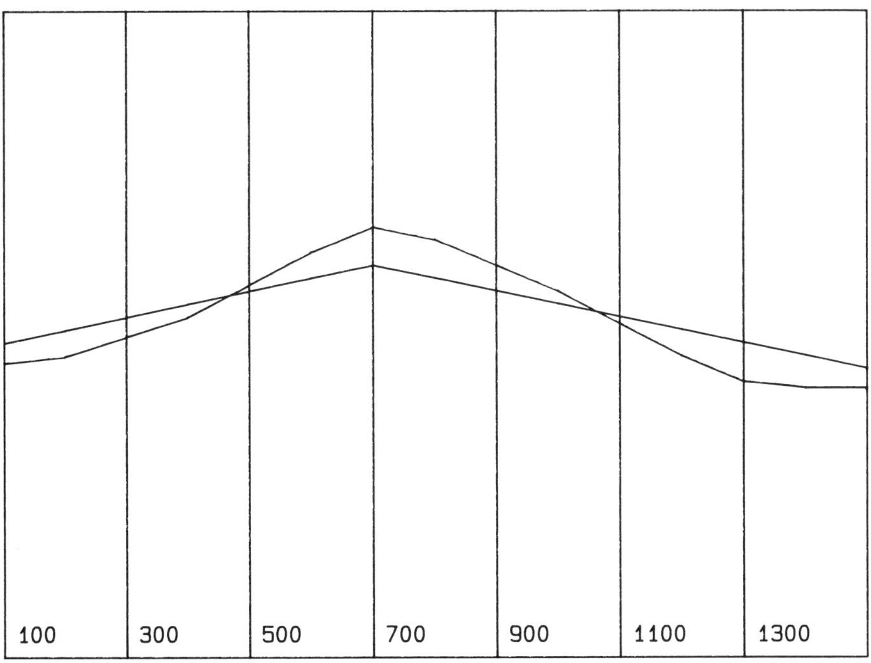

Fig. 2.12 Detailed plot of longitudinal section along centre line for earthworks shown in Fig. 2.10, for comparison with Fig. 2.11.

(ddd) Plotter routine for HPGL Line numbers 29000–29450

This routine is similar to (xx) where the frame is drawn by pen 1 (black) and the odd cross-sections by pen 4 (blue). The formation level is plotted by pen 2 (red) and the ground level by pen 3 (green). Pen 2 (red) is then chosen to label the plot as before (Fig. 2.12).

(eee) Plotter routine for Apple 410 Line numbers 29000–29450

This routine is similar to (ddd) with the plotter commands changed as shown in Table 2.1

(fff) Plotter routine for Epson HI-80 Line numbers 29500–29950

The same applies to this routine as to (eee).

2.3.2 PC-Earthworks – numeric variables

A	= Area of cross-section
AØ	= Type of cross-section selection
A1	= Area of fill
A2	= Area of cut
A6	= Flag for program A6
AC	= Area of cut
AF	= Area of fill
B	= Formation breadth of road
BYTE(3)	= Array for digitizer output
C	= Counter in error trace routine
C1,C2	= Cut areas used in extrapolation
C(N–1)	= Cut area array
CH(N–1)	= Cross-section chainage array
DE	= Direct entry volumes for mass-haul
F1,F2	= Fill areas used in extrapolation
F(N–1)	= Fill area array
FL(N–1)	= Array of cross-section formation levels
FWCX	= *X*-coordinate at centre of formation width
FWCY	= *Y*-coordinate at centre of formation width
FWLX	= *X*-coordinate at LHS of formation width
FWLY	= *Y*-coordinate at LHS of formation width
FWRX	= *X*-coordinate at RHS of formation width
FWRY	= *Y*-coordinate at RHS of formation width
GRIDX(I)	= *X*-coordinate of odd cross-sections for mass-haul
GX	= *X*-coordinate at lower left-hand corner of plot
GY	= *Y*-coordinate at lower left-hand corner of plot
H	= Level difference, formation and point
H(N–1)	= Array of formation depths/heights
HP%	= Integer variable to identify plotter
I	= Integer loop variable for number of cross-section

J,J2 = Integer loop variables
K,K2 = Ground surface slope
L = Distance between cross-sections
L0 = Chainage of first cross-section
L1 = Extrapolated chainage length
LSMAX = Maximum level of longitudinal section
LSMIN = Minimum level of longitudinal section
LSRNG = Level range of longitudinal section
LSV = Longitudinal section vertical scale factor
LSX = Longitudinal section horizontal scale factor
M,M2 = Slope of cut and fill side slopes
MC = Mass-haul cut volume total
MC(N–1) = Mass-haul cut volume array
MF = Mass-haul fill volume total
MF(N–1) = Mass-haul fill volume array
MG = Multiplication factor for plotted display
MGD = Magnification factor for digitizer
MHMAX = Mass-haul maximum value
MHMIN = Mass-haul minimum value
MHRNG = Mass-haul range
MHV = Mass-haul vertical scale factor
MHX = Mass-haul horizontal scale factor
MX = Multiplication factor for Epson HI-80 plotter
N = Number of cross-sections
NUM = Number of cross-sections for mass-haul
P = Number of points in cross-section
PC = Total cut volume, prismoidal rule
PC(N–1) = Cut volume array, prismoidal rule
PF(N–1) = Fill volume array, prismoidal rule
PP = Screen display variable
PX = *X*-coordinate at upper right-hand corner of plot
PY = *Y*-coordinate at upper right-hand corner of plot
PLOT% = Integer variable for plotter selection
RGB% = Integer variable to select screen colour
S = Step for loop (either 1 or 2)
SCF = Screen factor for cut or fill
SX = Horizontal display width in user units
SY = Vertical display height in user units
SCR% = Integer variable to select screen or plotter
SR% = Integer variable for screen resolution
T = Number of prismoidal rule cross-sections
TØ = Number of cross-sections
T1,T2 = Storage variables used in area calculations
TC = Total cut volume, end areas rule
TF = Total fill volume, end areas rule
U = Program selection from menu
VØ = Type of volume calculation selection

VC	= Total volume cut, end areas rule
VC(N–1)	= Cut volume array, end areas rule
VF	= Total volume fill, end areas rule
VF(N–1)	= Fill volume array, end areas rule
W–W2	= Widths of cut or fill from centre line of cross-section
X(P–1)	= Point offset array in cross-section
XCEN	= *X*-coordinate of cross-section centre line
XD	= Variable used to store *X*-location of plotted points
XD(P–1)	= Array for *X*-coordinates of digitized points
XF(N–1)	= Array for *X*-coordinates of LS formation levels
XP(P–1)	= Array for *X*-coordinates of plotted points
Y(P–1)	= Point height array in cross-section
YCEN	= *Y*-coordinate of formation level and zero of mass-haul diagram
YD	= Variable used to store *Y*-location of plotted points
YD(P–1)	= Array for *Y*-coordinates of digitized points
YF(N–1)	= Array for *Y*-coordinates of LS formation levels
YP(P–1)	= Array for *Y*-coordinates of plotted points

2.3.3 PC-Earthworks – string variables

A$		Storage for location of earthworks
B$		Storage for operator's name
C$		Storage for date of calculations
DATE$		Variable used to return the date of the computer's calendar/ clock: US format 'MM–DD–YYYY'
EO$	=	**"SORRY, DATA ERROR . . . PLEASE RE-ENTER"**
F$		Storage for cut **"C"**, fill **"F"** or both **"B"**
G$	=	**"IS CURVATURE CORRECTION REQUIRED (Y/N)"**
GD$		Entry for graphics dump on printer
H$		Selection of data for mass-haul computations
M$	=	**"PC-EARTHWORKS PROGRAM – MASS-HAUL"**
P$	=	**"DO YOU WISH PRINTOUT OF ABOVE DATA (Y/N)"**
Q$		Question response (Y/N)
SL$		Storage for positive/negative (P/N) slope to left of centre line
SR$		Storage for positive/negative (P/N) slope to right of centre line
T$	=	**"PC-GRAPHICS"**
TIME$		Variable used to return the time of the built-in clock in HH:MM:SS format

2.3.4 PC-Earthworks – BASIC program

```
1Ø  REM  <PC-EARTH> Earthworks program for IBM PC or Compatible
12  REM  BASIC Programs for Land Surveying, Chapter 4
13  REM  + Computer Graphics for Surveying, Chapter 2
14  REM  (C) COPYRIGHT 1984-87 P.H.Milne
16  REM  ALL RIGHTS RESERVED
18  REM  Version PC-1.Ø, Ø1-Ø1-87
2Ø CLEAR:KEY OFF
25 DEFDBL A,C,D,F,M,P,T,V
3Ø E = Ø:I = Ø:J = Ø:K = Ø:N = Ø:P1 = Ø:T = Ø:C = Ø:SCR%=Ø:HP%=Ø:PLOT
%=Ø:A$=" "
35 AØ = Ø:A6 = Ø:B = Ø:H = Ø:K2 = Ø:L = Ø:LØ = Ø:M = Ø:M2 = Ø:P = Ø:T
1 = Ø:T2 = Ø:W1 = Ø:W2 = Ø:SR%=1:PP=Ø:FRML=Ø
4Ø EØ$ = "SORRY, DATA ERROR ... PLEASE RE-ENTER."
45 P$ = "DO YOU WISH PRINTOUT OF DATA (Y/N)"
5Ø PCF$="PCSCREEN"
52 OPEN "I",#3,PCF$
54 INPUT#3,SCR%
56 INPUT#3,SCD%
58 CLOSE#3
6Ø ON (SCR%+1) GOTO 62,64,66
62 SCREEN Ø:WIDTH 4Ø:GOTO 8Ø
64 SCREEN 1:GOTO 68
66 SCREEN 1:GOTO 8Ø
68 IF SCD%=1 THEN 8Ø
7Ø COLOR 8,Ø
8Ø  DIM C(3Ø),F(3Ø),FL(3Ø),GL(3Ø),H(3Ø),X(3Ø),Y(3Ø),XD(3Ø),YD(3Ø),XF(
3Ø),YF(3Ø),XP(3Ø),YP(3Ø),CH(3Ø),GRIDX(3Ø)
85  DIM VC(3Ø),VF(3Ø),PC(3Ø),PF(3Ø)
9Ø  DIM MC(3Ø),MF(3Ø),MH(3Ø),BYTE(3)
95 G$ = "IS CURVATURE CORRECTION REQUIRED(Y/N)"
1ØØ  GOSUB 9ØØØ
11Ø  PRINT
3ØØ  PRINT " SELECT PROGRAM FROM MENU"
33Ø  PRINT
34Ø  PRINT "    <1> AREAS OF CROSS-SECTIONS"
35Ø  PRINT "    <2> VOLUMES FROM CROSS-SECTIONS"
36Ø  PRINT "    <3> MASS-HAUL FROM VOLUMES"
37Ø  PRINT "    <4> QUIT                     ";
38Ø  INPUT U
39Ø IF U<1 OR U>4 THEN 1ØØ
4ØØ IF U=4 THEN 95ØØ
5ØØ  GOSUB 9ØØØ
54Ø  PRINT
55Ø  PRINT "ENTER LOCATION OF WORK, ETC. WHEN       REQUESTED, AND PR
ESS <Enter>"
56Ø  PRINT:PRINT "ENTER LOCATION OF EARTHWORKS :-"
57Ø  INPUT A$
58Ø  PRINT "ENTER OPERATOR'S NAME :-"
59Ø  INPUT B$
6ØØ  PRINT "ENTER DATE OF CALCULATIONS AS DD/MM/YY"
61Ø  INPUT C$
64Ø  PRINT
65Ø  PRINT "DO YOU WISH PRINTOUT OF LOCATION ETC.,  (Y/N) ";
66Ø  INPUT Q$
67Ø  IF Q$ = "Y" THEN 6ØØØ
68Ø  IF Q$ <  > "N" THEN 65Ø
693  REM  ****************************************
695  REM  ENTER CROSS-SECTION DATA
7ØØ  GOSUB 9ØØØ
71Ø  PRINT "EARTHWORKS PROGRAM"
72Ø  PRINT "*************************************"
73Ø  PRINT : PRINT "ENTER NUMBER OF CROSS-SECTIONS ";
74Ø  INPUT N: IF N < 2 THEN 73Ø
```

```
750  PRINT : PRINT "ENTER CHAINAGE OF FIRST CROSS-SECTION"
760  INPUT L0: IF L0 < 0 THEN 750
770  PRINT : PRINT "ENTER DISTANCE BETWEEN CROSS-SECTIONS"
780  INPUT L: IF L <  = 0 THEN 770
790 ON U GOTO 800,2800,5000
793  REM  ****************************************
795  REM  START AREAS OF CROSS-SECTION DATA
800  GOSUB 9000
810  PRINT "EARTHWORKS PROGRAM - AREAS"
820  PRINT "*****************************************"
830 I=0:REM START LOOP FOR I=0 TO N-1
835 IF I=0 THEN 850
840 GOSUB 9000
850  PRINT : PRINT "SELECT TYPE OF CROSS-SECTION FROM MENU :"
855  PRINT
860  PRINT "    <1>   ONE-LEVEL  (CUT OR FILL)"
870  PRINT "    <2>   TWO-LEVEL  (CUT OR FILL)"
880  PRINT "    <3>   TWO-LEVEL  (CUT ++ FILL)"
890  PRINT "    <4> THREE-LEVEL  (CUT OR FILL)"
900  PRINT "    <5>   IRREGULAR  (CUT OR FILL)"
910  PRINT "    <6>   IRREGULAR  (CUT ++ FILL)"
920  INPUT A0
930  IF A0 < 1 OR A0 > 6 THEN 840
940 F$ = " ": REM  EMPTY F$
950  ON A0 GOTO 1000,1200,1450,1700,2000,2500
960  LOCATE 24,1: PRINT "SUMMARY OF CUT/FILL (Y/N)";
970  INPUT Q$: IF Q$ = "Y" THEN 6600
980  IF Q$ <  > "N" THEN 960
990 GOTO 2800
993  REM  ****************************************
995  REM  START OF CROSS-SECTION AREA ROUTINES
1000  GOSUB 9000:SL$="":SR$=""
1010 X$="<A1> ONE-LEVEL":PRINT X$
1020  PRINT "*****************************************"
1030  PRINT : PRINT "CROSS-SECTION NO. ";I + 1
1040  PRINT "CHAINAGE          = ";L0 + I * L
1050  PRINT : PRINT "ENTER CUT OR FILL (C/F)";: INPUT F$
1060  IF F$ <  > "C" AND F$ <  > "F" THEN 1050
1070  PRINT : PRINT "ENTER FORMATION WIDTH ";: INPUT B
1075  IF B <  = 0 THEN 1070
1080  IF F$ = "C" THEN  PRINT "ENTER FORMATION DEPTH ";: INPUT H
1090  IF F$ = "F" THEN  PRINT "ENTER FORMATION HEIGHT ";: INPUT H
1095  IF H <  = 0 THEN 1080
1096 IF F$="C" THEN H(I)=H
1097 IF F$="F" THEN H(I)=-H
1100  PRINT "ENTER SIDE SLOPE 1:";: INPUT M
1105  IF M <  = 0 THEN 1100
1110 A =  INT ((H * (B + M * H)) * 1000 + .5) / 1000:W =  INT ((B / 2
 + M * H) * 1000 + .5) / 1000
1120  PRINT : PRINT "WIDTH W = ";W;" : AREA = ";A
1130  IF F$ = "C" THEN 1150
1140 F(I) = A:C(I) = 0: GOTO 1160
1150 C(I) = A:F(I) = 0:
1160 W1 = W:W2 = W
1161 PRINT "Do you wish Plot of DATA (y/n)";
1162 INPUT Q$
1163 IF Q$="y" THEN 20000
1164 IF Q$<>"n" THEN 1161
1165 LOCATE 23,1:PRINT P$;:INPUT Q$
1170  IF Q$ = "Y" THEN 6100
1180  IF Q$ <  > "N" THEN 1165
1190 I=I+1:IF I>N-1 THEN 960
1192 GOTO 840
1193  REM  ****************************************
1200  GOSUB 9000:R = 0:SL$="":SR$=""
```

```
1210 X$="<A2> TWO-LEVEL":PRINT X$
1220  PRINT "****************************************"
1230  PRINT : PRINT "CROSS-SECTION NO. ";I + 1
1240  PRINT "CHAINAGE          = ";LØ + I * L
1250  PRINT : PRINT "ENTER CUT OR FILL (C/F)";: INPUT F$
1260  IF F$ <  > "C" AND F$ <  > "F" THEN 125Ø
1270  PRINT : PRINT "ENTER FORMATION WIDTH ";: INPUT B
1275  IF B <  = Ø THEN 127Ø
1280  IF F$ = "C" THEN  PRINT "ENTER FORMATION DEPTH ";: INPUT H
1290  IF F$ = "F" THEN  PRINT "ENTER FORMATION HEIGHT ";: INPUT H
1295  IF H <  = Ø THEN 128Ø
1296 IF F$="C" THEN H(I)=H
1297 IF F$="F" THEN H(I)=-H
13ØØ  PRINT "ENTER CROSSFALL  1:";: INPUT K:IF K <=Ø THEN 13ØØ
13Ø2  PRINT "IS SLOPE +VE/-VE FROM L-R (P/N) ";: INPUT SL$
13Ø5  IF SL$ <  > "P" AND SL$ <  > "N" THEN 13Ø2
131Ø  PRINT "ENTER SIDE SLOPE 1:";: INPUT M
1315  IF M <  = Ø THEN 131Ø
132Ø W1 = (B / 2 + M * H) * (K / (K - M))
133Ø W2 = (B / 2 + M * H) * (K / (K + M))
134Ø A = ((B / 2 + M * H) * (W1 + W2) - B * B / 2) / (2 * M)
135Ø A =  INT (A * 1ØØØ + .5) / 1ØØØ:W1 =  INT (W1 * 1ØØØ + .5) / 1ØØ
Ø:W2 =  INT (W2 * 1ØØØ + .5) / 1ØØØ
136Ø  PRINT : PRINT "WIDTH W1 = ";W1;" WIDTH W2 = ";W2: PRINT "AREA
   = ";A
137Ø  PRINT : PRINT G$;: INPUT Q$: IF Q$ = "Y" THEN 265Ø
138Ø  IF Q$ <  > "N" THEN 137Ø
139Ø  IF F$ = "C" THEN 141Ø
14ØØ F(I) = A:C(I) = Ø: GOTO 142Ø
141Ø C(I) = A:F(I) = Ø
142Ø PRINT "Do you wish Plot of DATA (y/n)";
1422 INPUT Q$
1423 IF Q$="y" THEN 2ØØØØ
1424 IF Q$<>"n" THEN 142Ø
1425  LOCATE 23,1: PRINT P$;: INPUT Q$: IF Q$ = "Y" THEN 61ØØ
143Ø  IF Q$ <  > "N" THEN 1425
144Ø I=I+1:IF I>N-1 THEN 96Ø
1442 GOTO 84Ø
1443  REM ****************************************
145Ø  GOSUB 9ØØØ:R = Ø:SL$="":SR$=""
146Ø X$="<A3> TWO-LEVEL (CUT ++ FILL)":PRINT X$
147Ø  PRINT "****************************************"
148Ø  PRINT : PRINT "CROSS-SECTION NO. ";I + 1
149Ø  PRINT "CHAINAGE          = ";LØ + I * L
15ØØ  PRINT : PRINT "ENTER WHETHER CUT OR FILL (C/F)": PRINT "AT CENT
RE LINE ";: INPUT F$
151Ø  IF F$ <  > "C" AND F$ <  > "F" THEN 15ØØ
152Ø  PRINT "ENTER FORMATION WIDTH ";: INPUT B
1525  IF B <  = Ø THEN 152Ø
153Ø  IF F$ = "C" THEN  PRINT "ENTER FORMATION DEPTH ";: INPUT H
154Ø  IF F$ = "F" THEN  PRINT "ENTER FORMATION HEIGHT ";: INPUT H
1545  IF H < Ø THEN 153Ø
1546 IF F$="C" THEN H(I)=H
1547 IF F$="F" THEN H(I)=-H
155Ø  PRINT "ENTER CROSSFALL        1:";: INPUT K:IF K <= Ø THEN 155Ø
1552  PRINT "IS SLOPE +VE/-VE FROM L-R (P/N) ";: INPUT SL$
1555  IF SL$ <  > "P" AND SL$ <  > "N" THEN 1552
156Ø  PRINT "ENTER CUT  SIDE SLOPE 1:";: INPUT M2
1565  IF M2 <  = Ø THEN 156Ø
157Ø  PRINT "ENTER FILL SIDE SLOPE 1:";: INPUT M
1575  IF M <  = Ø THEN 157Ø
158Ø  IF F$ = "F" THEN H = ( - 1) * H
1585  REM  CALCULATE AREA OF FILL (A1)
159Ø A1 = ((B / 2 - K * H) ^ 2) / (2 * (K - M))
1595 W1 =  INT ((K / (K - M2)) * (B / 2 + M2 * H) * 1ØØØ + .5) / 1ØØØ
```

```
1600 F(I) =  INT (A1 * 1000 + .5) / 1000
1605  REM  CALCULATE AREA OF CUT (A2)
1610 A2 = ((B / 2 + K * H) ^ 2) / (2 * (K - M2))
1615 W2 =  INT ((K / (K - M)) * (B / 2 - M * H) * 1000 + .5) / 1000
1620 C(I) =  INT (A2 * 1000 + .5) / 1000
1630  PRINT : PRINT "CUT WIDTH = ";W1;" : AREA = ";C(I)
1640  PRINT "FILL WIDTH= ";W2;" : AREA = ";F(I)
1650  PRINT : PRINT G$;: INPUT Q$: IF Q$ = "Y" THEN 2700
1660  IF Q$ <  > "N" THEN 1650
1670 PRINT "Do you wish Plot of DATA (y/n)";
1672 INPUT Q$
1673 IF Q$="y" THEN 20000
1674 IF Q$<>"n" THEN 1420
1675  LOCATE 23,1: PRINT P$;: INPUT Q$: IF Q$ = "Y" THEN 6100
1680  IF Q$ <  > "N" THEN 1675
1690 I=I+1:IF I>N-1 THEN 960
1692 GOTO 840
1693  REM  ***************************************
1700  GOSUB 9000
1710 X$="<A4> THREE-LEVEL":PRINT X$
1720  PRINT "***************************************"
1730  PRINT : PRINT "CROSS-SECTION NO. ";I + 1
1740  PRINT "CHAINAGE          = ";L0 + I * L
1750  PRINT : PRINT "ENTER WHETHER CUT OR FILL (C/F)": PRINT "AT CENT
RE LINE ";: INPUT F$
1760  IF F$ <  > "C" AND F$ <  > "F" THEN 1750
1770  PRINT "ENTER FORMATION WIDTH ";: INPUT B
1775  IF B <  = 0 THEN 1770
1780  IF F$ = "C" THEN  PRINT "ENTER FORMATION DEPTH ";: INPUT H
1790  IF F$ = "F" THEN  PRINT "ENTER FORMATION HEIGHT ";: INPUT H
1795  IF H < 0 THEN 1780
1796 IF F$="C" THEN H(I)=H
1797 IF F$="F" THEN H(I)=-H
1800  PRINT : PRINT "ENTER SIDE SLOPE    1:";: INPUT M
1805  IF M <  = 0 THEN 1800
1810  PRINT : PRINT "ENTER CROSSFALL-LHS 1:";: INPUT K
1815  IF K <  = 0 THEN 1810
1820  PRINT "IS SLOPE +VE/-VE FROM C.L. (P/N) ";: INPUT SL$
1830  IF SL$ <  > "P" AND SL$ <  > "N" THEN 1820
1835 IF F$="F" THEN 1850
1840  IF SL$ = "N" THEN W1 = (B / 2 + M * H) * (K / (K + M))
1842  IF SL$ = "P" THEN W1 = (B / 2 + M * H) * (K / (K - M))
1845 GOTO 1860
1850  IF SL$ = "P" THEN W1 = (B / 2 + M * H) * (K / (K + M))
1855  IF SL$ = "N" THEN W1 = (B / 2 + M * H) * (K / (K - M))
1860  PRINT : PRINT "ENTER CROSSFALL-RHS 1:";: INPUT K2
1865  IF K2 <  = 0 THEN 1860
1870  PRINT "IS SLOPE +VE/-VE FROM C.L. (P/N) ";: INPUT SR$
1880  IF SR$ <  > "P" AND SR$ <  > "N" THEN 1870
1885  IF F$="F" THEN 1900
1890  IF SR$ = "N" THEN W2 = (B / 2 + M * H) * (K2 / (K2 + M))
1892  IF SR$ = "P" THEN W2 = (B / 2 + M * H) * (K2 / (K2 - M))
1895 GOTO 1910
1900  IF SR$ = "P" THEN W2 = (B / 2 + M * H) * (K2 / (K2 + M))
1905  IF SR$ = "N" THEN W2 = (B / 2 + M * H) * (K2 / (K2 - M))
1910 A = ((W1 + W2) * (M * H + B / 2) - B * B / 2) / (2 * M):A =  INT
 (A * 1000 + .5) / 1000
1915 W1 =  INT (W1 * 1000 + .5) / 1000:W2 =  INT (W2 * 1000 + .5) / 1
000
1920  PRINT : PRINT "WIDTH W1 = ";W1;" : WIDTH W2 = ";W2: PRINT "AREA
     = ";A
1930  IF F$ = "C" THEN 1950
1940 F(I) = A:C(I) = 0: GOTO 1960
1950 C(I) = A:F(I) = 0
1960 PRINT "Do you wish Plot of DATA (y/n)";
```

```
1962 INPUT Q$
1963 IF Q$="y" THEN 20000
1964 IF Q$<>"n" THEN 1960
1965  LOCATE 23,1: PRINT P$;: INPUT Q$
1970  IF Q$ = "Y" THEN 6100
1980  IF Q$ <  > "N" THEN 1960
1990 I=I+1:IF I>N-1 THEN 960
1992 GOTO 840
1993  REM  ****************************************
2000  GOSUB 9000
2010 A6=0:X$="<A5> IRREGULAR":PRINT X$
2020  PRINT "****************************************"
2030  PRINT : PRINT "CROSS-SECTION NO. ";I + 1
2040  PRINT "CHAINAGE            = ";L0 + I * L: IF A6 = 2 THEN 2080
2050  PRINT : PRINT "ENTER CUT OR FILL (C/F) ";: INPUT F$
2060  IF F$ <  > "C" AND F$ <  > "F" THEN 2050
2070  PRINT "ENTER FORMATION WIDTH ";: INPUT B
2071  IF B <  = 0 THEN 2070
2075  IF F$ = "C" THEN  PRINT "ENTER FORMATION DEPTH ";: INPUT H
2076  IF F$ = "F" THEN  PRINT "ENTER FORMATION HEIGHT ";: INPUT H
2077  IF H < 0 THEN 2075
2078 IF F$="C" THEN H(I)=H
2079 IF F$="F" THEN H(I)=-H
2080  PRINT : PRINT "ENTER NO. OF POINTS IN CROSS-SECTION";
2090  INPUT P
2100  IF P =  > 3 THEN 2131
2110  PRINT  CHR$ (7): GOSUB 9100
2120  PRINT "THREE OR MORE POINTS ARE REQUIRED !"
2130  GOTO 2080
2131  PRINT:PRINT " Select DATA Input method :-"
2132  PRINT:PRINT "    <1> Keyboard"
2133  PRINT:PRINT "    <2> Digitiser ";
2134  INPUT SEL%
2135  IF SEL% <1 OR SEL% >2 THEN 2131
2136  IF SEL% = 2 THEN 12000
2138 GOSUB 9000
2140  PRINT : PRINT "ENTER OFFSET,HEIGHT (X,Y) OF EACH POINT IN CLOCK
WISE DIRECTION"
2150  PRINT "LHS OF C.L. AND BELOW FORMATION IS -VE"
2160  PRINT "RHS OF C.L. AND ABOVE FORMATION IS +VE"
2170  FOR J = 0 TO P - 1
2180  PRINT : PRINT "ENTER X,Y OF PT. NO. ";J + 1;: INPUT X(J),Y(J)
2190  NEXT J
2195  REM  CALCULATE AREA FROM VARIABLES T1,T2
2200 T1 = 0:T2 = 0: REM  SET VARIABLES TO ZERO
2210  FOR J2 = 0 TO P - 2
2220 T1 = T1 + (Y(J2)) * (X(J2 + 1))
2230 T2 = T2 + (X(J2)) * (Y(J2 + 1)): NEXT J2
2240 T1 = T1 + (Y(P - 1)) * (X(0))
2250 T2 = T2 + (X(P - 1)) * (Y(0))
2260 A = (T1 - T2) / 2:A =  INT (A * 1000 + .5) / 1000
2270  PRINT : PRINT " AREA = ";A
2271 PRINT " Do you wish plot of DATA (y/n)";
2272 INPUT Q$
2273 IF Q$="y" THEN 20000
2274 IF Q$<>"n" THEN 2271
2280  IF A6 <  > 0 THEN 2530
2290  IF F$ = "C" THEN 2310
2300 F(I) = A:C(I) = 0: GOTO 2320
2310 C(I) = A:F(I) = 0
2320 LOCATE 23,1: PRINT P$;: INPUT Q$
2330  IF Q$ = "Y" THEN 6400
2340  IF Q$ <  > "N" THEN 2320
2350 I=I+1:IF I>N-1 THEN 960
2352 GOTO 840
```

```
2493  REM  ****************************************
25ØØ  GOSUB 9ØØØ
251Ø X$="<A6> IRREGULAR (CUT ++ FILL)":PRINT X$
252Ø A6 = 1: GOTO 2Ø2Ø
253Ø  IF F$ = "C" THEN C(I) = A
254Ø  IF F$ = "F" THEN F(I) = A
255Ø  LOCATE 23,1: PRINT P$;: INPUT Q$
256Ø  IF Q$ = "Y" THEN 64ØØ
257Ø  IF Q$ <  > "N" THEN 255Ø
258Ø  IF A6 = 2 THEN 263Ø
259Ø A6 = 2: REM  COLLECT SECOND AREA
2595  REM  CHECK IF F$="C" OR "F"
26ØØ  GOSUB 9ØØØ:IF F$ = "C" THEN 262Ø
26Ø5  REM  I.E. F$="F" FOR FILL; CHANGE F$
261Ø F$ = "C": PRINT "NOW ENTER CUT :-": GOTO 2Ø3Ø
262Ø F$ = "F": PRINT "NOW ENTER FILL :- ": GOTO 2Ø3Ø
2625  REM  ARRIVE HERE AT END OF CALCULATIONS
263Ø I=I+1:IF I>N-1 THEN 96Ø
2632 GOTO 84Ø
2643  REM  ****************************************
2645  REM  EQUIVALENT AREA (CURVATURE CORRECTION) FOR <A2>
265Ø EC = (W1 * W2 * (W1 + W2)) / (3 * K * A)
266Ø  PRINT : PRINT "ENTER RADIUS OF CURVATURE ";: INPUT R
2665  IF R <  = Ø THEN 266Ø
267Ø  PRINT : PRINT "IS CENTRE OF CURVE ON LOWER (L) OR      HIGHER (
H) SIDE OF SLOPE ";: INPUT L$
2675  IF L$ <  > "L" AND L$ <  > "H" THEN 267Ø
268Ø  IF L$ = "L" THEN A =  INT (A * (1 + EC / R) * 1ØØØ + .5) / 1ØØØ
: GOTO 139Ø
269Ø  IF L$ = "H" THEN A =  INT (A * (1 - EC / R) * 1ØØØ + .5) / 1ØØØ
: GOTO 139Ø
2693  REM  ****************************************
2695  REM  EQUIVALENT AREA (CURVATURE CORRECTION) FOR <A3>
27ØØ E1 = (W1 + B / 2 - K * H) / 3
271Ø E2 = (W2 + B / 2 + K * H) / 3
272Ø  PRINT : PRINT "ENTER RADIUS OF CURVATURE ";: INPUT R
2725  IF R <  = Ø THEN 272Ø
273Ø  PRINT : PRINT "IS CENTRE OF CURVE ON LOWER(L) OR       HIGHER(H
) SIDE OF SLOPE ";: INPUT L$: IF L$ <  > "L" AND L$ <  > "H" THEN 273
Ø
274Ø  IF L$ = "H" THEN 277Ø
2745  REM  I.E. L$="L"
275Ø C(I) =  INT ((C(I)) * (1 + E1 / R) * 1ØØØ + .5) / 1ØØØ
276Ø F(I) =  INT ((F(I)) * (1 - E2 / R) * 1ØØØ + .5) / 1ØØØ: GOTO 167
Ø
277Ø C(I) =  INT ((C(I)) * (1 - E1 / R) * 1ØØØ + .5) / 1ØØØ
278Ø F(I) =  INT ((F(I)) * (1 + E2 / R) * 1ØØØ + .5) / 1ØØØ: GOTO 167
Ø
2793  REM  ****************************************
2795  REM  START VOLUMES ROUTINE
28ØØ  GOSUB 9ØØØ
281Ø  PRINT "EARTHWORKS PROGRAMS - VOLUMES"
282Ø  PRINT "****************************************"
283Ø IF U=1 THEN 3ØØØ
284Ø  FOR I = Ø TO N - 1
285Ø  PRINT : PRINT "CROSS-SECTION ";I + 1;
286Ø  PRINT " : CHAINAGE ";LØ + I * L
287Ø C(I) = Ø:F(I) = Ø
288Ø  PRINT "ENTER CUT/FILL/BOTH (C/F/B) ";
289Ø  INPUT F$
29ØØ  IF F$ = "C" THEN 293Ø
291Ø  IF F$ = "F" THEN 294Ø
292Ø  IF F$ <  > "B" THEN 288Ø
293Ø  PRINT "ENTER CUT AREA ";: INPUT C(I): IF C(I) < Ø THEN 293Ø
2935  IF F$ = "C" THEN 295Ø
```

```
294Ø  PRINT "ENTER FILL AREA ";: INPUT F(I): IF F(I) < Ø THEN 294Ø
295Ø  NEXT I
296Ø  PRINT : PRINT "DO YOU WISH PRINTOUT SUMMARY OF CUT/FILL (Y/N)";
: INPUT Q$
297Ø  IF Q$ = "Y" THEN 66ØØ
298Ø  IF Q$ < > "N" THEN 296Ø
2993  REM ***************************************
2995  REM  VOLUME COMPUTATIONS
3ØØØ  PRINT : PRINT "SELECT TYPE OF VOLUME CALCULATIONS :-"
3Ø1Ø  PRINT
3Ø2Ø  PRINT "    <1> MEAN AREA RULE "
3Ø3Ø  PRINT "    <2> END AREAS RULE "
3Ø4Ø  PRINT "    <3> PRISMOIDAL RULE"
3Ø5Ø  PRINT "    <4> MASS-HAUL CALC."
3Ø6Ø  PRINT "    <5> QUIT            "
3Ø7Ø  INPUT VØ: IF VØ < 1 OR VØ > 5 THEN 3ØØØ
3Ø8Ø  ON VØ GOTO 32ØØ,34ØØ,4ØØØ,5ØØØ,95ØØ
3193  REM ***************************************
32ØØ  GOSUB 9ØØØ
321Ø  PRINT "EARTHWORKS PROGRAM - VOLUMES"
322Ø V$ = "<V1> MEAN AREA RULE":AC = Ø:AF = Ø
323Ø  PRINT : PRINT V$
324Ø  PRINT "***************************************"
325Ø  FOR I = Ø TO N - 1
3255  REM  COLLECT TOTAL CUT AREA IN AC
326Ø AC = AC + C(I)
3265  REM  COLLECT TOTAL FILL AREA IN AF
327Ø AF = AF + F(I)
328Ø  NEXT I
3285  REM  TOTAL VOLUME CUT = VC
329Ø VC = AC * L * (N - 1) / N
33ØØ VC =  INT (VC * 1ØØØ + .5) / 1ØØØ
33Ø5  REM  TOTAL VOLUME FILL = VF
331Ø VF = AF * L * (N - 1) / N
332Ø VF =  INT (VF * 1ØØØ + .5) / 1ØØØ
333Ø  PRINT : PRINT "TOTAL VOLUME CUT  = ";VC
334Ø  PRINT "TOTAL VOLUME FILL = ";VF
335Ø  PRINT : PRINT P$;: INPUT Q$
336Ø  IF Q$ = "Y" THEN 68ØØ
337Ø  IF Q$ < > "N" THEN 335Ø
338Ø  GOTO 3ØØØ
3393  REM ***************************************
3395  REM  VOLUME ROUTINE VØ=2
34ØØ  GOSUB 9ØØØ
341Ø  PRINT "EARTHWORKS PROGRAM - VOLUMES"
342Ø V$ = "<V2> END AREAS RULE":TC = Ø:TF = Ø
343Ø  PRINT : PRINT V$
344Ø  PRINT "***************************************"
345Ø  FOR I = Ø TO N - 2: IF C(I) < > Ø AND C(I + 1) < > Ø THEN 346
Ø
3452  IF C(I) = Ø AND C(I + 1) = Ø THEN 346Ø
3455  GOTO 37ØØ
346Ø VC(I) = (C(I) + C(I + 1)) * L / 2
3465  REM  COLLECT TOTAL VOLUMES IN TC AND TF
347Ø TC = TC + VC(I)
348Ø VC(I) =  INT (VC(I) * 1ØØØ + .5) / 1ØØØ
349Ø  IF F(I) < > Ø AND F(I + 1) < > Ø THEN 35ØØ
3492  IF F(I) = Ø AND F(I + 1) = Ø THEN 35ØØ
3495  GOTO 38ØØ
35ØØ VF(I) = (F(I) + F(I + 1)) * L / 2
351Ø TF = TF + VF(I):VF(I) =  INT (VF(I) * 1ØØØ + .5) / 1ØØØ
352Ø  NEXT I:TC =  INT (TC * 1ØØØ + .5) / 1ØØØ:TF =  INT (TF * 1ØØØ +
 .5) / 1ØØØ
353Ø  PRINT "CRXS.    CHAINAGE  CUT VOL.  FILL VOL."
354Ø  PRINT "======================================="
```

```
3550  FOR J = Ø TO N - 2
3560  PRINT J + 1; TAB( 1Ø);LØ + J * L
3570  PRINT  TAB( 2Ø);VC(J); TAB( 3Ø);VF(J)
3580  NEXT J
3590  PRINT N; TAB( 1Ø);LØ + (N - 1) * L
3600  PRINT "======================================="
3610  PRINT "TOTAL VOLUMES = "; TAB( 2Ø);TC; TAB( 3Ø);TF
3620  PRINT "***************************************"
3630  PRINT : PRINT P$;: INPUT Q$
3640  IF Q$ = "Y" THEN 7ØØØ
3650  IF Q$ <  > "N" THEN 363Ø
3660  GOTO 3ØØØ
3693  REM  ****************************************
3695  REM  CALCULATE CUT LENGTH
37ØØ  PRINT "CROSS-SECTION ";I + 1;" CUT AREA = ";C(I)
371Ø  PRINT "CROSS-SECTION ";I + 2;" CUT AREA = ";C(I + 1)
372Ø  PRINT : PRINT "ARE THESE VALUES TO BE USED (Y/N)";: INPUT Q$
373Ø  IF Q$ = "Y" THEN 346Ø
3735  REM  I.E. Q$="N" - COMPUTE CORRECT VOLUME
374Ø  PRINT "ENTER PREVIOUS/NEXT CROSS-SECTION AREA": INPUT C2
375Ø  IF C(I) = Ø THEN C1 = C(I + 1)
376Ø  IF C(I + 1) = Ø THEN C1 = C(I)
3765  REM  CALCULATE LENGTH OF CUT
377Ø L1 = L * C1 / (C2 - C1)
378Ø VC(I) = C1 * L1 / 2
379Ø  GOTO 347Ø
3793  REM  ****************************************
3795  REM  CALCULATE FILL DISTANCE L1
38ØØ  PRINT "CROSS-SECTION ";I + 1;" : FILL AREA = ";F(I)
381Ø  PRINT "CROSS-SECTION ";I + 2;" : FILL AREA = ";F(I + 1)
382Ø  PRINT : PRINT "ARE THESE VALUES TO BE USED (Y/N)";: INPUT Q$
383Ø  IF Q$ = "Y" THEN 35ØØ
3835  REM  I.E. Q$="N"-COMPUTE CORRECT VOLUME
384Ø  PRINT "ENTER PREVIOUS/NEXT CROSS-SECTION AREA": INPUT F2
3845  IF F2 <  = Ø THEN 384Ø
385Ø  IF F(I) = Ø THEN F1 = F(I + 1)
386Ø  IF F(I + 1) = Ø THEN F1 = F(I)
3865  REM  CALCULATE LENGTH OF FILL
387Ø L1 = L * F1 / (F2 - F1)
388Ø VF(I) = F1 * L1 / 2
389Ø  GOTO 351Ø
3993  REM  ****************************************
3995  REM  PRISMOIDAL CALCULATIONS
4ØØØ  GOSUB 9ØØØ
4Ø1Ø  PRINT "EARTHWOKS PROGRAM - VOLUMES"
4Ø2Ø V$ = "<V3> PRISMOIDAL RULE":T = N - 1:PC = Ø:PF = Ø
4Ø3Ø  PRINT : PRINT V$
4Ø4Ø  PRINT "***************************************"
4Ø45  REM  CHECK THAT N IS ODD
4Ø5Ø  IF (N - 1) / 2 <> INT((N - 1) / 2) THEN T = N - 2
4Ø6Ø  FOR I = Ø TO T - 2 STEP 2
4Ø65  REM  CALCULATE CUT VOL BETWEEN I,I+1,I+2
4Ø7Ø PC(I) = (C(I) + 4 * C(I + 1) + C(I + 2)) * L / 3
4Ø8Ø PC = PC + PC(I)
4Ø9Ø PC(I) =  INT (PC(I) * 1ØØØ + .5) / 1ØØØ
4Ø95  REM  CALCULATE FILL VOL. BETWEEN I,I+1,I+2
41ØØ PF(I) = (F(I) + 4 * F(I + 1) + F(I + 2)) * L / 3
411Ø PF = PF + PF(I)
412Ø PF(I) =  INT (PF(I) * 1ØØØ + .5) / 1ØØØ
413Ø  NEXT I
414Ø PC =INT(PC * 1ØØØ + .5)/1ØØØ:PF =INT(PF * 1ØØØ + .5)/1ØØØ
415Ø  PRINT "CRXS.    CHAINAGE  CUT VOL.  FILL VOL."
416Ø  PRINT "======================================="
417Ø  FOR J = Ø TO T - 2 STEP 2
418Ø  PRINT J + 1; TAB( 1Ø);LØ + J * L
```

```
4190  PRINT  TAB( 20);PC(J); TAB( 30);PF(J)
4200  NEXT J
4210  PRINT T + 1; TAB( 10);LØ + T * L
4220  PRINT "========================================"
4230  PRINT "TOTAL VOLUMES"; TAB( 20);PC; TAB( 30);PF
4240  PRINT "****************************************"
4250  PRINT : PRINT P$;: INPUT Q$
4260  IF Q$ = "Y" THEN 7200
4270  IF Q$ <  > "N" THEN 4250
4280  GOTO 3000
4993  REM  ****************************************
4995  REM  MASS-HAUL COMPUTATIONS
5000  GOSUB 9000:D = Ø:MC = Ø:MF = Ø
5010 M$ = "PC-EARTHWORKS PROGRAM - MASS-HAUL": PRINT M$
5020  PRINT "****************************************"
5030  IF U <  > 3 THEN 5200
5035  REM  I.E. U=3, ENTER CRXS AND VOLUME
5040 D = 1: FOR I = Ø TO N - 2
5050  PRINT : PRINT "VOLUME CRXS ";I + 1;" TO ";I + 2
5060  PRINT "CHAINAGE ";LØ + I * L;" TO ";LØ + (I + 1) * L
5070 MC(I) = Ø:MF(I) = Ø
5080  PRINT : PRINT "ENTER CUT/FILL/BOTH (C/F/B)";
5090  INPUT F$
5100  IF F$ = "C" THEN 5130
5110  IF F$ = "F" THEN 5140
5120  IF F$ <  > "B" THEN 5080
5130  PRINT "ENTER CUT  VOL. ";: INPUT MC(I): IF MC(I) < Ø THEN 5130
5135  IF F$ = "C" THEN 5150
5140  PRINT "ENTER FILL VOL. ";: INPUT MF(I): IF MF(I) < Ø THEN 5140
5150 MC = MC + MC(I):MF = MF + MF(I): NEXT I
5155 MC=INT(MC*1ØØØ+.5)/1ØØØ:MF=INT(MF*1ØØØ+.5)/1ØØØ
5160  PRINT : PRINT "DO YOU WISH A PRINTOUT SUMMARY OF       VOLUMES
(Y/N) ";
5170  INPUT Q$: IF Q$ = "Y" THEN 7400
5180  IF Q$ <  > "N" THEN 5160
5193  REM  ****************************************
5195  REM  MASS-HAUL ROUTINE FOR VOLUMES
5200  GOSUB 9000:H$ = " "
5210  PRINT M$
5220  PRINT "****************************************": IF D = 1 THEN
 5270
5230  PRINT : PRINT "DO YOU WISH TO USE END AREA (E)          PRISMOID
AL (P) VOLUMES, OR QUIT(Q) ";: INPUT H$
5240  IF H$ = "Q" THEN 5480
5250  IF H$ = "P" THEN 5500
5260  IF H$ <  > "E" THEN 5230
5265  REM  I.E. END AREA VOLS VC(I) AND VF(I)
5270 MH = Ø: FOR I = Ø TO N - 2: IF D = 1 THEN 5290
5280 MC(I) = VC(I):MF(I) = VF(I):MC = TC:MF = TF
5290 MH = MH + MC(I) - MF(I)
5300 MH(I)=INT(MH*1ØØØ+.5)/1ØØØ
5305 MC(I)=INT(MC(I)*1ØØØ+.5)/1ØØØ:MF(I)=INT(MF(I)*1ØØØ+.5)/1ØØØ
5310  NEXT I: IF D = 1 THEN D = 2
5315 MH=INT(MH*1ØØØ+.5)/1ØØØ:MC=INT(MC*1ØØØ+.5)/1ØØØ:MF=INT(MF*1ØØØ+.
5)/1ØØØ
5320  PRINT : PRINT "CHAIN. CUT VOL.  FILL VOL.  AGG.VOL."
5330  PRINT "========================================"
5340  FOR J = Ø TO N - 2
5350  PRINT LØ + J * L
5360  PRINT  TAB( 8);MC(J); TAB( 18);MF(J); TAB( 29);MH(J)
5370  NEXT J
5380  PRINT LØ + (N - 1) * L
5390  PRINT "========================================"
5400  PRINT "TOTAL"; TAB( 8);MC; TAB( 18);MF; TAB( 29);MH
5410  PRINT "****************************************"
```

```
5420  PRINT : PRINT P$;: INPUT Q$
5430  IF Q$ = "Y" THEN 7600
5440  IF Q$ <  > "N" THEN 5420
5450 PRINT:PRINT "Do you wish plot of Mass-Haul Diagram":PRINT "(y/n)
";
5460 INPUT Q$:IF Q$="y" THEN 24000
5470 IF Q$ <> "n" THEN 5450
5480 PRINT:PRINT " Plot of longitudinal section (y/n)";:INPUT Q$:IF Q
$="y" THEN 27000
5485 IF Q$ <> "n" THEN 5480
5490  GOTO 9500
5493  REM  ****************************************
5495  REM  PRISMOIDAL RULE VOLS PC(I) AND PF(I)
5500 MH = 0: FOR I = 0 TO T - 2 STEP 2
5510 MC(I) = PC(I):MF(I) = PF(I)
5520 MH = MH + MC(I) - MF(I)
5530 MH(I) = MH
5540  NEXT I:MC = PC:MF = PF
5550  PRINT : PRINT "CHAIN. CUT VOL.  FILL VOL.  AGG.VOL."
5560  PRINT "======================================="
5570  FOR J = 0 TO T - 2 STEP 2
5580  PRINT L0 + J * L
5590  PRINT  TAB( 8);MC(J); TAB( 18);MF(J); TAB( 29);MH(J)
5600  NEXT J
5610  PRINT L0 + T * L
5620  PRINT "======================================="
5630  PRINT "TOTAL"; TAB( 8);MC; TAB( 18);MF; TAB( 29);MH
5640  PRINT "***************************************"
5650  PRINT : PRINT P$;: INPUT Q$
5660  IF Q$ = "Y" THEN 7600
5670  IF Q$ <  > "N" THEN 5650
5675 PRINT:PRINT "Do you wish plot of Mass-Haul Diagram":PRINT "(y/n)
";
5680 INPUT Q$:IF Q$="y" THEN 24000
5685 IF Q$ <> "n" THEN 5675
5690  GOTO 5200
5985  REM  ****************************************
5990  REM  PRINTOUT OF LOCATION, ETC
6000 LPRINT "LOCATION OF EARTHWORKS :- ";A$
6010 LPRINT "*************************************************************
************"
6020 LPRINT "OPERATOR'S NAME      :- ";B$
6030 LPRINT "DATE OF CALCULATIONS :- ";C$
6060 LPRINT "*************************************************************
************"
6070 LPRINT
6080  GOTO 700
6090 :
6093  REM  ****************************************
6095  REM  PRINTOUT FOR CROSS-SECTIONS A1-A4
6100 LPRINT "PC-EARTHWORKS PROGRAM :-"
6120 LPRINT X$
6130 LPRINT "****************************************"
6140 LPRINT "CROSS-SECTION NO. ";I+1
6150 LPRINT "CHAINAGE          = ";L0+I*L
6160 LPRINT "========================================"
6170 LPRINT "FORMATION WIDTH = ";B
6180 IF F$="C" THEN LPRINT "FORMATION DEPTH  = ";H
6190 IF F$="F" THEN LPRINT "FORMATION HEIGHT = ";H
6200  IF A0 = 3 THEN 6290
6210 LPRINT "SIDE SLOPES 1:";M
6220  IF A0 = 1 THEN 6340
6230  IF A0 = 4 THEN 6260
6240 LPRINT "CROSSFALL    1:";K:IF R<>0 THEN LPRINT "RADIUS OF CURVATU
RE = ";R
```

```
6250  GOTO 6340
6260 LPRINT "CROSSFALL-LHS 1:";K;" ";SL$
6270 LPRINT "CROSSFALL-RHS 1:";K2;" ";SR$
6280  GOTO 6340
6290 LPRINT "CUT  SIDE SLOPE 1:";M2
6300 LPRINT "FILL SIDE SLOPE 1:";M
6310 LPRINT "CROSSFALL       1:";K:IF R<>0 THEN LPRINT "RADIUS OF CUR
VATURE = ";R
6320 LPRINT "CUT  WIDTH = ";W1;" : AREA = ";C(I)
6330 LPRINT "FILL WIDTH = ";W2;" : AREA = ";F(I):GOTO 6370
6340 LPRINT "WIDTH W1 = ";W1;" : WIDTH W2 = ";W2
6350 IF F$="C" THEN LPRINT "AREA OF CUT = ";C(I)
6360 IF F$="F" THEN LPRINT "AREA OF FILL = ";F(I)
6370 LPRINT "***************************************"
6380 LPRINT
6390 I=I+1:IF I>N-1 THEN 960
6392 GOTO 840
6393  REM  ***************************************
6395  REM  PRINTOUT FOR CROSS-SECTIONS A5-A6
6400 LPRINT "PC-EARTHWORKS PROGRAM :-"
6410 LPRINT X$
6420 LPRINT "***************************************"
6430 LPRINT "CROSS-SECTION NO. ";I+1
6440 LPRINT "CHAINAGE           = ";L0+I*L
6450 LPRINT "======================================="
6460 LPRINT "FORMATION WIDTH = ";B
6470 LPRINT :LPRINT "PT. NO.  OFFSET    HEIGHT"
6480 LPRINT "======================================="
6490  FOR J = 0 TO P - 1
6495 X(J)=INT(X(J)*1000+.5)/1000:Y(J)=INT(Y(J)*1000+.5)/1000
6500 LPRINT J+1;TAB(10);X(J);TAB(20);Y(J):NEXT J
6510  IF A0 = 6 AND F$ = "C" THEN 6550
6515  IF A0 = 6 AND F$ = "F" THEN 6560
6520 LPRINT:IF F$="C" THEN LPRINT "AREA OF CUT = ";C(I)
6530 IF F$="F" THEN LPRINT "AREA OF FILL = ";F(I)
6540  GOTO 6570
6550 LPRINT:LPRINT "AREA OF CUT  = ";C(I):GOTO 6570
6560 LPRINT:LPRINT "AREA OF FILL = ";F(I)
6570 LPRINT "***************************************"
6580 LPRINT
6585  IF A6 = 1 THEN 2590
6590 I=I+1:IF I>N-1 THEN 960
6592 GOTO 840
6593  REM  ***************************************
6595  REM  PRINTOUT SUMMARY OF AREAS
6600 LPRINT "PC-EARTHWORKS PROGRAM :-"
6620 LPRINT "***************************************"
6630 LPRINT "SUMMARY OF CROSS-SECTIONAL AREAS"
6640 LPRINT:LPRINT "CRXS.    CHAINAGE  CUT AREA  FILL AREA"
6650 LPRINT "======================================="
6660  FOR I = 0 TO N - 1
6670 LPRINT I+1;TAB(10);L0+I*L;TAB(20);C(I);TAB(30);F(I)
6680  NEXT I
6690 LPRINT "***************************************"
6700 LPRINT
6710 IF U=1 THEN 2800
6720 IF U=2 THEN 3000
6793  REM  ***************************************
6795  REM  PRINTOUT OF <V1> RESULTS - TOTAL VOLUME
6800 LPRINT "PC-EARTHWORKS PROGRAM - VOLUMES"
6820 LPRINT "***************************************"
6830 LPRINT V$
6840 LPRINT:LPRINT "TOTAL VOLUME OF CUT  = ";VC
6850 LPRINT "TOTAL VOLUME OF FILL = ";VF
6860 LPRINT "***************************************"
```

```
6870 LPRINT
6880  GOTO 3000
6993  REM  ***************************************
6995  REM  PRINTOUT OF <V2> RESULTS - TOTAL VOLUMES
7000 LPRINT "PC-EARTHWORKS PROGRAM - VOLUMES"
7020 LPRINT "****************************************"
7030 LPRINT V$
7040 LPRINT:LPRINT "CRXS.    CHAINAGE  CUT VOL.  FILL VOL."
7050 LPRINT "========================================"
7060  FOR J = 0 TO N - 2
7070 LPRINT J+1;TAB(10);L0+J*L
7080 LPRINT TAB(20);VC(J);TAB(30);VF(J)
7090  NEXT J
7100 LPRINT N;TAB(10);L0+(N-1)*L
7110 LPRINT "========================================"
7120 LPRINT "TOTAL VOLUMES = ";TAB(20);TC;TAB(30);TF
7130 LPRINT "****************************************"
7140 LPRINT
7150  GOTO 3000
7193  REM  ***************************************
7195  REM  PRINTOUT OF <V3> RESULTS - TOTAL VOLUMES
7200 LPRINT "PC-EARTHWORKS PROGRAM - VOLUMES"
7220 LPRINT "****************************************"
7230 LPRINT V$
7240 LPRINT:LPRINT "CRXS.    CHAINAGE  CUT VOL.  FILL VOL."
7250 LPRINT "========================================"
7260  FOR J = 0 TO T - 2 STEP 2
7270 LPRINT J+1;TAB(10);L0+J*L
7280 LPRINT TAB(20);PC(J);TAB(30);PF(J)
7290  NEXT J
7300 LPRINT T+1;TAB(10);L0+T*L
7310 LPRINT "========================================"
7320 LPRINT "TOTAL VOLUMES = ";TAB(20);PC;TAB(30);PF
7330 LPRINT "****************************************"
7340 LPRINT
7350  GOTO 3000
7393  REM  ***************************************
7395  REM  SUMMARY PRINTOUT OF VOLUMES
7400 LPRINT M$
7420 LPRINT "****************************************"
7430 LPRINT "SUMMARY OF CUT/FILL VOLUMES"
7440 LPRINT:LPRINT "CRXS.    CHAINAGE  CUT VOL.  FILL VOL."
7450 LPRINT "========================================"
7460  FOR J = 0 TO N - 2
7470 LPRINT J+1;TAB(10);L0+J*L
7475 MC(J)=INT(MC(J)*1000+.5)/1000:MF(J)=INT(MF(J)*1000+.5)/1000
7480 LPRINT TAB(20);MC(J);TAB(30);MF(J)
7490  NEXT J
7500 LPRINT N;TAB(10);L0+(N-1)*L
7510 LPRINT "========================================"
7520 LPRINT "TOTAL VOLUMES = ";TAB(20);MC;TAB(30);MF
7530 LPRINT "****************************************"
7540 LPRINT
7550  GOTO 5200
7593  REM  ***************************************
7595  REM  MASS-HAUL PRINTOUT FOR BOTH END AREAS AND PRISMOIDAL
7600 LPRINT M$
7620 LPRINT "****************************************"
7630 LPRINT:LPRINT "CHAIN. CUT VOL.  FILL VOL.  AGG.VOL."
7640 LPRINT "========================================"
7650  IF H$ = "P" THEN 7680
7655  REM  I.E. H$="E" OR DIRECT
7660 T0 = N - 1
7670 FOR J=0 TO N-2
7672 LPRINT L0+J*L
```

```
7674 LPRINT TAB(8);MC(J);TAB(18);MF(J);TAB(30);MH(J)
7676 NEXT J
7678 GOTO 7730
7680 T0 = T
7690  FOR J = 0 TO T - 2 STEP 2
7700 LPRINT L0+J*L
7710 LPRINT TAB(8);MC(J);TAB(18);MF(J);TAB(30);MH(J)
7720  NEXT J
7730 LPRINT L0+T0*L
7740 LPRINT "========================================"
7750 LPRINT "TOTAL";TAB(8);MC;TAB(18);MF;TAB(30);MH
7760 LPRINT "****************************************"
7770 LPRINT
7780  IF H$="E" THEN 5450
7790  IF H$="P" THEN 5675
7800 H$ = "E":GOTO 5450
8985  REM  ****************************************
8990  REM  GOSUB ROUTINE TO CLEAR SCREEN
9000  CLS:T$="PC-Graphics"
9005 IF A$ <> " " THEN T$=A$
9010 LOCATE 1,2:PRINT T$
9020 LOCATE 1,30:PRINT DATE$
9030 LOCATE 2,2:PRINT "PC-Earthworks"
9040 LOCATE 2,30:PRINT TIME$
9050 PRINT "======================================="
9060 PRINT
9070  RETURN
9093  REM  ***********************************************
9095  REM  GOSUB ROUTINE FOR ERRORS
9100 BEEP
9110 PRINT:PRINT E0$
9120  PRINT
9130  RETURN
9493  REM  ***********************************************
9495  REM  END OF PROGRAM
9500  GOSUB 9000
9510  PRINT
9520  PRINT "  END EARTHWORKS PROGRAM"
9530  PRINT
9540  PRINT "**************************"
9542  FOR I = 1 TO 500
9544  NEXT I
9546  REM RETURN TO HELLO MENU
9548 STOP
9550  REM  PROGRAM PREPARED BY
9560  REM  DR. P.H. MILNE
9610  REM  ***************************
9620  END
11955 REM ***** Digitiser Input *****
12000 GOSUB 9000
12010 PRINT:PRINT " Place drawing on Digitiser Tablet"
12020 GOSUB 50500
12030 PRINT:PRINT " Click cursor on LHS of formation width"
12040 FOR K=1 TO 3
12050 INPUT#5,BYTE(K)
12060 NEXT K
12070 FWLX=BYTE(1):FWLY=BYTE(2):CLOSE#5
12080 GOSUB 50500
12090 PRINT " Click cursor on RHS of formation width"
12100 FOR K=1 TO 3
12110 INPUT#5,BYTE(K)
12120 NEXT K
12130 FWRX=BYTE(1):FWRY=BYTE(2):CLOSE#5
12140 MGD=B/(FWRX-FWLX)
12150 FWCX=(FWLX+FWRX)/2
```

```
12160 FWCY=(FWLY+FWRY)/2
12170 PRINT
12200 FOR J=Ø TO P-1
12210 GOSUB 50500
12220 PRINT " Click cursor button on Pt.";J+1
12230 FOR K=1 TO 3
12240 INPUT#5,BYTE(K)
12250 NEXT K
12260 XD(J)=BYTE(1):YD(J)=BYTE(2):CLOSE#5
12270 X(J)=(XD(J)-FWCX)*MGD
12280 Y(J)=(YD(J)-FWCY)*MGD
12290 NEXT J
12300 GOTO 2200
19995 REM ***** Plot Output AØ *****
20000 GOSUB 9000
20010 PRINT:PRINT " Select Display - ";TAB(20);"<1> Graphics Screen"
20020 PRINT:PRINT TAB(20);"<2> Plotter"
20030 PRINT:PRINT TAB(20);"<3> Return  ";
20040 INPUT SCR%
20050 IF SCR% <1 OR SCR% >3 THEN 20000
20060 ON SCR% GOTO 20100,22000,20090
20090 ON AØ GOTO 1165,1425,1675,1965,2280,2280
20095 REM ***** Screen Plot A1 *****
20100 PRINT:PRINT " Select Maximum Offset ";:INPUT MAXOFF
20110 GX=5*SR%:GY=25:SX=240*SR%:SY=150:XD=Ø:YD=Ø:MG=SX/(2*MAXOFF)
20120 GOSUB 9000:WINDOW (Ø,Ø) - (320*SR%,200)
20130 IF SR%=1 THEN RGB%=2 ELSE RGB%=3
20140 PX=GX+SX:PY=GY+SY:XCEN=GX+SX/2:YCEN=GY+SY/2
20150 LINE (GX,GY) - (PX,PY),RGB%,B
20160 IF SR%=1 THEN 20180
20170 LINE (GX-1,GY) - (PX-1,PY),RGB%,B
20180 LINE (XCEN,GY)-(XCEN,PY),RGB%
20190 ON AØ GOTO 20200,20500,20700,21000,21200,21200
20195 REM ***** Screen Details A1 *****
20200 P=6
20210 IF F$="C" THEN SCF=1
20220 IF F$="F" THEN SCF=-1
20230 X(Ø)=Ø:Y(Ø)=Ø
20240 X(1)=(-(B/2)):Y(1)=Ø
20250 X(2)=(-W):Y(2)=(H*SCF)
20260 X(3)=Ø:Y(3)=(H*SCF)
20270 X(4)=(W):Y(4)=(H*SCF)
20280 X(5)=(B/2):Y(5)=Ø
20290 IF SCR% = 2 THEN 22200
20295 REM ***** Screen Plot AØ *****
20300 FOR J=Ø TO P-1
20310 XP(J)=XCEN+(X(J)*MG):YP(J)=YCEN+(Y(J)*MG)
20320 IF XD=Ø THEN XD=XP(J)*SR%
20330 IF YD=Ø THEN YD=YP(J)
20340 PSET (XD,YD)
20350 LINE -(XP(J)*SR%,YP(J))
20360 XD=XP(J)*SR%:YD=YP(J)
20370 NEXT J
20380 LINE (XD,YD)-(XP(Ø)*SR%,YP(Ø))
20390 LOCATE 5,32*SR%:PRINT "Cross"
20395 LOCATE 6,32*SR%:PRINT "Section"
20400 LOCATE 7,32*SR%:PRINT "No.";I+1
20410 LOCATE 9,32*SR%:PRINT "Ch.";LØ+I*L
20420 LOCATE 12,32*SR%:PRINT "Screen"
20430 LOCATE 13,32*SR%:PRINT "Dump(y/n)"
20440 LOCATE 14,32*SR%:PRINT "         "
20450 BEEP:LOCATE 14,32*SR%:INPUT GD$
20460 IF GD$="n" THEN 20490
20470 IF GD$<>"y" THEN 20440
20480 LCOPY
```

```
20490 GOTO 20000
20495 REM ***** Screen Details A2 *****
20500 P=6
20510 IF F$="C" THEN SCF=1
20520 IF F$="F" THEN SCF=-1
20530 X(0)=0:Y(0)=0
20540 X(1)=(-(B/2)):Y(1)=0
20550 X(3)=0:Y(3)=(H*SCF)
20560 X(5)=(B/2):Y(5)=0
20570 HW1=W1/ABS(K):HW2=W2/ABS(K)
20580 IF SL$="N" THEN 20640
20590 IF F$="C" THEN X(2)=-W2 ELSE X(2)=-W1
20600 IF F$="C" THEN X(4)=W1 ELSE X(4)=W2
20610 IF F$="C" THEN Y(2)=Y(3)-HW2 ELSE Y(2)=Y(3)-HW1
20620 IF F$="C" THEN Y(4)=Y(3)+HW1 ELSE Y(4)=Y(3)+HW2
20630 GOTO 20680
20640 IF F$="C" THEN X(2)=-W1 ELSE X(2)=-W2
20650 IF F$="C" THEN X(4)=W2 ELSE X(4)=W1
20660 IF F$="C" THEN Y(2)=Y(3)+HW1 ELSE Y(2)=Y(3)+HW2
20670 IF F$="C" THEN Y(4)=Y(3)-HW2 ELSE Y(4)=Y(3)-HW1
20680 IF SCR% = 2 THEN 22200
20690 GOTO 20300
20695 REM ***** Screen Details A3 *****
20700 P=6
20710 IF F$="C" THEN SCF=1
20720 IF F$="F" THEN SCF=+1
20730 X(0)=0:Y(0)=0
20740 X(1)=(-(B/2)):Y(1)=0
20750 X(3)=0:Y(3)=(H*SCF)
20760 X(5)=(B/2):Y(5)=0
20770 HW1=W1/ABS(K):HW2=W2/ABS(K)
20780 IF SL$="N" THEN 20840
20790 X(2)=-W2
20800 X(4)=W1
20810 Y(2)=Y(3)-HW2
20820 Y(4)=Y(3)+HW1
20830 GOTO 20880
20840 X(2)=-W1
20850 X(4)=W2
20860 Y(2)=Y(3)+HW1
20870 Y(4)=Y(3)-HW2
20880 IF SCR% = 2 THEN 22200
20890 GOTO 20300
20995 REM ***** Screen Details A4 *****
21000 P=6
21010 IF F$="C" THEN SCF=1
21020 IF F$="F" THEN SCF=-1
21030 X(0)=0:Y(0)=0
21040 X(1)=(-(B/2)):Y(1)=0
21050 X(3)=0:Y(3)=(H*SCF)
21060 X(5)=(B/2):Y(5)=0
21070 HW1=W1/ABS(K):HW2=W2/ABS(K2)
21080 X(2)=-W1
21090 X(4)=W2
21100 IF SL$="P" THEN Y(2)=Y(3)+HW1
21110 IF SL$="N" THEN Y(2)=Y(3)-HW1
21120 IF SR$="P" THEN Y(4)=Y(3)+HW2
21130 IF SR$="N" THEN Y(4)=Y(3)-HW2
21140 IF SCR% = 2 THEN 22200
21150 GOTO 20300
21195 REM ***** Plot Output A5 + A6 *****
21200 FOR J=0 TO P-1
21210 XP(J)=XCEN+(X(J)*MG):YP(J)=YCEN+(Y(J)*MG)
21220 IF XD=0 THEN XD=XP(J)*SR%
21230 IF YD=0 THEN YD=YP(J)
```

```
21240 PSET (XD,YD)
21250 LINE -(XP(J)*SR%,YP(J))
21260 XD=XP(J)*SR%:YD=YP(J)
21270 NEXT J
21280 LINE (XD,YD)-(XP(0)*SR%,YP(0))
21290 GOTO 20390
21995 REM ***** CS Plotter Output *****
22000 GOSUB 9000:IF HP% <> 0 THEN 22090
22010 PRINT:PRINT " Select Plotter";TAB(18);"<1> Apple 410":PRINT:PRI
NT TAB(18);"<2> Epson HI-80"
22020 PRINT:PRINT TAB(18);"<3> Hewlett-Packard":PRINT:PRINT TAB(18);"
<4> Return         ";
22030 INPUT PLOT%
22040 IF PLOT% < 1 OR PLOT% > 4 THEN 22000
22050 IF PLOT% = 4 THEN 20000
22060 IF PLOT% = 3 THEN HP% = 1:REM HP-GL
22070 IF PLOT% = 2 THEN HP% = 4:REM Epson
22080 IF PLOT% = 1 THEN HP% = 3:REM Apple
22090 PRINT:PRINT " Select Maximum Offset ";:INPUT MAXOFF
22100 PRINT:PRINT " Plot on new sheet of paper (y/n)";:INPUT Q$:IF Q$
="n" THEN 22110
22102 IF Q$ <> "y" THEN 22100
22105 PP=0
22110 GX=0:GY=0:SX=180:SY=120:XD=0:YD=0:MG=SX/(2*MAXOFF):MX=2670/SX
22140 PX=GX+SX:PY=GY+SY:XCEN=GX+SX/2:YCEN=GY+SY/2
22190 ON A0 GOTO 20200,20500,20700,21000,22200,22200
22195 REM ***** Plotter Routines *****
22200 ON HP% GOTO 22250,22250,22500,22750
22245 REM ***** HP-GL Plotter (HP-7475A) *****
22250 GOSUB 50000
22260 IF PP=1 THEN 22390
22270 GOSUB 51000
22360 PRINT#2,"SP3;"
22370 PRINT#2,"PU;PA";XCEN;",";GY;";"
22380 PRINT#2,"PD;PA";XCEN;",";PY;";":PP=1
22390 PRINT#2,"SP4;"
22400 FOR J=0 TO P-1
22410 XP(J)=XCEN+(X(J)*MG):YP(J)=YCEN+(Y(J)*MG)
22420 IF XD=0 THEN XD=XP(J)
22430 IF YD=0 THEN YD=YP(J)
22440 PRINT#2,"PU;PA";XD;",";YD;";"
22450 PRINT#2,"PD;PA";XP(J);",";YP(J);";"
22460 XD=XP(J):YD=YP(J)
22470 NEXT J
22480 PRINT#2,"PA";XP(0);",";YP(0);";"
22490 GOTO 23000
22495 REM ***** Apple 410 Plotter *****
22500 GOSUB 50000
22510 IF PP=1 THEN 22640
22520 GOSUB 53000
22610 PRINT#2,"PS3;"
22620 PRINT#2,"MA";XCEN;",";GY;";"
22630 PRINT#2,"DA";XCEN;",";PY;";":PP=1
22640 PRINT#2,"PS4;"
22650 FOR J=0 TO P-1
22660 XP(J)=XCEN+(X(J)*MG):YP(J)=YCEN+(Y(J)*MG)
22670 IF XD=0 THEN XD=XP(J)
22680 IF YD=0 THEN YD=YP(J)
22690 PRINT#2,"MA";XD;",";YD;";"
22700 PRINT#2,"DA";XP(J);",";YP(J);";"
22710 XD=XP(J):YD=YP(J)
22720 NEXT J
22730 PRINT#2,"DA";XP(0);",";YP(0);";"
22740 GOTO 23000
22745 REM ***** Epson HI-80 Plotter *****
```

```
22750 GOSUB 50000
22760 GY=5:PY=125:YCEN=65
22765 IF PP=1 THEN 22890
22770 GOSUB 54000
22860 PRINT#2,"SP3"
22870 PRINT#2,"MA";XCEN*MX;",";GY*MX
22880 PRINT#2,"DA";XCEN*MX;",";PY*MX:PP=1
22890 PRINT#2,"SP4"
22900 FOR J=0 TO P-1
22910 XP(J)=(XCEN+(X(J)*MG))*MX:YP(J)=(YCEN+(Y(J)*MG))*MX
22920 IF XD=0 THEN XD=XP(J)
22930 IF YD=0 THEN YD=YP(J)
22940 PRINT#2,"MA";XD;",";YD
22950 PRINT#2,"DA";XP(J);",";YP(J)
22960 XD=XP(J):YD=YP(J)
22970 NEXT J
22980 PRINT#2,"DA";XP(0);",";YP(0)
22990 GOTO 23000
22995 REM ***** Plotter Title *****
23000 ON HP% GOTO 23100,23100,23300,23400
23095 REM ***** HP-GL *****
23100 IF PTXT=1 THEN 23140
23110 PRINT#2,"SP2;":PRINT#2,"SI0.2,0.3;"
23120 PRINT#2,"PU;PA5,-3;"
23130 PRINT#2,"LB";A$;" - Cross-Section No. ";I+1;" - Chainage ";L0+I
*L;" : Date :";C$:PTXT=1
23135 PRINT#2, CHR$(3)
23140 CLOSE#2
23150 GOTO 20000
23295 REM ***** Apple 410 *****
23300 IF PTXT=1 THEN 23340
23310 PRINT#2,"PS2;":PRINT#2,"LS2;"
23320 PRINT#2,"MA5,-3;"
23330 PRINT#2,"PL";A$;" - Cross-Section No. ";I+1;" - Chainage ";L0+I
*L;" : Date :";C$:PTXT=1
23335 PRINT#2,CHR$(13) CHR$(10) CHR$(3)
23340 CLOSE#2
23350 GOTO 20000
23395 REM ***** Epson HI-80 *****
23400 IF PTXT=1 THEN 23440
23410 PRINT#2,"SP2":PRINT#2,"SI30,20"
23420 PRINT#2,"MA";5*MX;",";+2*MX
23430 PRINT#2,"LA";A$;" - Cross-Section No. ";I+1;" - Chainage ";L0+I
*L;" : Date :";C$:PTXT=1
23440 CLOSE#2
23450 GOTO 20000
23995 REM ***** Select Mass-Haul Display *****
24000 GOSUB 9000:IF H$=" " THEN H$ = "E"
24010 PRINT:PRINT " Select Display - ";TAB(20);"<1> Graphics Screen"
24020 PRINT:PRINT TAB(20);"<2> Plotter"
24030 PRINT:PRINT TAB(20);"<3> Return  ";
24040 INPUT SCR%
24050 IF SCR% <1 OR SCR% >3 THEN 24000
24060 IF SCR% <> 3 THEN 24100
24070 IF H$ = "E" THEN 5480
24080 IF H$ = "P" THEN 5690
24095 REM ***** Set up Variables *****
24100 IF H$ = "E" THEN NUM = N-1
24110 IF H$ = "E" THEN S=1
24120 IF H$ = "P" THEN NUM = T
24130 IF H$ = "P" THEN S=2
24140 FOR I = 0 TO NUM
24150 CH(I) = L0 + I*L
24160 NEXT I
24200 MHMAX = -30000
```

```
24210 MHMIN = 30000
24220 FOR J=0 TO (NUM-1) STEP S
24230 IF MH(J) > MHMAX THEN MHMAX = MH(J)
24240 IF MH(J) < MHMIN THEN MHMIN = MH(J)
24250 NEXT J
24260 MHRNG = (MHMAX - MHMIN)*1.2
24270 IF SCR% = 1 THEN 24500
24280 IF SCR% = 2 THEN 25000
24500 PRINT
24510 GX=5*SR%:GY=25:SX=240*SR%:SY=150:XD=0:YD=0
24520 GOSUB 9000:WINDOW (0,0) - (320*SR%,200)
24530 IF SR%=1 THEN RGB%=2 ELSE RGB%=3
24540 PX=GX+SX:PY=GY+SY
24550 LINE (GX,GY) - (PX,PY),RGB%,B
24560 IF SR%=1 THEN 24580
24570 LINE (GX-1,GY) - (PX-1,PY),RGB%,B
24580 MHV = SY/MHRNG:MHX = SX/(CH(NUM)-CH(0)):YCEN = GY + (ABS(MHMIN)
*1.2*MHV)
24590 LINE (GX,YCEN) - (PX,YCEN),RGB%
24595 REM ***** Draw Vertical Grid *****
24600 FOR I = 2 TO NUM-2 STEP 2
24610 GRIDX(I) = GX + (I*L*MHX)*SR%
24620 LINE (GRIDX(I),GY) - (GRIDX(I),PY),3
24630 NEXT I
24700 XD=GX:YD=YCEN
24710 FOR K = 0 TO (NUM-1) STEP S
24720 XP(K) = GX + (S*L + K*L)*MHX*SR%
24730 YP(K) = YCEN + MH(K)*MHV
24740 LINE (XD,YD) - (XP(K),YP(K)),1
24750 XD=XP(K):YD=YP(K)
24760 NEXT K
24800 LOCATE 5,32*SR%:PRINT "Mass-"
24810 LOCATE 6,32*SR%:PRINT "Haul"
24820 LOCATE 7,32*SR%:PRINT "Diagram"
24830 LOCATE 10,32*SR%:PRINT "Screen"
24840 LOCATE 11,32*SR%:PRINT "Dump(y/n)"
24850 LOCATE 12,32*SR%:PRINT "         "
24860 BEEP:LOCATE 12,32*SR%:INPUT GD$
24870 IF GD$="n" THEN 24900
24880 IF GD$<>"y" THEN 24850
24890 LCOPY
24900 GOTO 24000
24995 REM ***** M-H Plotter Output *****
25000 GOSUB 9000:IF HP% <> 0 THEN 25100
25010 PRINT:PRINT " Select Plotter";TAB(18);"<1> Apple 410":PRINT:PRI
NT TAB(18);"<2> Epson HI-80"
25020 PRINT:PRINT TAB(18);"<3> Hewlett-Packard":PRINT:PRINT TAB(18);"
<4> Return          ";
25030 INPUT PLOT%
25040 IF PLOT% < 1 OR PLOT% > 4 THEN 25000
25050 IF PLOT% = 4 THEN 24000
25060 IF PLOT% = 3 THEN HP% = 1:REM HP-GL
25070 IF PLOT% = 2 THEN HP% = 4:REM Epson
25080 IF PLOT% = 1 THEN HP% = 3:REM Apple
25090 PRINT:PRINT " Place paper on plotter and switch on":PRINT:PRINT
 " Ready to plot (y/n) ";:INPUT Q$:IF Q$ = "y" THEN 25110
25095 IF Q$ <> "n" THEN 25090
25100 GOTO 25000
25110 GX=0:GY=0:SX=180:SY=120:XD=0:YD=0:MX=2670/SX
25140 PX=GX+SX:PY=GY+SY
25150 MHV = SY/MHRNG:MHX = SX/(CH(NUM)-CH(0)):YCEN = GY + (ABS(MHMIN)
*1.2*MHV)
25200 ON HP% GOTO 25500,25500,26000,26500
25495 REM ***** HP-GL Plotter (HP-7475A) *****
25500 GOSUB 50000
```

```
25510 GOSUB 51000
25610 PRINT#2,"SP3;"
25620 PRINT#2,"PU;PA";GX;",";YCEN;";"
25630 PRINT#2,"PD;PA";PX;",";YCEN;";"
25640 PRINT#2,"SP4;"
25650 FOR I = 2 TO NUM-2 STEP 2
25660 GRIDX(I) = GX + (I*L*MHX)
25670 PRINT#2,"PU;PA";GRIDX(I);",";GY;";"
25680 PRINT#2,"PD;PA";GRIDX(I);",";PY;";"
25690 NEXT I
25700 PRINT#2,"SP2;"
25710 XD=GX:YD=YCEN
25720 FOR K = 0 TO (NUM-1) STEP S
25730 XP(K) = GX + (S*L + K*L)*MHX
25740 YP(K) = YCEN + MH(K)*MHV
25750 PRINT#2,"PU;PA";XD;",";YD;";"
25760 PRINT#2,"PD;PA";XP(K);",";YP(K);";"
25770 XD=XP(K):YD=YP(K)
25780 NEXT K
25800 PRINT#2,"SI0.15,0.22;"
25810 FOR J=0 TO (NUM-2) STEP 2
25820 PRINT#2,"PU;PA";GRIDX(J);",";(YCEN-2);";"
25830 PRINT#2,"LB";CH(J)
25840 PRINT#2, CHR$(3)
25850 NEXT J
25900 PRINT#2,"SI0.2,0.3;"
25910 PRINT#2,"PA5,-3;"
25920 PRINT#2,"LB";A$;" - Mass-Haul Diagram : Date :";C$
25930 PRINT#2, CHR$(3)
25940 CLOSE#2
25950 GOTO 24000
25995 REM ***** Apple 410 Plotter *****
26000 GOSUB 50000
26010 GOSUB 53000
26110 PRINT#2,"PS3;"
26120 PRINT#2,"MA";GX;",";YCEN;";"
26130 PRINT#2,"DA";PX;",";YCEN;";"
26140 PRINT#2,"PS4;"
26150 FOR I = 2 TO NUM-2 STEP 2
26160 GRIDX(I) = GX + (I*L*MHX)
26170 PRINT#2,"MA";GRIDX(I);",";GY;";"
26180 PRINT#2,"DA";GRIDX(I);",";PY;";"
26190 NEXT I
26200 PRINT#2,"PS2;"
26210 XD=GX:YD=YCEN
26220 FOR K = 0 TO (NUM-1) STEP S
26230 XP(K) = GX + (S*L + K*L)*MHX
26240 YP(K) = YCEN + MH(K)*MHV
26250 PRINT#2,"MA";XD;",";YD;";"
26260 PRINT#2,"DA";XP(K);",";YP(K);";"
26270 XD=XP(K):YD=YP(K)
26280 NEXT K
26300 PRINT#2,"LS1;"
26310 FOR J=0 TO (NUM-2) STEP 2
26320 PRINT#2,"MA";GRIDX(J);",";(GY+3);";"
26330 PRINT#2,"PL";CH(J)
26340 PRINT#2,CHR$(13) CHR$(10) CHR$(3)
26350 NEXT J
26400 PRINT#2,"LS2;"
26410 PRINT#2,"MA5,-3;"
26420 PRINT#2,"PL";A$;" - Mass-Haul Diagram : Date :";C$
26430 PRINT#2,CHR$(13) CHR$(10) CHR$(3)
26440 CLOSE#2
26450 GOTO 24000
26495 REM ***** Epson HI-80 Plotter *****
```

```
26500 GOSUB 50000
26510 GY=5:PY=125
26520 GOSUB 54000
26610 PRINT#2,"SP3"
26620 PRINT#2,"MA";GX*MX;",";YCEN*MX
26630 PRINT#2,"DA";PX*MX;",";YCEN*MX
26640 PRINT#2,"SP4"
26650 FOR I = 2 TO NUM-2 STEP 2
26660 GRIDX(I) = (GX + (I*L*MHX))*MX
26670 PRINT#2,"MA";GRIDX(I);",";GY*MX
26680 PRINT#2,"DA";GRIDX(I);",";PY*MX
26690 NEXT I
26700 PRINT#2,"SP2"
26710 XD=GX*MX:YD=YCEN*MX
26720 FOR K = 0 TO (NUM-1) STEP S
26730 XP(K) = (GX + (S*L + K*L)*MHX)*MX
26740 YP(K) = (YCEN + MH(K)*MHV)*MX
26750 PRINT#2,"MA";XD;",";YD
26760 PRINT#2,"DA";XP(K);",";YP(K)
26770 XD=XP(K):YD=YP(K)
26780 NEXT K
26800 PRINT#2,"SI25,20"
26810 FOR J=0 TO (NUM-2) STEP 2
26820 PRINT#2,"MA";GRIDX(J);",";(YCEN-2)*MX
26830 PRINT#2,"LA";CH(J)
26850 NEXT J
26900 PRINT#2,"SI30,20"
26910 PRINT#2,"MA"5*MX;",";+2*MX
26920 PRINT#2,"LA";A$;" - Mass-Haul Diagram : Date :";C$
26940 CLOSE#2
26950 GOTO 24000
26995 REM ***** Long. Section Display *****
27000 GOSUB 9000:NUM=N-1:IF U <> 1 THEN 27080
27010 PRINT:PRINT " Select Display - ";TAB(20);"<1> Graphics Screen"
27020 PRINT:PRINT TAB(20);"<2> Plotter"
27030 PRINT:PRINT TAB(20);"<3> Quit     ";
27040 INPUT SCR%
27050 IF SCR% <1 OR SCR% >3 THEN 27000
27060 IF SCR% <> 3 THEN 27100
27070 GOTO 9500
27080 BEEP:PRINT:PRINT " This option only available if cross-    sect
ions and formation depth known"
27085 FOR J=1 TO 1500:NEXT J
27090 GOTO 9500
27100 GOSUB 9000:IF FRML=1 THEN 27270
27110 PRINT:PRINT " Enter formation level at each cross     section -
"
27120 FOR K=0 TO NUM
27130 PRINT:PRINT " C-S No.";(K+1);" - Chainage";L0+K*L
27140 PRINT " Formation Level ";:INPUT FL(K)
27150 IF FL(K)<0 THEN 27130
27160 NEXT K
27170 FRML=1
27200 LSMAX = 0:LSMIN=10000
27210 FOR J=0 TO NUM
27220 CH(J)=L0+J*L
27225 GL(J)=FL(J)+H(J)
27230 IF GL(J) > LSMAX THEN LSMAX = GL(J)
27235 IF GL(J) < LSMIN THEN LSMIN = GL(J)
27240 IF FL(J) > LSMAX THEN LSMAX = FL(J)
27245 IF FL(J) < LSMIN THEN LSMIN = FL(J)
27250 NEXT J
27260 LSRNG = (LSMAX - LSMIN)*2
27270 IF SCR% = 1 THEN 27500
27280 IF SCR% = 2 THEN 28000
```

```
27500 PRINT
27510 GX=5*SR%:GY=25:SX=240*SR%:SY=150:XD=0:YD=0
27520 GOSUB 9000:WINDOW (0,0) - (320*SR%,200)
27530 IF SR%=1 THEN RGB%=2 ELSE RGB%=3
27540 PX=GX+SX:PY=GY+SY
27550 LINE (GX,GY) - (PX,PY),RGB%,B
27560 IF SR%=1 THEN 27580
27570 LINE (GX-1,GY) - (PX-1,PY),RGB%,B
27580 LSV=SY/LSRNG:LSX=SX/(CH(NUM)-CH(0))
27590 IF LSV > 2.4 THEN LSV=2.4
27600 FOR I = 2 TO NUM-2 STEP 2
27610 GRIDX(I) = GX + (I*L*LSX)*SR%
27620 LINE (GRIDX(I),GY) - (GRIDX(I),PY),3
27630 NEXT I
27695 REM ***** Draw formation level *****
27700 XD=GX:YD=GY+50+(FL(0)-LSMIN)*LSV
27710 FOR K = 0 TO NUM
27720 XF(K) = GX + (K*L*LSX*SR%)
27730 YF(K) = GY+50+(FL(K)-LSMIN)*LSV
27740 LINE (XD,YD) - (XF(K),YF(K)),RGB%
27750 XD=XF(K):YD=YF(K)
27760 NEXT K
27795 REM ***** Draw ground level *****
27800 XD=GX:YD=GY+50+(GL(0)-LSMIN)*LSV
27810 FOR J = 0 TO NUM
27820 XP(J) = GX + (J*L*LSX*SR%)
27830 YP(J) = GY+50+(GL(J)-LSMIN)*LSV
27840 LINE (XD,YD) - (XP(J),YP(J)),1
27850 XD=XP(J):YD=YP(J)
27860 NEXT J
27900 LOCATE 5,32*SR%:PRINT "Long."
27910 LOCATE 6,32*SR%:PRINT "Section"
27930 LOCATE 10,32*SR%:PRINT "Screen"
27940 LOCATE 11,32*SR%:PRINT "Dump(y/n)"
27950 LOCATE 12,32*SR%:PRINT "          "
27960 BEEP:LOCATE 12,32*SR%:INPUT GD$
27970 IF GD$="n" THEN 27000
27980 IF GD$<>"y" THEN 27950
27985 LCOPY
27990 GOTO 27000
27995 REM ***** Plotter Output for L-S *****
28000 GOSUB 9000:IF HP% <> 0 THEN 28110
28010 PRINT:PRINT " Select Plotter";TAB(18);"<1> Apple 410":PRINT:PRI
NT TAB(18);"<2> Epson HI-80"
28020 PRINT:PRINT TAB(18);"<3> Hewlett-Packard":PRINT:PRINT TAB(18);"
<4> Return         ";
28030 INPUT PLOT%
28040 IF PLOT% < 1 OR PLOT% > 4 THEN 28000
28050 IF PLOT% = 4 THEN 27000
28060 IF PLOT% = 3 THEN HP% = 1:REM HP-GL
28070 IF PLOT% = 2 THEN HP% = 4:REM Epson
28080 IF PLOT% = 1 THEN HP% = 3:REM Apple
28090 PRINT:PRINT " Place paper on plotter and switch on":PRINT:PRINT
 " Ready to plot (y/n) ";:INPUT Q$:IF Q$ = "y" THEN 28110
28095 IF Q$ <> "n" THEN 28090
28100 GOTO 28000
28110 GX=0:GY=0:SX=180:SY=120:XD=0:YD=0:MX=2670/SX
28140 PX=GX+SX:PY=GY+SY
28150 LSV = SY/LSRNG:LSX = SX/(CH(NUM)-CH(0))
28160 IF LSV > 2.4 THEN LSV=2.4
28200 ON HP% GOTO 28500,28500,29000,29500
28495 REM ***** HP-GL Plotter (HP-7475A) *****
28500 GOSUB 50000
28520 GOSUB 51000
28530 PRINT#2,"SP4;"
```

```
28540 FOR I = 2 TO NUM-2 STEP 2
28550 GRIDX(I) = GX + (I*L*LSX)
28560 PRINT#2,"PU;PA";GRIDX(I);",";GY;";"
28570 PRINT#2,"PD;PA";GRIDX(I);",";PY;";"
28580 NEXT I
28590 PRINT#2,"SP2;"
28600 XD=GX:YD=GY+50+(FL(0)-LSMIN)*LSV
28610 FOR K = 0 TO NUM
28620 XF(K) = GX + (K*L*LSX)
28630 YF(K) = GY+50+(FL(K)-LSMIN)*LSV
28650 PRINT#2,"PU;PA";XD;",";YD;";"
28660 PRINT#2,"PD;PA";XF(K);",";YF(K);";"
28670 XD=XF(K):YD=YF(K)
28680 NEXT K
28690 PRINT#2,"SP3;"
28700 XD=GX:YD=GY+50+(GL(0)-LSMIN)*LSV
28710 FOR J = 0 TO NUM
28720 XP(J) = GX + (J*L*LSX)
28730 YP(J) = GY+50+(GL(J)-LSMIN)*LSV
28750 PRINT#2,"PU;PA";XD;",";YD;";"
28760 PRINT#2,"PD;PA";XP(J);",";YP(J);";"
28770 XD=XP(J):YD=YP(J)
28780 NEXT J
28790 PRINT#2,"SP2;"
28800 PRINT#2,"SI0.15,0.22;"
28810 FOR J=0 TO (NUM-2) STEP 2
28820 PRINT#2,"PU;PA";GRIDX(J);",";(GY+3);";"
28830 PRINT#2,"LB";CH(J)
28840 PRINT#2, CHR$(3)
28850 NEXT J
28900 PRINT#2,"SI0.2,0.3;"
28910 PRINT#2,"PA5,-3;"
28920 PRINT#2,"LB";A$;" - Longitudinal Section : Date :";C$
28930 PRINT#2, CHR$(3)
28940 CLOSE#2
28950 GOTO 27000
28995 REM ***** Apple 410 Plotter *****
29000 GOSUB 50000
29020 GOSUB 53000
29030 PRINT#2,"PS4;"
29040 FOR I = 2 TO NUM-2 STEP 2
29050 GRIDX(I) = GX + (I*L*LSX)
29060 PRINT#2,"MA";GRIDX(I);",";GY;";"
29070 PRINT#2,"DA";GRIDX(I);",";PY;";"
29080 NEXT I
29090 PRINT#2,"PS2;"
29100 XD=GX:YD=GY+50+(FL(0)-LSMIN)*LSV
29110 FOR K = 0 TO NUM
29120 XF(K) = GX + (K*L*LSX)
29130 YF(K) = GY+50+(FL(K)-LSMIN)*LSV
29150 PRINT#2,"MA";XD;",";YD;";"
29160 PRINT#2,"DA";XF(K);",";YF(K);";"
29170 XD=XF(K):YD=YF(K)
29180 NEXT K
29190 PRINT#2,"PS3;"
29200 XD=GX:YD=GY+50+(GL(0)-LSMIN)*LSV
29210 FOR J = 0 TO NUM
29220 XP(J) = GX + (J*L*LSX)
29230 YP(J) = GY+50+(GL(J)-LSMIN)*LSV
29250 PRINT#2,"MA";XD;",";YD;";"
29260 PRINT#2,"DA";XP(J);",";YP(J);";"
29270 XD=XP(J):YD=YP(J)
29280 NEXT J
29290 PRINT#2,"PS2;"
29300 PRINT#2,"LS1;"
```

```
29310 FOR J=0 TO (NUM-2) STEP 2
29320 PRINT#2,"MA";GRIDX(J);",";(GY+3);";"
29330 PRINT#2,"PL";CH(J)
29340 PRINT#2,CHR$(13) CHR$(10) CHR$(3)
29350 NEXT J
29400 PRINT#2,"LS2;"
29410 PRINT#2,"MA5,-3;"
29420 PRINT#2,"PL";A$;" - Longitudinal Section : Date :";C$
29430 PRINT#2,CHR$(13) CHR$(10) CHR$(3)
29440 CLOSE#2
29450 GOTO 27000
29495 REM ***** Epson HI-80 Plotter *****
29500 GOSUB 50000
29510 GY=5:PY=125
29520 GOSUB 54000
29530 PRINT#2,"SP4"
29540 FOR I = 2 TO NUM-2 STEP 2
29550 GRIDX(I) = (GX + (I*L*LSX))*MX
29560 PRINT#2,"MA";GRIDX(I);",";GY*MX
29570 PRINT#2,"DA";GRIDX(I);",";PY*MX
29580 NEXT I
29590 PRINT#2,"SP2"
29600 XD=GX:YD=(GY+50+(FL(0)-LSMIN)*LSV)*MX
29610 FOR K = 0 TO NUM
29620 XF(K) = (GX + (K*L*LSX))*MX
29630 YF(K) = (GY+50+(FL(K)-LSMIN)*LSV)*MX
29650 PRINT#2,"MA";XD;",";YD
29660 PRINT#2,"DA";XF(K);",";YF(K)
29670 XD=XF(K):YD=YF(K)
29680 NEXT K
29690 PRINT#2,"SP3"
29700 XD=GX:YD=(GY+50+(GL(0)-LSMIN)*LSV)*MX
29710 FOR J = 0 TO NUM
29720 XP(J) = (GX + (J*L*LSX))*MX
29730 YP(J) = (GY+50+(GL(J)-LSMIN)*LSV)*MX
29750 PRINT#2,"MA";XD;",";YD
29760 PRINT#2,"DA";XP(J);",";YP(J)
29770 XD=XP(J):YD=YP(J)
29780 NEXT J
29790 PRINT#2,"SP2"
29800 PRINT#2,"SI25,20"
29810 FOR J=0 TO (NUM-2) STEP 2
29820 PRINT#2,"MA";GRIDX(J);",";(GY+3)*MX
29830 PRINT#2,"LA";CH(J)
29850 NEXT J
29900 PRINT#2,"SI30,20"
29910 PRINT#2,"MA"5*MX;",";+2*MX
29920 PRINT#2,"LA";A$;" - Longitudinal Section : Date :";C$
29940 CLOSE#2
29950 GOTO 27000
```

2.3.5 PC-Earthworks – computer printouts

Cross-section details for Figs 2.3–2.6

```
LOCATION OF EARTHWORKS :- Townhead
********************************************************************
OPERATOR'S NAME        :- A.N. Other
DATE OF CALCULATIONS :- Ø1/Ø1/87
********************************************************************

PC-EARTHWORKS PROGRAM :-
<A1> ONE-LEVEL
****************************************
CROSS-SECTION NO.  1
CHAINAGE           =  1ØØ
========================================
FORMATION WIDTH =  12
FORMATION DEPTH  =  3
SIDE SLOPES 1: 3
WIDTH W1 =  15  : WIDTH W2 =  15
AREA OF CUT =  63
****************************************

PC-EARTHWORKS PROGRAM :-
<A2> TWO-LEVEL
****************************************
CROSS-SECTION NO.  2
CHAINAGE           =  13Ø
========================================
FORMATION WIDTH =  12
FORMATION DEPTH  =  1.5
SIDE SLOPES 1: 3
CROSSFALL   1: 2Ø
WIDTH W1 =  12.353  : WIDTH W2 =  9.13
AREA OF CUT =  25.596
****************************************

PC-EARTHWORKS PROGRAM :-
<A3> TWO-LEVEL (CUT ++ FILL)
****************************************
CROSS-SECTION NO.  3
CHAINAGE           =  16Ø
========================================
FORMATION WIDTH =  12
FORMATION HEIGHT = -.5
CUT  SIDE SLOPE 1: 2
FILL SIDE SLOPE 1: 2
CROSSFALL        1: 3
CUT  WIDTH =  15  : AREA =  1Ø.125
FILL WIDTH =  21  : AREA =  28.125
****************************************

PC-EARTHWORKS PROGRAM :-
<A4> THREE-LEVEL
****************************************
CROSS-SECTION NO.  4
CHAINAGE           =  19Ø
========================================
FORMATION WIDTH =  12
FORMATION HEIGHT =  .5
SIDE SLOPES 1: 2
CROSSFALL-LHS 1: 12  N
CROSSFALL-RHS 1: 1Ø  N
WIDTH W1 =  8.399999  : WIDTH W2 =  8.75
AREA OF FILL =  12.Ø12
****************************************
```

```
PC-EARTHWORKS PROGRAM :-
***************************************
SUMMARY OF CROSS-SECTIONAL AREAS

CRXS.     CHAINAGE  CUT AREA  FILL AREA
=======================================
 1         1ØØ       63         Ø
 2         13Ø       25.596     Ø
 3         16Ø       1Ø.125     28.125
 4         19Ø       Ø          12.Ø12
***************************************

EARTHWORKS PROGRAM - VOLUMES
***************************************
<V2> END AREAS RULE

CRXS.     CHAINAGE  CUT VOL.  FILL VOL.
=======================================
 1         1ØØ
                     1328.94    Ø
 2         13Ø
                     535.815    421.875
 3         16Ø
                     151.875    6Ø2.Ø55
 4         19Ø
=======================================
TOTAL VOLUMES =      2Ø16.63    1Ø23.93
***************************************
```

Cross-section details for Figs 1.7 and 2.7

```
PC-EARTHWORKS PROGRAM :-
<A5> IRREGULAR
***************************************
CROSS-SECTION NO.  1
CHAINAGE        =  1ØØ
=======================================
FORMATION WIDTH =  6

PT. NO.  OFFSET    HEIGHT
=======================================
 1        -3         Ø
 2        -3          .3
 3        -4          .3
 4        -4         -.3
 5        -4.5       -.3
 6        -4.5       Ø
 7        -1Ø.5      4
 8        -8         5
 9        -5         4.5
 1Ø        4         6
 11        8         5
 12        12        5
 13        4.5       Ø
 14        4.5       -.3
 15        4         -.3
 16        4          .3
 17        3          .3
 18        3         Ø

AREA OF CUT =  83.7
***************************************
```

Cross-section details for Fig. 2.8

```
PC-EARTHWORKS PROGRAM :-
<A6> IRREGULAR (CUT ++ FILL)
****************************************
CROSS-SECTION NO.  1
CHAINAGE          =  600
========================================
FORMATION WIDTH =  12

PT. NO.  OFFSET    HEIGHT
========================================
 1        -2         0
 2         2         3
 3         6         4
 4         13        4.5
 5         6         0

AREA OF CUT  =  34
****************************************

PC-EARTHWORKS PROGRAM :-
<A6> IRREGULAR (CUT ++ FILL)
****************************************
CROSS-SECTION NO.  1
CHAINAGE          =  600
========================================
FORMATION WIDTH =  12

PT. NO.  OFFSET    HEIGHT
========================================
 1        -2         0
 2        -4        -2
 3        -8        -4
 4        -14       -4
 5        -6         0

AREA OF FILL =  22
****************************************
```

Earthwork volume computations for Figs 2.9–2.12

```
LOCATION OF EARTHWORKS :- Renfrew
****************************************************************
OPERATOR'S NAME       :- John Smith
DATE OF CALCULATIONS :- 01/01/87
****************************************************************

PC-EARTHWORKS PROGRAM :-
****************************************
SUMMARY OF CROSS-SECTIONAL AREAS

CRXS.     CHAINAGE  CUT AREA  FILL AREA
========================================
 1         100       0          28.385
 2         200       0          40.747
 3         300       0          28.385
 4         400       0          17.67
 5         500       8.604      0
 6         600       40.747     0
 7         700       70.418     0
 8         800       70.418     0
 9         900       40.747     0
 10        1000      17.67      0
 11        1100      0          8.604
```

```
 12       1200      0         40.747
 13       1300      0         70.418
 14       1400      0         54.758
 15       1500      0         28.385
****************************************

PC-EARTHWORKS PROGRAM - VOLUMES
****************************************
<V3> PRISMOIDAL RULE

CRXS.    CHAINAGE  CUT VOL.  FILL VOL.
========================================
 1        100
                    0         7325.267
 3        300
                    286.8     3302.167
 5        500
                    8067      0
 7        700
                    13094.567
                              0
 9        900
                    3714.233  286.8
 11       1100
                    0         8067
 13       1300
                    0         10594.5
 15       1500
========================================
TOTAL VOLUMES =     25162.6   29575.733
****************************************

PC-EARTHWORKS PROGRAM - MASS-HAUL
****************************************

CHAIN. CUT VOL.  FILL VOL.  AGG.VOL.
========================================
 100
        0         7325.267   -7325.267
 300
        286.8     3302.167   -10340.634
 500
        8067      0          -2273.634
 700
        13094.567
                  0           10820.933
 900
        3714.233  286.8       14248.366
 1100
        0         8067        6181.366
 1300
        0         10594.5    -4413.134
 1500
========================================
TOTAL   25162.6   29575.733  -4413.134
****************************************
```

3

Surface modelling

3.1 INTRODUCTION

Traditionally, the map has been regarded as an invaluable source of information to both the public and industry in general. Historically, the cartographic map sheet was, and still is, a work of art requiring input from modern surveying and stereo instruments. Maps can generally be divided into two groups: (1) planimetric maps, which in plan represent such natural and artificial features as streams, lakes, site boundaries and public works; and (2) topographic maps, which may include some of the above, but essentially represent the relief or contours of the ground.

This chapter deals initially with topographic mapping which is the basis of all other forms of mapping and is not necessarily confined to land surveying. The latter part of this chapter covers surface modelling, often referred to as Digital Ground Modelling (DGM), Digital Terrain Modelling (DTM) or Digital Elevation Modelling (DEM).

Other users, such as market researchers, engineers and scientists often require the two-dimensional representation of three-dimensional data. In a normal three-dimensional system, using conventional survey methods, the digital information is often lost because the map sheet only gives a visual picture. If the recovery of information is required for position and distances or areas, surveys often have to be repeated. However, if an organized computer database is established with a three-dimensional data grid which defines a set of points in three-dimensional (X, Y, Z) space, then this database can be altered and viewed at will. These data points may lie in a regular or irregular (random) grid where the grid data values define the 'height' (Z-coordinate) of each point above or below the X–Y plane. Such topographic or contour mapping programs, for the drawing of isolines along which there is a constant value, have many users as shown in Table 3.1.

Applications include contours in topographic mapping (equal elevations), isohyetals for rainfall analysis, isobars for pressure distribution, isophotes for light intensity, and many others.

In addition to producing conventional two-dimensional contour maps, there is often a wish to display isometric projections (sometimes called mesh or fishnet diagrams) of the spatial data, preferably with hidden line removal to present a smooth uncluttered view.

Table 3.1 Contour applications

User	*Uses*
Architects	Contour plans for land developments, quarries, reservoirs, etc.
Engineers (Civil, Structural, Marine, Environmental, etc.)	Ground contouring, digital ground modelling, stress and noise contours
Geographers	Topographic mapping, contouring, population distribution, etc.
Geologists	Interpolating and contouring the earth's surface and subterranean formations
Geophysicists	Interpolating and contouring the earth's gravitational and magnetic fields, mine modelling etc.
Hydrologists	Run-off and rainfall distribution
Meteorologists	Temperature and barometric pressure studies
Scientists	Temperature and pressure studies, statistical data analysis
Surveyors	Contouring from tacheometric or aerial surveys, digital ground models, topographic mapping, etc.

The programs described in this chapter allow three-dimensional data, either on a regular grid (PC-Grid) or at irregular or random spacing (PC-Random) to be entered and stored on disc for subsequent plotting, either as contours (PC-Contour) or as an isometric projection (PC-Surfplot). In all cases the user has a choice of plotting on the screen or directly to a plotter. In fact it is often best to preview the plot on the graphics screen before dumping the graphics image on the printer or plotter to ensure the correct orientation and magnification. Two demonstration data files are also provided (Glapwell and Glenmore) to check the user's output with that shown in the illustrations.

3.2 PC-GRID

This program is for the creation of a data file from regularly spaced data, i.e. data at the corners of a square or rectangular grid, which can then be accessed by either PC-Contour to draw contours or by PC-Surfplot to display an isometric projection on screen or plotter.

The initial computer listing in this section, PC-Grid/Data is a blank data file for entering your own data and should be saved under the title **"GRIDRDAT"**. The second computer listing is PC-Grid/Demo and includes data for a spoil-heap at Glapwell, and should be saved under the title **"GRIDDEMO"** as it can be used as a demonstration file to check your program and the output to screen or plotter.

3.2.1 PC-Grid/Data – subroutine index

	Line numbers	*Function*
(a)	10–80	Initialization and control
(b)	100–560	Selection of application
(c)	1000–1410	Reading and storing of data
(d)	2000–2140	Presentation of data file information
(e)	3000–8990	Data file space
(f)	9000–9050	Routine to clear screen
(g)	9500–9540	Termination of program

(a) Initialization and control Line numbers 10–80

The program is initialized and the installation file checked for the availability of graphics screens and colour for subsequent display.

(b) Selection of application Line numbers 100–560

The user is given a choice of three applications:

⟨1⟩ Read DATA and present details
⟨2⟩ Read DATA and store on disc
⟨3⟩ Prepare DATA file.

Thus this program saved under the title **"GRIDRDAT"** can be used initially to prepare a data file, then to read the data to check it is spatially correct and finally to store the data on disc for access by PC-Contour and PC-Surfplot.

A note on data file preparation is included on the screen to save the user constantly referring back to the book, and the user is reminded that on completion of entering the data, the file should be saved under another name for reference and subsequent recall. Some programs require data to be entered from screen prompts, but should an error occur, the data often has to be re-entered. To overcome this problem, this program uses **DATA** statements so that the data file can be built up in several keyboard sessions if required.

(c) Reading and storing of data Line numbers 1000–1410

The data file is then read and, depending on the number of rows (**NR**) and number of columns (**NC**) in the grid, an array **Z(NR,NC)** is then dimensioned to store the data values. These are subsequently stored on disc for quick access by PC-Contour and PC-Surfplot. Before running this program, the user should ensure that there is space on disc to store the data file, and that the disc is not 'write-protected'!

(d) Presentation of data file information Line numbers 2000–2140

This routine is provided as a check to allow the user to see if the data information has been entered correctly before saving on disc.

(e) Data file space Line numbers 3000–8990

Line numbers from 3000 to 8990 are available for the insertion of data as suggested in the data file preparation notes at lines 500–560.

(f) Routine to clear screen Line numbers 9000–9050

This routine is used to clear the screen and provide lines of text at the top of the screen giving the title of the software program and the date.

(g) Termination of program Line numbers 9500–9540

This is the final display at the completion of the program, and the user is given an option to run PC-Contour or return to BASIC command level.

3.2.2 PC-Grid/Data – numeric variables

DX	= *X*-distance of grid
DY	= *Y*-distance of grid
GRID%	= Application selection
I	= Integer variable used in loops
J	= Integer variable used in loops
MAX	= Maximum *Z*-value
MIN	= Minimum *Z*-value
NC	= Number of columns in grid
NR	= Number of rows in grid
SCD%	= Monitor display, monochrome/colour.
SCR%	= Screen resolution
Z(NR,NC)	= Array to hold *Z*-values

3.2.3 PG-Grid/Data – string variables

DATE$	= Variable used to return the date of the computer's calendar/clock in US format MM–DD–YYYY
F$	= File name for data
Q$	= Question response (Y/N)
T$	= **"PC-Grid/Data"**

3.2.4 PC-Grid/Data – BASIC program

```
10 REM  <GRIDRDAT> Program for IBM PC and Compatibles
12 REM  Version PC-1.0, 01-01-87
14 REM  (C)  Copyright  1987  P.H. Milne
16 REM  All rights reserved
20 REM  Creates Data File for Access by PC-Contour
25 REM  Enter Name of Data File Here for Reference
30  CLEAR
35 PCF$="PCSCREEN"
40 OPEN "I",#3,PCF$
45 INPUT#3,SCR%
50 INPUT#3,SCD%
55 CLOSE#3
60 ON (SCR%+1) GOTO 62,64,66
62 SCREEN 0:WIDTH 40:GOTO 80
64 SCREEN 1:GOTO 68
66 SCREEN 1:GOTO 80
68 IF SCD%=1 THEN 80
70 COLOR 8,0
80 KEY OFF
100 CLS
110 GOSUB 9000
120 PRINT
130 PRINT "  THIS PROGRAM READS INFORMATION FROM "
140 PRINT "  REGULAR SQUARE GRID POINTS INTO A "
150 PRINT "  DATA FILE FOR PLOTTING BY PC-Contour"
160 PRINT
170 PRINT " Select application required :-"
180 PRINT
190 PRINT " <1> Read DATA and present details"
200 PRINT
210 PRINT " <2> Read DATA and store on disc"
220 PRINT
230 PRINT " <3> Prepare DATA file"
240 PRINT
250 PRINT " <4> Quit ";
260 INPUT GRID%:IF GRID%<1 OR GRID%>4 THEN 100
270 IF GRID%>2 THEN 290
280 PRINT:PRINT:PRINT " One moment ... reading DATA file "
290 ON GRID% GOTO 1030,1030,300,9500
300 CLS
310 GOSUB 9000
320 PRINT " To check on DATA file preparation"
330 PRINT " list lines 500-560 on screen after"
340 PRINT " the Ok prompt."
350 PRINT
370 PRINT " After entering your own data, either"
380 PRINT " RUN or SAVE your program immediately."
390 PRINT:PRINT
400 END
500 REM ***** Data File Preparation Notes *****
510 REM Line 3000 : Data File Name
520 REM Line 3010 : No. of Rows, No. of Columns for grid
530 REM Line 3020 : X-distance, Y-distance
540 REM Line 3030 : Maximum reading, Minimum reading
550 REM LINE 3040- onwards, Z-data values
560 REM To generate new data file delete Lines 3000-8990
995 REM ***** Data File Read Routine *****
1000 CLS:GOSUB 9000
1020 INPUT " Press <Enter> when Data File Ready";R$
1030  READ F$
1040  READ NR,NC
1050  READ DX,DY
```

```
1060  READ MAX,MIN
1070  DIM Z((NR - 1),(NC - 1))
1080  FOR I = 0 TO (NR - 1)
1090  FOR J = 0 TO (NC - 1)
1100  READ Z(I,J)
1110  NEXT J
1120  NEXT I
1130 NR =  INT (NR)
1140 NC =  INT (NC)
1190 PRINT
1200  PRINT " PC Data File is now being generated..."
1210  OPEN "O",#1,F$
1220  PRINT#1, NR
1230  PRINT#1, NC
1240  PRINT#1, MAX
1250  PRINT#1, MIN
1260  PRINT#1, DX
1270  PRINT#1, DY
1280  FOR I = 0 TO (NR - 1)
1290  FOR J = 0 TO (NC - 1)
1300 H = Z(I,J)
1310  PRINT#1, H
1320  NEXT J
1330  NEXT I
1340  CLOSE#1
1350 PRINT
1360 BEEP:PRINT " PC Data File ";F$;" stored on disc."
1370 PRINT
1380 PRINT " Do you wish to check Data File (y/n)";:INPUT Q$
1390 IF Q$="n" THEN 9500
1400 IF Q$<>"y" THEN 1370
1410 GOTO 2000
1995 REM ***** Data File Information *****
2000 CLS
2010 GOSUB 9000
2020 PRINT " THE DATA FILE ";F$;" HAS BEEN READ"
2030 PRINT
2040 PRINT " No.  Rows = ";NR;"   No.  Cols.= ";NC
2050 PRINT " Minimum value = ";MIN
2060 PRINT " Maximum value = ";MAX
2070 PRINT " Range         = ";(MAX-MIN)
2080 PRINT " 1 Map row     = ";DY/(NR-1);" Units"
2090 PRINT " 1 Map column  = ";DX/(NC-1);" Units"
2100 PRINT
2110 PRINT " Do you wish to run PC-Contour now":PRINT " (y/n) ";:INPU
T Q$
2120 IF Q$="n" THEN 9500
2130 IF Q$<>"y" THEN 2100
2140 LOAD "CONTOUR",R
2995 REM ***** Data Statements start here *****
3000 REM Lines 3000 onwards are for your File Data
8995 REM ***** PC-Grid Heading *****
9000 T$="PC-Grid/Data"
9010 LOCATE 1,2:PRINT T$
9020 LOCATE 1,30:PRINT DATE$
9030 PRINT "======================================="
9040 PRINT
9050 RETURN
9495 REM ***** Program End *****
9500 CLS
9510 GOSUB 9000
9520 PRINT:PRINT "  Program Complete"
9530 PRINT:PRINT
9540 END
```

3.2.5 PC-Grid/Demo

To be able to run PC-Grid/Data it is necessary to prepare a data file. It is suggested that once the following **DATA** statements have been entered into the file PC-Grid/Data, it is resaved under the title **"GRIDDEMO"** with changed remarks in lines 10 and 25. The data file presented is for a spoil-heap at Glapwell, and the resulting data file will be automatically saved on disc as **"GLAPWELL"** for subsequent access by PC-Contour and PC-Surfplot.

The initial data statements give the file name **"GLAPWELL"** to be stored in **F$**, and the number of rows (**NR**) as 31 and number of columns (**NC**) as 41. The Total *X*-distance (**DX**) is 400 metres and the *Y*-distance (**DY**) is 300 metres, that is *Z*-values are given on a 10 metre square grid. The maximum (**MAX**) and minimum (**MIN**) *Z*-values are noted as 131.5 m and 74.1 m for reference and advice in later selecting the contour values required to be plotted. Since the zero (0, 0) of the IBM PC screen is in the top left-hand corner, this has been taken as the zero for data generation purposes. Thus the data in the top left-hand corner of the plotted area is stored in **Z(Ø,Ø)** and the data for the lower right-hand corner stored in **Z((NR–1),(NC–1))**. For maximum screen usage **NR** should be 75% of **NC**, to give a 1:1 natural grid plot on screen or plotter.

3.2.6 PC-Grid/Demo – data

```
3ØØØ  DATA  "GLAPWELL"
3Ø1Ø  DATA  31,41
3Ø2Ø  DATA  4ØØ,3ØØ
3Ø3Ø  DATA  131.5,74.1
3Ø4Ø  DATA  78.5,81,83,86.4,88.5,92.3,93.7,95.4,96.1,96.2,95.9,95.1,9
3.2,92.8,92.6,92.7,93.2,93.7,93.2,91
3Ø5Ø  DATA  87.6,87.2,9Ø.3,92.6,94.8,97.3,99.9,1Ø2.4,1Ø4.9,1Ø7.2,1Ø8.
3,1Ø9.8,11Ø.9,111.8,112.3,112.3,112.2,112,111.5,111.Ø,11Ø.5
3Ø6Ø  DATA  78.7,81.2,84.7,87.6,9Ø.4,93.6,95.6,97.5,98.4,98.3,98.2,97
.7,96.8,95.4,95.2,95.4,96.8,96.8,94.9,92.8
3Ø7Ø  DATA  9Ø.1,89.5,9Ø.9,93.7,96.5,99.2,1Ø2.2,1Ø5.3,1Ø7.8,11Ø.1,112
.6,114.3,115.7,117.2,117.9,118,117.9,117.2,116.8,115,113.2
3Ø8Ø  DATA  78.3,81.1,84.7,87.4,9Ø.2,93.4,95.9,98.2,99.8,1ØØ.7,1ØØ.6,
1ØØ.1,99.2,98.7,98.5,98.5,98.7,98.5,97.3,95.1
3Ø9Ø  DATA  92.3,89.7,9Ø.8,94,97.2,1ØØ.6,1Ø3.5,1Ø7.1,11Ø.1,114.3,116.
2,118,12Ø.1,12Ø.9,121.2,121.5,121.8,121.5,121.2,12Ø.1,119
31ØØ  DATA  77.9,8Ø.2,83.6,86.1,89.3,92.7,96.Ø,98.1,1ØØ.3,1Ø1.8,1Ø2.4
,1Ø2.2,1Ø1.7,1Ø1,1ØØ.8,1Ø1,1ØØ.7,1ØØ.1,99.2,97.4
311Ø  DATA  93.7,91,9Ø.2,93.8,97.4,1Ø1.2,1Ø5.1,1Ø8.5,112.3,117.4,12Ø.
4,121.8,122.4,123.1,123.7,123.9,124,123.9,123.3,122.6,121.6
312Ø  DATA  77.3,79.4,82.4,84.3,87.5,92.3,95.8,98.1,1ØØ.4,1Ø2.9,1Ø3.7
,1Ø3.6,1Ø3.5,1Ø3.1,1Ø2.8,1Ø2.7,1Ø2.6,1Ø2.4,1Ø1.2,1Ø1
313Ø  DATA  96.2,92.1,89.3,93.2,97.4,1Ø1.9,1Ø5.7,1Ø9.9,115.2,12Ø.3,12
2.3,123.7,124.9,125.6,125.8,126.5,126.8,126.5,126.2,125.3,124.3
314Ø  DATA  76.5,78.7,8Ø.2,83.1,86.2,9Ø.6,95.7,97.9,1ØØ.4,1Ø3,1Ø3.9,1
Ø4.9,1Ø5.2,1Ø5.4,1Ø5.2,1Ø5.1,1Ø4.9,1Ø4.7,1Ø3.5,1Ø2.4
315Ø  DATA  99.8,93.7,9Ø.1,92.3,96.5,1Ø1.6,1Ø5.9,11Ø.2,116.1,12Ø.5,12
2.1,122.9,124.8,126.2,127.6,128.8,129.9,129.7,129,128.5,128
316Ø  DATA  75.3,77.8,79.1,81.7,85.1,9Ø.1,95.2,97.8,1ØØ.4,1Ø2.9,1Ø4,1
Ø5.4,1Ø6.2,1Ø6.6,1Ø6.7,1Ø6.9,1Ø6.7,1Ø6.3,1Ø5.7,1Ø4.5
```

```
3170  DATA  102.4,97.3,91.4,90.6,95.8,101.4,105.8,110.1,115.3,119.8,1
21,121.5,122.4,124.6,126.4,128.2,131.3,131.5,131.5,131,130.5
3180  DATA  74.9,77.3,78.8,80.6,85.1,90.2,95.1,97.9,100.3,102.8,104.5
,105.9,107,107.9,108.2,108.4,108.,107.9,107.2,106.1
3190  DATA  103.9,100.5,93.2,90.8,95.1,100.1,104.2,109.4,113,116.5,11
8.6,119.8,120.9,122.4,124.6,126.5,129,131.3,131.5,131,130.5
3200  DATA  74.5,77,78.3,80.3,84.9,90.7,94.9,97.9,100.3,102.7,104.5,1
05.9,107,108.2,109,109.9,110.2,109.5,108.6,107.2
3210  DATA  105.2,102.9,97.3,91.2,94.4,98.7,102.2,106.8,110.5,113.4,1
15.3,117.2,118.7,120.3,122.4,124.8,127.6,130.8,131,129,127
3220  DATA  74.1,77,78.5,80.5,85.4,92.3,94.8,98,100.5,102.8,104.4,105
.6,107.2,108.5,110.1,111.2,111.5,111.2,110.4,108.6
3230  DATA  106.8,104.2,100.3,93.7,92.3,97.3,100.3,103.8,107.3,110.1,
112.4,114,115.9,118.6,120.5,123,126.5,130.1,129.6,127.2,125
3240  DATA  75.4,77.2,78.9,80.9,86.8,92.6,95.2,98.5,100.7,103.1,104.9
,105.8,107.4,109,110.8,112.4,112.9,112.7,112.3,111.1
3250  DATA  107.9,104.9,102.1,96.2,92.2,95.3,98,100.4,104.6,107.2,109
.2,111.8,114.2,117.4,119,122.1,125.3,128.4,127,124,122
3260  DATA  76.2,77.7,79.8,82.4,87.6,92.7,95.6,98.6,100.9,103.2,105.1
,105.9,107.6,109.5,111.3,112.9,114.1,114.6,114.8,113.3
3270  DATA  110.2,107.2,102.9,98.4,91.5,94.2,96.8,98.8,101.2,104.3,10
7.1,108.9,112.6,115.8,118.7,121.2,125.1,127,124.8,123,121.2
3280  DATA  77.1,78.2,80.2,84.6,88.1,93.2,96.1,99.3,101.4,103.6,105.3
,106.8,108,109.9,112,113.2,115.7,116.2,116.2,115.7
3290  DATA  112.8,108.1,104.6,100.7,92.6,93.2,95.2,97.6,98.8,102.1,10
4.8,107.8,111.1,114.9,117.8,120.7,125.1,125.1,123.8,121.5,119.3
3300  DATA  77.8,79.6,82.1,85.7,89.8,94.1,97.4,100.2,102.4,104.3,105.
7,107.1,108.3,110.8,112.2,114.3,116.4,117.7,117.9,116.9
3310  DATA  114.9,109.8,105.7,102.3,94.9,93,94.6,96.9,98.2,100.3,103.
7,106.9,110.1,113.8,117.6,120.3,123.8,123.8,122.6,120.2,118
3320  DATA  78.3,80.2,83.1,87.4,92.1,95.2,98.3,101.2,103.1,104.9,106.
2,107.6,108.9,111.2,112.7,114.8,117.4,119.3,120.2,118.7
3330  DATA  116,112,107,102.8,97.1,92.9,94,96,97.9,99.8,102.7,105.3,1
09.6,113.2,117,119.8,122.6,122.6,120.5,118.7,116.5
3340  DATA  78.8,82.6,85.6,89.9,93,97.4,100.2,102.4,104.2,105.6,106.9
,108.1,109.5,111.4,113.1,115.2,118.3,120.8,121.7,120.3
3350  DATA  116.4,112.4,107.4,103.6,98.8,92.8,93.8,95.1,97.6,99.6,102
.3,104.8,108.2,112.3,115.8,118.6,121.5,120.8,118.3,116.4,114.3
3360  DATA  79.9,82.7,87.9,92.6,95.9,99.8,101.8,103.3,105.1,106.3,107
.6,108.8,109.9,112,114.2,116.8,119.1,121.7,123,120.9
3370  DATA  117.2,113.2,108.2,104.3,100.4,92.7,93.8,94.7,97.2,99.1,10
1.5,104,107.6,111.5,115.1,117.3,119.5,118,116,112.6,109.5
3380  DATA  81.2,85.7,90.8,94.8,97.9,100.9,102.8,104.8,105.8,107.1,10
8.2,109.4,110.5,112.9,115.2,117.7,119.7,122.2,123.3,121.1
3390  DATA  117.4,113.5,109.3,105.2,101.8,93.5,94,94.6,96.5,98.5,101,
103.2,107,110.1,113.7,115.9,117.2,115.1,112.6,110.2,108
3400  DATA  82.4,87.6,92.6,96.1,98.7,102.1,103.3,105.6,106.7,108.3,10
8.8,109.9,111.6,113.7,116.3,118.2,120.1,122.7,123.3,120.8
3410  DATA  117.4,114.1,109.8,105.9,102.4,94.7,94.2,94.6,95.9,98,100.
1,102.6,105.9,109.8,112.3,113.6,113.8,112.7,110.5,109,107.5
3420  DATA  84.9,89.5,93.2,97.3,100.4,102.9,104.8,106.2,107.4,108.4,1
09.4,110.6,112.4,114.3,116.9,118.8,120.6,122.9,123,120.4
3430  DATA  117.2,114.8,110.1,106.2,102.4,95.2,94.3,94.5,95.8,97.8,99
.6,101.8,105.1,108,110.8,111.3,110.7,109.7,108.3,107.2,106.1
3440  DATA  87,90.6,94.8,98.8,101.6,103.9,106,107.3,108.1,109,109.9,1
11.7,113,115.8,117.4,119,120.8,123.2,122.6,119.8
3450  DATA  117.1,114.9,110.2,107,103,96.3,94.5,94.5,95.1,96.8,97.9,9
9.3,102.2,104.8,106.3,106.9,106.9,106.3,106.3,105.1,103.9
3460  DATA  88.1,92.6,97.2,100.3,102.9,105.3,106.9,108.2,108.9,109.7,
110.8,112.1,114.2,116.3,118.1,119.2,121.4,123.8,121.7,118.7
3470  DATA  116.8,115.1,111.3,107.6,103.7,99.8,95.1,94.6,94.7,95.5,96
.8,97.4,99.3,100.7,101.9,102.6,102.9,103.7,103.8,103,102.2
3480  DATA  89.9,93.5,97.6,101.2,103.7,106.1,107.4,108.4,109.4,110.3,
111.2,112.7,114.9,116.6,118.7,119.5,121.7,123.4,120.6,118.2
```

```
3490  DATA  116.2,114.8,111.9,108.5,104.6,100.7,97,95.1,94.8,95.3,96.
1,96.7,97.2,97.8,98.9,99.7,100.1,101.9,102.8,103,103.5
3500  DATA  92,95.1,99.2,102.4,104.8,106.6,107.9,108.8,109.7,110.6,11
2.1,113.3,115.2,116.8,117.8,119.5,121.7,121.7,119.6,117.4
3510  DATA  116,114.6,112.1,109.7,105.2,101.3,99.2,97.1,94.9,95.1,95.
4,95.8,96.2,96.6,97,97.2,97.4,99.8,101.6,102.7,103.8
3520  DATA  93.2,96.8,100.7,103.1,105.8,107.2,108.2,109,110.1,111.2,1
12.4,114.2,115.9,116.9,117.4,119.2,120.8,120.1,118.3,116.8
3530  DATA  115.9,114.3,112.1,110.8,107.4,102,100.1,98.2,96.5,97.6,97
.2,97.3,97.4,97.2,97,96.5,96,95.5,97.8,100.2,102.6
3540  DATA  93.7,97.4,101.8,104.2,106.1,107.4,108.5,109.5,110.5,111.6
,112.7,114.5,116.1,117,117.3,118.5,119.2,118.5,117.4,116.3
3550  DATA  115.6,113.8,111.8,110.8,108,102.7,100.2,97.6,97.3,100.1,9
9.8,99.7,99.5,98,97.6,97.2,96,96,97.2,98.8,100.4
3560  DATA  93.7,97.6,102.3,104.8,106.3,107.8,108.8,109.6,110.7,111.9
,113.1,115.1,116.3,117.3,117.3,117.6,118.6,118,117,116
3570  DATA  114.9,112,110.9,110.2,108,105.1,102.4,99.3,99.3,100.5,100
.8,100.8,100.7,100.4,99.5,98.8,97.6,96.2,96.3,97.4,98.5
3580  DATA  93.7,97.7,102.3,104.8,106.6,108,108.9,109.7,111.2,112.2,1
13.5,115.1,117.4,118.1,118,117.2,118.3,117.9,117,115.8
3590  DATA  112.1,109.6,109.2,108.5,107.9,104.9,102.7,101.8,101.8,102
.1,102,102,101.9,101.7,100.8,99.9,98.7,97.3,96.4,96.5,96.6
3600  DATA  92.6,97.4,102.1,104.6,106.9,107.9,108.8,109.8,111.4,112.3
,113.7,116,119,120.7,121,120.4,117.3,117.8,117,115.6
3610  DATA  110.9,109.4,108.8,108.2,107.3,104,103.5,103.3,103,102.9,1
02.8,102.7,102.6,102.2,101.9,101.7,99.8,98.2,97.3,96.8,96.3
3620  DATA  92.3,95.7,101.4,104.1,106.5,107.9,108.8,109.9,111.4,112.4
,114.1,116.8,119.8,121.5,121.8,121.4,117.9,117.9,117.4,114.8
3630  DATA  111.2,109.5,108.5,107.4,105.2,104.6,103.6,103.5,103.5,103
.5,103.5,103.3,103.1,102.8,102.3,101.5,100.8,99.1,98.2,97.1,96
3640  DATA  92,96,100.7,103.6,106.1,107.7,108.6,109.7,111.6,112.5,114
.5,117.5,118.9,121.5,123,122.4,118.5,118,117.8,115.2
3650  DATA  111.5,109.6,108.3,106.6,103.2,104.3,103.7,103.6,104,104.1
,104.2,103.9,103.6,103.4,103.7,101.3,101.3,99.7,99.1,97.4,96.2
```

3.3 PC-RANDOM

Not all spatial data is collected or sampled on a regular grid basis and this program allows contours to be drawn from irregularly or random spaced data. The data has first to be read by the program PC-Random/RData and then a disc data file must be prepared for access by PC-Contour or PC-Surfplot.

The initial computer listing in this section, PC-Random/RData is a blank data file for entering your own data and should be saved under the title **"RANDRDAT"**. The second computer listing is PC-Random/Demo and includes data for a tacheometry survey at Glenmore in Scotland; it should be saved on disc under the title **"RANDDEMO"** as it can be used as a demonstration file to check your program and the output to screen or plotter.

3.3.1 PC-Random/RData – subroutine index

	Line numbers	*Function*
(a)	10–90	Initialization and control
(b)	100–560	Selection of application
(c)	1000–1540	Reading and storing of data
(d)	2000–2440	Presentation of data file information
(e)	2500–2990	PC-Random iteration routine
(f)	3000–3490	Graphics screen display
(g)	4000–4850	Plotter routines
(h)	6000–6270	Printout of data file details
(i)	9000–9050	Routine to clear screen
(j)	9500–9650	Termination of program

(a) Initialization and control Line numbers 10–90

The program is initialized and the installation file checked for the availability of graphics screens and colour for subsequent display of the random data points.

(b) Selection of application Line numbers 100–560

The user is given a choice of three applications:

⟨1⟩ Read DATA and present details
⟨2⟩ Read DATA and store on disc
⟨3⟩ Prepare DATA file

Thus this program stored under the title **"RANDRDAT"** can be used initially to prepare a data file, then to read the data to check it is spatially correct, and finally to store the data on disc for access by PC-Contour or PC-Surfplot.

To save the user constantly referring back to the book, a note on data file preparation is included on the screen, and the user reminded that on completion of entering the data, the file should be saved under another name for reference and subsequent recall. Data is entered in data statements as described in 3.2.1(b).

(c) Reading and storing of data Line numbers 1000–1540

The data file is then read, first its name, **F$** and then the *X*-coordinates of the left-hand side and right-hand side of the plot, followed by the bottom and top coordinates, i.e. the plotting area is defined by *X*-min, *X*-max, *Y*-min, *Y*-max. The number of data points, *N*, is then read and arrays set up to store the *X*, *Y* and *Z*-coordinates of the points. There is no need to enter the maximum and minimum *Z*-values in this program as this routine

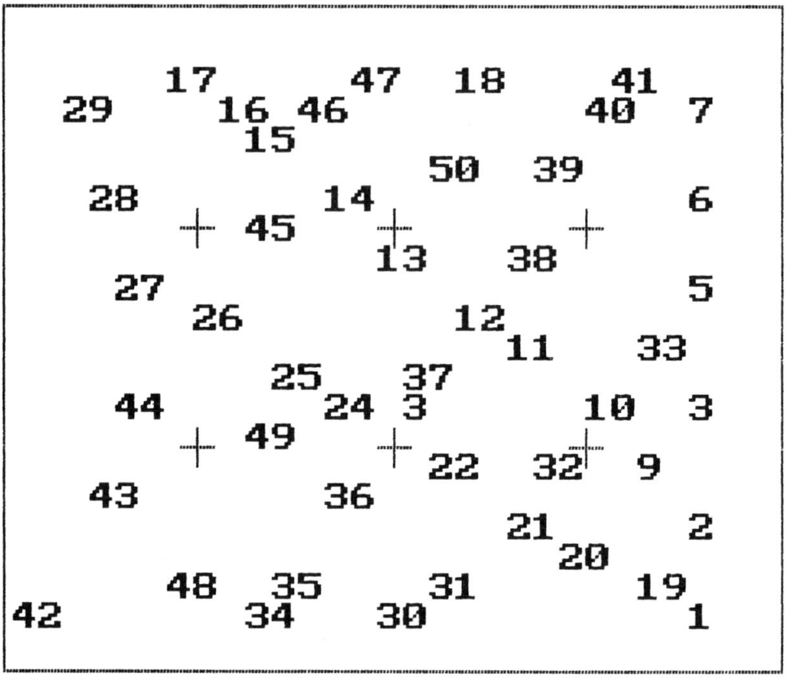

Fig. 3.1 PC-Random/Demo plot of data points for Glenmore tacheometry survey on Screen 1 (medium-resolution).

automatically locates them and stores them in the variables **MAX** and **MIN** respectively.

The plotted area is then split up into a grid; the maximum number of columns is taken as 41 and the maximum number of rows is taken as 31. If, for example, a square area was entered, the number of columns would be automatically reduced to 31 to ensure the correct scaling of the data on the screen or plotter. As PC-Contour and PC-Surfplot require grid points for the interpolation of contours, these grid points are calculated in the routine at line 2500 before being stored on disc. The user should ensure before running this program that there is space on disc to store the data file, and that the disc is not 'write-protected'. The user should also be patient during this interpolation process, as with a microcomputer running at 4.77 MHz, the time for the data file generation can take 45–50 minutes. Faster machines will take less time.

(d) Presentation of data file information Line numbers 2000–2440

This routine is provided as a check to allow the user to see if the data information has been entered correctly before storing on disc. The user is

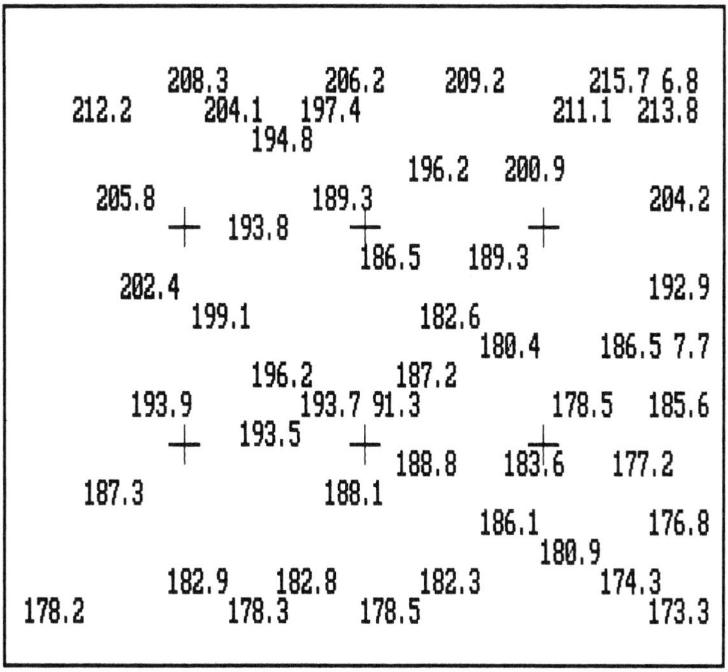

Fig. 3.2 PC-Random/Demo plot of ground levels for Glenmore tacheometry survey on Screen 2 (high-resolution).

also given an opportunity to display the data on the graphics screen or plotter.

(e) PC-Random iteration routine Line numbers 2500–2990

As mentioned in (c) this iteration routine is used to scan the PC-Random data and calculate the value of each grid point from the five nearest data points using an inverse square distance weighted interpolation. As in the normal hand drawing of contours, the final contours will only be as good as the selection of the data points to accurately represent the ground surface. Students should resist the temptation to reduce data coverage on fairly level ground as this will impair the accuracy of the final contours.

(f) Graphics screen display Line numbers 3000–3490

The display screen chosen for the display of either the data points or their *Z*-value is shown in Fig. 1.4. The screen is first set up to cover *X*-min, *X*-max, *Y*-min, *Y*-max and then the position of each data point calculated and labelled in text mode, using the variables **CROW** and **CCOL** for the rows and columns. On completion, the user can obtain a graphics printer screen dump or plot. Hard copies of both the graphics screen and plotter can be obtained through this menu of either the data points, Fig. 3.1, or their *Z*-value, Fig. 3.2.

(g) Plotter routines Line numbers 4000–4850

Three plotter routines are provided for the Apple 410, Epson HI-80 and Hewlett-Packard HP-7475A colour plotters, as discussed in Chapter 2. The user has a choice of plotting either the data points or their *Z*-values. On completion the plot is labelled with the data file name and date of plotting.

(h) Printout of data file details Line numbers 6000–6270

This printout routine is provided to keep a record of the data information entered, *X*-min, *X*-max, *Y*-min, *Y*-max, the number of points and their *X*, *Y*, *Z*-coordinates for future reference.

(i) Routine to clear screen Line numbers 9000–9050

This routine is provided to clear the screen before each plot and provide lines of text at the top of the screen to identify the program title and data file presentation, together with the current date.

(j) Termination of program Line numbers 9500–9650

This is the final display at the completion of the program, and the user is given an option to run PC-Contour or return to BASIC command level.

3.3.2 PC-Random/RData – numeric variables

CCOL	=	Column number for text display on graphics screen
CROW	=	Row number for text display on graphics screen
DX	=	*X*-distance of plotted area
DY	=	*Y*-distance of plotted area
FX	=	*X*-distance between grid points
FY	=	*Y*-distance between grid points
GPXØ–2	=	Variables to plot grid nodes on screen
GPYØ–2	=	Variables to plot grid nodes on screen
GX	=	*X*-coordinate at lower left-hand corner of plot
GY	=	*Y*-coordinate at lower left-hand corner of plot
HP%	=	Integer variable to identify plotter
I	=	Integer variable used in loops
J	=	Integer variable used in loops
K	=	Integer variable used in loops
L1–L5	=	Variables to identify level points in grid
MAX	=	Maximum *Z*-value in data
MF	=	Ratio of **DX** to **DY**
MG	=	Multiplication factor for plotted display

MH	=	Multiplication factor for horizontal resolution
MIN	=	Minimum *Z*-value in data
MV	=	Multiplication factor for vertical resolution
MX	=	Multiplication factor for Epson HI-80 plotter
N	=	Number of data points
NC	=	Number of columns in grid
NR	=	Number of rows in grid
PP	=	Screen display variable
PX	=	*X*-coordinate at upper right-hand corner of plot
PY	=	*Y*-coordinate at upper right-hand corner of plot
P(N)	=	Array for data point identification
PLOT%	=	Integer variable for plotter selection
RL	=	Reduced level of data output
R(N)	=	Array to store reduced levels
RAND%	=	Integer variable to select application
RGB%	=	Integer variable to select line colour
S1,S2	=	Variables to calculate grid node levels
SCD%	=	Integer variable to define screen display
SCR%	=	Integer variable to select screen or plotter
SD%	=	Select display, identification number or level of data point
SEL%	=	Integer variable to select screen or plotter
SR%	=	Integer variable for screen resolution
S(N)	=	Array to store identification number or level of data point
SX	=	Horizontal display width in user units
SY	=	Vertical display height in user units
X	=	*X*-coordinate of grid point
XL	=	*X*-coordinate of left-hand side of plot
XR	=	*X*-coordinate of right-hand side of plot
X(N)	=	Array for *X*-coordinates of data points
Y	=	*Y*-coordinate of grid point
YB	=	*Y*-coordinate at foot of plot
YT	=	*Y*-coordinate at top of plot
Y(N)	=	Array for *Y*-coordinates of data points
ZP	=	*Z*-coordinate of interpolated grid points
Z(N)	=	Array for *Z*-coordinates of data points

3.3.3 PC-Random/RData – string variables

DATE$		Variable used to return the date of the computer's calendar/clock in US format MM–DD–YYYY.
F$		File name
GD$		Entry for graphics dump or printer
P$	=	**"PC-Random"**
PCF$	=	**"PCSCREEN"**, installation file for graphics
Q$		Question response, (y/n) or (Y/N)
T$	=	**P$ + F$**

3.3.4 PC-Random/RData – BASIC program

```
10 REM  <RANDRDAT> Program for IBM PC and Compatibles
12 REM  Version PC-1.0, 01-01-1987
14 REM  (C)   Copyright  1987  P.H. Milne
16 REM  All rights reserved
18 REM  Creates Data File for Access by PC-Contour
20 REM  Plot available of DATA on Apple, Epson or Hewlett-Packard Plotter
22 REM  Setting in Line 4090 "COM1:1200,N,7,1,CS2000,DS15000"
25 REM  Storage space for DATA in lines 5000-onwards
30 CLEAR
35 PCF$="PCSCREEN"
40 OPEN "I",#3,PCF$
45 INPUT#3,SCR%
50 INPUT#3,SCD%
55 CLOSE#3
60 ON (SCR%+1) GOTO 62,64,66
62 SCREEN 0:WIDTH 40:GOTO 80
64 SCREEN 1:GOTO 68
66 SCREEN 1:GOTO 80
68 IF SCD%=1 THEN 80
70 COLOR 8,0
80 KEY OFF
90 F$=" "
100 CLS
110 GOSUB 9000
120  PRINT
130  PRINT "  THIS PROGRAM READS INFORMATION FROM"
140  PRINT "  RANDOMLY SPACED POINTS INTO A"
150  PRINT "  DATA FILE FOR PLOTTING BY PC-Contour"
160  PRINT
170 PRINT "  Select application required :-"
180 PRINT
190 PRINT "  <1> Read DATA and present details"
200 PRINT
210 PRINT "  <2> Read DATA and store on disc"
220 PRINT
230 PRINT "  <3> Prepare DATA file"
240 PRINT
250 PRINT "  <4> Quit ";
260 INPUT RAND%:IF RAND%<1 OR RAND%>4 THEN 100
290 ON RAND% GOTO 1030,1030,300,9500
300 CLS
310 GOSUB 9000
320 PRINT " To examine Data File preparation"
330 PRINT " list lines 500-560 on screen after"
340 PRINT " the Ok prompt"
350 PRINT
370 PRINT " After entering your own data, either"
380 PRINT " RUN or SAVE your program immediately."
390 PRINT:PRINT
400 END
500 REM ***** Data File Preparation Notes *****
510 REM Line 5000 : Data File Name
520 REM Line 5010 : X-coordinates LHS, RHS
530 REM Line 5020 : Y-coordinates bottom,top
540 REM Line 5030 : Number of data points
550 REM Line 5040 - onwards : DATA  Point No, X-coord, Y-coord, Z-coord
560 REM To create own data file start at line 5000
1000 CLS:GOSUB 9000
1020 INPUT "  Press <Enter> when Data File Ready ";R$
1030 READ F$
1040 READ XL,XR
1050 READ YB,YT
```

```
1060 READ N
1070  DIM P(N),X(N),Y(N),Z(N),R(N),S(N)
1080 PRINT:PRINT
1090 PRINT "  DATA now being read from file"
1100 MAX=-25000
1110 MIN=25000
1120 FOR I=1 TO N
1130 READ P(I),X(I),Y(I),Z(I)
1140  IF Z(I) > MAX THEN MAX = Z(I)
1150  IF Z(I) < MIN THEN MIN = Z(I)
1160  NEXT I
1170 DX = XR - XL
1180 DY = YT - YB
1190 MF = DX / DY
1200 NC = 41
1210 NR=(NC-1)/MF + 1
1220  IF NR<=31 THEN 1250
1230 NR = 31
1240 NC=(NR-1)*MF + 1
1250 NC =  INT (NC+.1)
1260 NR =  INT (NR+.1)
1270 PRINT:IF RAND%=1 THEN 1510
1280 PRINT "  Has Data been stored on Disc already":INPUT "  (y/n) "
Q$: IF Q$="y" THEN 1510
1290 IF Q$<>"n" THEN 1270
1300 PRINT:PRINT:PRINT "  PC-Contour Data File :";F$:PRINT "  IS NOW
BEING GENERATED ....."
1310  OPEN "O",#1,F$
1320  PRINT#1, NR
1330  PRINT#1, NC
1340  PRINT#1, MAX
1350  PRINT#1, MIN
1360  PRINT#1, DX
1370  PRINT#1, DY
1380 FX = (DX / (NC - 1))
1390 FY = (DY / (NR - 1))
1400  FOR I = 0 TO (NR - 1)
1410 Y = YT - I * FY
1420  FOR J = 0 TO (NC - 1)
1430 X = XL + J * FX
1450  GOSUB 2500
1460  NEXT J
1470  NEXT I
1480  CLOSE#1
1500 BEEP:PRINT:PRINT "  Data File -";F$;" - Closed"
1510 BEEP:PRINT:PRINT "  Do you wish Printout of Data (y/n)";
1520 INPUT Q$
1530 IF Q$="y" THEN 6000
1540 IF Q$<>"n" THEN 1510
1995 REM ***** Data Information and Plot Selection *****
2000 CLS
2002 SCREEN SCR%:IF SCR%=0 THEN 2010
2004 IF SCR%=1 THEN 2008
2006 SCREEN 1:GOTO 2010
2008 COLOR 8,0
2010 GOSUB 9000
2020 PRINT " THE DATA FILE ";F$;" HAS BEEN READ"
2030 PRINT
2040 PRINT " No. Rows = ";NR; " No. Columns = ";NC
2050 PRINT " Minimum Value  = ";MIN
2060 PRINT " Maximum Value  = ";MAX
2070 PRINT " Range          = ";(MAX-MIN)
2080 PRINT " 1 Map Row      = ";DY/(NR-1);"  Units"
2090 PRINT " 1 Map Column   = ";DX/(NC-1);"  Units"
2100 BEEP
```

```
2110 PRINT:PRINT " Select Display :-"
2120 PRINT:PRINT " <1> SCREEN, <2> PLOTTER or <3> QUIT"
2130 IF SCR%<>0 THEN 2150
2140 PRINT:PRINT " N.B. Option <1> not available"
2150 INPUT SEL%
2160 IF SEL%<1 OR SEL%>3 THEN 2100
2170 IF SCR%=0 THEN IF SEL%=1 THEN 2000
2180 ON SEL% GOTO 2300,4000,9500
2295 REM  Choose Screen for Plotting
2300 CLS
2302 SCREEN SCR%
2304 IF SCR%=1 THEN 2308
2306 SCREEN 1:GOTO 2310
2308 COLOR 8,0
2310 GOSUB 9000
2320 PRINT:PRINT TAB(3);"Select SCREEN";TAB(20);"<1> Colour -"
2330 PRINT TAB(20);"    Low resolution"
2340 PRINT:PRINT TAB(20);"<2> Black & White -"
2350 PRINT TAB(20);"    High Resolution"
2360 PRINT:PRINT TAB(20);:INPUT SR%
2370 IF SR%<1 OR SR%>2 THEN 2320
2380 PRINT:PRINT:PRINT TAB(3);"Select Data for Screen :-"
2390 PRINT:PRINT TAB(20);"<1> Point No. or"
2400 PRINT:PRINT TAB(20);"<2> Z-Value ";:INPUT SD%
2410 IF SD%<1 OR SD%>2 THEN 2380
2420 IF SR%=1 THEN 3000
2430 SCREEN 2
2440 GOTO 3000
2495  REM  ***** PC-Random Iteration Subroutine *****
2500  FOR K = 1 TO N
2510 R(K) = (X - X(K)) ^ 2 + (Y - Y(K)) ^ 2
2520  IF R(K) = 0 GOTO 2980
2530  NEXT K
2540 RL = 25000
2550  FOR K = 1 TO N
2560  IF R(K) > RL GOTO 2590
2570 L1 = K
2580 RL = R(K)
2590  NEXT K
2600 RL = 25000
2610  FOR K = 1 TO N
2620  IF K = L1 GOTO 2660
2630  IF R(K) > RL GOTO 2660
2640 L2 = K
2650 RL = R(K)
2660  NEXT K
2670 RL = 25000
2680  FOR K = 1 TO N
2690  IF K = L1 GOTO 2740
2700  IF K = L2 GOTO 2740
2710  IF R(K) > RL GOTO 2740
2720 L3 = K
2730 RL = R(K)
2740  NEXT K
2750 RL = 25000
2760  FOR K = 1 TO N
2770  IF K = L1 GOTO 2830
2780  IF K = L2 GOTO 2830
2790  IF K = L3 GOTO 2830
2800  IF R(K) > RL GOTO 2830
2810 L4 = K
2820 RL = R(K)
2830  NEXT K
2840 RL = 25000
2850  FOR K = 1 TO N
```

```
2860  IF K = L1 GOTO 2930
2870  IF K = L2 GOTO 2930
2880  IF K = L3 GOTO 2930
2890  IF K = L4 GOTO 2930
2900  IF R(K) > RL GOTO 2930
2910 L5 = K
2920 RL = R(K)
2930  NEXT K
2940 S1 = Z(L1) / R(L1) + Z(L2) / R(L2) + Z(L3) / R(L3) + Z(L4) / R(L
4) + Z(L5) / R(L5)
2950 S2 = 1 / R(L1) + 1 / R(L2) + 1 / R(L3) + 1 / R(L4) + 1 / R(L5)
2960 ZP = S1 / S2
2970  GOTO 2990
2980 ZP = Z(K)
2990  PRINT#1, ZP:RETURN
2995 REM ***** Screen Plot *****
3000 CLS
3010 GX=5*SR%:GY=5:MG=0:MH=0:MV=0:SX=240*SR%:SY=180:PP=0
3020 MH=SX/(DX*SR%):MV=SY/DY
3030 IF MH<=MV THEN MG=MH ELSE MG=MV
3040 WINDOW (0,0) - (320*SR%,200)
3050 IF SR%=1 THEN RGB%=2 ELSE RGB%=3
3060 PX=GX+MG*DX*SR%:PY=GY+MG*DY
3070 LINE (GX,GY)-(PX,PY),RGB%,B
3080 IF SR%=1 THEN 3100
3090 LINE (GX-1,GY)-(PX-1,PY),RGB%,B
3100 FOR I=10 TO (NC-10) STEP 10
3110 FOR J=10 TO (NR-10) STEP 10
3120 GPX0=GX+(6*I*SR%)
3130 GPX1=GPX0-5*SR%:GPX2=GPX0+5*SR%
3140 GPY0=GY+(6*J)
3150 GPY1=GPY0-5:GPY2=GPY0+5
3160 LINE (GPX1,GPY0) - (GPX2,GPY0),RGB%
3170 LINE (GPX0,GPY1) - (GPX0,GPY2),RGB%
3180 NEXT J
3190 NEXT I
3195 IF PP=1 THEN 3300
3200 ' Start of Screen Plot
3210 LOCATE 1,2:PRINT "PC-Random/RData - ";F$
3215 LOCATE 1,30*SR%:PRINT DATE$
3220 FOR J=1 TO N
3230 IF SD%=1 THEN S(J)=P(J)
3240 IF SD%=2 THEN S(J)=Z(J)
3250 CCOL=(GX+((X(J)-XL)*MG)*SR%)/8:CROW=(GY+(Y(J)-YB)*MG)/8
3255 IF CCOL<2 THEN CCOL=2
3256 IF CCOL>28*SR% THEN CCOL=28*SR%
3260 IF CROW<=3 THEN CROW=3
3265 IF CROW>22 THEN CROW=22
3270 LOCATE (26-CROW),CCOL-1
3280 PRINT S(J)
3290 NEXT J
3295 PP=1:GOTO 3070
3300 BEEP
3310 LOCATE 3,32*SR%:PRINT "Screen"
3320 LOCATE 4,32*SR%:PRINT "Dump(y/n)"
3330 LOCATE 5,32*SR%:PRINT "          "
3340 LOCATE 5,32*SR%:INPUT GD$
3350 IF GD$="n" THEN 3400
3360 IF GD$<>"y" THEN 3300
3370 LCOPY
3400 BEEP
3410 LOCATE 9,32*SR%:PRINT "Select"
3420 LOCATE 10,32*SR%:PRINT "Next"
3430 LOCATE 11,32*SR%:PRINT "Choice :"
3440 LOCATE 13,32*SR%:PRINT "<1>Screen"
```

```
3450 LOCATE 14,32*SR%:PRINT "<2>Plot"
3460 LOCATE 15,32*SR%:PRINT "<3>Quit"
3465 LOCATE 16,32*SR%:PRINT "          "
3470 LOCATE 16,32*SR%:INPUT SELECTION
3480 IF SELECTION <1 OR SELECTION>3 THEN 3400
3490 ON SELECTION GOTO 2300,4000,9500
3995 REM ***** Colour Plotter Display *****
4000 GX=0:GY=0:SX=192:SY=120
4010 MG=0:MH=0:MV=0
4020 MH=SX/DX:MV=SY/DY:MX=2670/SX
4030 IF MH<=MV THEN MG=MH ELSE MG=MV
4040 PX=MG*DX:PY=MG*DY
4050 CLS:GOSUB 9000
4051 PRINT:PRINT " Select Plotter";TAB(18);"<1> Apple 410"
4052 PRINT:PRINT TAB(18);"<2> Epson HI-80":PRINT:PRINT TAB(18);"<3> H
ewlett-Packard";:INPUT PLOT%
4054 IF PLOT%<1 OR PLOT%>3 THEN 4050
4056 IF PLOT%=1 THEN HP=3:REM Apple
4058 IF PLOT%=2 THEN HP=4:REM Epson
4059 IF PLOT%=3 THEN HP=1:REM HP-GL
4060 PRINT:PRINT:PRINT TAB(3);"Select Data for Plotter :-"
4062 PRINT:PRINT TAB(18);"<1> Point No. or"
4064 PRINT:PRINT TAB(18);"<2> Z-Value ";
4066 INPUT SD%
4068 IF SD%<1 OR SD%>2 THEN 4060
4070 PRINT:PRINT:PRINT "  Place paper on Plotter and switch on"
4080 PRINT:INPUT "  To Continue Press <Enter> ";R$
4090 OPEN "COM1:1200,N,7,1,CS2000,DS15000" FOR OUTPUT AS #2
4100 ON HP GOTO 4250,4250,4105,4200
4103  REM  ***** Apple 410 *****
4105 PRINT#2,"SP3;"
4110 PRINT#2,"WD -5,-5,192,125;"
4120 PRINT#2,"PS1;"
4140 PRINT#2,"MA";GX;",";GY;";"
4141 PRINT#2,"DA";PX;",";GY;";"
4142 PRINT#2,"DA";PX;",";PY;";"
4143 PRINT#2,"DA";GX;",";PY;";"
4144 PRINT#2,"DA";GX;",";GY;";"
4150 FOR I=10 TO (NC-10) STEP 10
4155 FOR J=10 TO (NR-10) STEP 10
4160 PGX0=GX+I*PX/(NC-1)
4161 PGX1=PGX0-2:PGX2=PGX0+2
4162 PGY0=GY+J*PY/(NR-1)
4163 PGY1=PGY0-2:PGY2=PGY0+2
4170 PRINT#2,"MA";PGX1;",";PGY0;";":PRINT#2,"DA";PGX2;",";PGY0;";"
4175 PRINT#2,"MA";PGX0;",";PGY1;";":PRINT#2,"DA";PGX0;",";PGY2;";"
4180 NEXT J
4185 NEXT I
4190 GOTO 4300
4195  REM  ***** Epson HI-80 *****
4200 PRINT#2,"IW0,0,2670,1920"
4201 PRINT#2,"MA";GX;",";GY
4202 PRINT#2,"OR"
4203 PRINT#2,"SP1":PRINT#2,"LT0"
4204 PRINT#2,"MA";GX;",";GY
4205 PRINT#2,"DA";PX*MX;",";GY;",";PX*MX;",";PY*MX;",";GX;",";PY*MX;"
,";GX;",";GY
4210 FOR I=10 TO (NC-10) STEP 10
4211 FOR J=10 TO (NR-10) STEP 10
4212 PGX0=GX+I*PX*MX/(NC-1)
4213 PGX1=PGX0-2*MX:PGX2=PGX0+2*MX
4214 PGY0=GY+J*PY*MX/(NR-1)
4215 PGY1=PGY0-2*MX:PGY2=PGY0+2*MX
4216 PRINT#2,"MA";PGX1;",";PGY0:PRINT#2,"DA";PGX2;",";PGY0
4217 PRINT#2,"MA";PGX0;",";PGY1:PRINT#2,"DA";PGX0;",";PGY2
```

```
4218 NEXT J
4220 NEXT I
4230  GOTO 4300
4245  REM  ***** Hewlett-Packard (HP-GL) *****
4250 PRINT#2,"PS3;"
4252 PRINT#2,"IP 170,602,15370,10602;"
4254 PRINT#2,"SC -10,180,-5,120;"
4256 PRINT#2,"SP1;"
4258 '
4260 PRINT#2,"PU;PA";GX;",";GY;";"
4261 PRINT#2,"PD;PA";PX;",";GY;";":TSEC=2 :GOSUB 60000
4262 PRINT#2,"PA";PX;",";PY;";":GOSUB 60000
4263 PRINT#2,"PA";GX;",";PY;";":GOSUB 60000
4264 PRINT#2,"PA";GX;",";GY;";":GOSUB 60000
4270 FOR I=10 TO (NC-10) STEP 10
4271 FOR J=10 TO (NR-10) STEP 10
4272 PGX0=GX+I*PX/(NC-1)
4273 PGX1=PGX0-2:PGX2=PGX0+2
4274 PGY0=GY+J*PY/(NR-1)
4275 PGY1=PGY0-2:PGY2=PGY0+2
4280 PRINT#2,"PU;PA";PGX1;",";PGY0;";":PRINT#2,"PD;PA";PGX2;",";PGY0;
";"
4285 PRINT#2,"PU;PA";PGX0;",";PGY1;";":PRINT#2,"PD;PA";PGX0;",";PGY2;
";"
4290 NEXT J
4295 NEXT I
4300 ON HP GOTO 4700,4700,4310,4400
4305  REM  ***** Apple 410 *****
4310 FOR J=1 TO N
4312 IF SD%=1 THEN S(J)=P(J)
4314 IF SD%=2 THEN S(J)=Z(J)
4320 PRINT#2,"PS2;":PRINT#2,"LS2;"
4330 PRINT#2,"MA";(GX+(X(J)-XL)*MG);",";(GY+(Y(J)-YB)*MG);";"
4340 PRINT#2,"PM8;"
4350 PRINT#2,"LS1;"
4360 PRINT#2,"PL":PRINT#2,S(J)
4370 PRINT#2,CHR$(13) CHR$(10) CHR$(3)
4380 NEXT J
4390 GOTO 4500
4395  REM  ***** Epson HI-80 *****
4400 FOR J=1 TO N
4410 IF SD%=1 THEN S(J)=P(J)
4414 IF SD%=2 THEN S(J)=Z(J)
4420 PRINT#2,"SP2":PRINT#2,"SI40,30"
4430 PRINT#2,"MA";(GX+(X(J)-XL)*MG)*MX;",";(GY+(Y(J)-YB)*MG)*MX
4440 PRINT#2,"AM8"
4450 PRINT#2,"SI25,20"
4460 PRINT#2,"LA";S(J)
4480 NEXT J
4490 GOTO 4600
4495  REM  ***** Apple 410 *****
4500 PRINT#2,"LS2;"
4510 PRINT#2,"MA 10,-3;"
4520 PRINT#2,"PL":PRINT#2,"PC-Random/RData File - ";F$;" : Date :";DA
TE$
4530  PRINT #2,CHR$(13) CHR$(10) CHR$(3)
4540 CLOSE#2
4550 GOTO 2000
4595  REM  ***** Epson HI-80 *****
4600 PRINT#2,"SI40,30"
4610 PRINT#2,"MA";5*MX;",";2*MX
4620 PRINT#2,"LAPC-Random/RData File - ";F$;" : Date :";DATE$
4640 CLOSE#2
4650 GOTO 2000
4695  REM  ***** Hewlett-Packard (HP-GL) *****
```

```
4700 FOR J=1 TO N
4712 IF SD%=1 THEN S(J)=P(J)
4714 IF SD%=2 THEN S(J)=Z(J)
4720 PRINT#2,"SP3;"
4730 PRINT#2,"PU;PA";(GX+(X(J)-XL)*MG);",";(GY+(Y(J)-YB)*MG);";"
4750 PRINT#2,"SI0.18,0.22;"
4760 PRINT#2,"LB+";S(J)
4770 PRINT#2,CHR$(3)
4780 NEXT J
4800 PRINT#2,"SP2;SI0.2,0.3;"
4810 PRINT#2,"PU;PA 10,-3;"
4820 PRINT#2,"LB":PRINT#2,"PC-Random/RData File - ";F$;" : Date :";DATE$
4830 PRINT#2,CHR$(3)
4840 CLOSE#2
4850 GOTO 2000
4995  REM  ***** User DATA Storage area *****
5995  REM ***** Printout of Data File Details *****
6000 CLS
6010 GOSUB 9000
6020 PRINT "  Place paper in Printer and switch on"
6030 PRINT
6040 INPUT "  To Continue Press <Enter> ";R$
6100 LPRINT "PC-Random/RData File - ";F$;" - Date :";DATE$
6120 LPRINT "*******************************************************************"
6130 LPRINT
6140 LPRINT "X-coordinate Left hand side ";TAB(35);"=";XL
6150 LPRINT "X-coordinate Right hand side";TAB(35);"=";XR
6160 LPRINT "Y-coordinate at foot of plot";TAB(35);"=";YB
6170 LPRINT "Y-coordinate at top  of plot";TAB(35);"=";YT
6180 LPRINT "Number of data points";TAB(35);"=";N
6190 LPRINT
6200 LPRINT "Point No.";TAB(15);"X-coord";TAB(30);"Y-coord";TAB(45);"Z-coord"
6210 LPRINT "======================================================="
6220 FOR I=1 TO N
6230 LPRINT P(I);TAB(14);X(I);TAB(29);Y(I);TAB(44);Z(I)
6240 NEXT I
6250 LPRINT "*******************************************************"
6260 LPRINT
6270 SR%=1:GOTO 2000
8995 REM ***** PC-Random Heading *****
9000 P$="PC-Random : ":T$=P$+F$
9010 LOCATE 1,2:PRINT T$
9020 LOCATE 1,30:PRINT DATE$
9030 PRINT " ========================================"
9040 PRINT
9050 RETURN
9500 CLS
9502 SCREEN SCR%:IF SCR%=0 THEN 9510
9504 IF SCR%=1 THEN 9508
9506 SCREEN 1:GOTO 9510
9508 COLOR 8,0
9510 GOSUB 9000
9520 PRINT:PRINT "  Program Complete  "
9530 PRINT
9540 PRINT:PRINT " Do you wish to run PC-Contour now (y/n)"
9550 INPUT Q$:IF Q$="n" THEN 9600
9560 IF Q$<>"y" THEN 9540
9570 PRINT:PRINT:PRINT "  Loading PC-Contour from disc ....."
9580 RUN "CONTOUR"
9590 PRINT
9600 PRINT:PRINT
9650 END
```

3.3.5 PC-Random/Demo

To be able to run PC-Random/RData it is necessary to prepare a data file. It is suggested that once the following **DATA** statements have been entered into the file PC-Random/RData that it is resaved under the title **"RANDDEMO"** with changed remarks in lines 10 and 25. The data file presented is for a tacheometric survey at Glenmore, and the resulting data file will be saved automatically on disc as **"GLENMORE"** for subsequent access by PC-Contour and PC-Surfplot.

Initial data statements give the file name **"Glenmore"** followed by the X-min, X-max, Y-min, Y-max coordinates of the plotted area. This is then followed by the number of data points (30), and the X, Y, Z-coordinates of each point. For simplicity, it is suggested that each data point is entered in a separate **DATA** statement for ease of checking and to save confusion with the data points. There is no need to identify the maximum and minimum Z-values as these will be located in the **READ** routine.

3.3.6 PC-Random/Demo – data

```
5000 DATA "Glenmore"
5010 DATA 0,400
5020 DATA 0,300
5030 DATA 50
5040 DATA 1,372,24,173.3:REM Line 1
5050 DATA 2,370,74,176.8
5060 DATA 3,368,122,185.6
5070 DATA 4,366,148,187.7
5080 DATA 5,364,185,192.9
5090 DATA 6,362,218,204.2
5100 DATA 7,360,254,213.8
5110 DATA 8,358,273,216.8
5120 DATA 9,342,97,177.2:REM Line 2
5130 DATA 10,309,122,178.5
5140 DATA 11,272,148,180.4
5150 DATA 12,241,171,182.6
5160 DATA 13,205,196,186.5
5170 DATA 14,176,219,189.3
5180 DATA 15,144,242,194.8
5190 DATA 16,121,261,204.1
5200 DATA 17,99,278,208.3
5210 DATA 18,251,276,209.2
5220 DATA 19,336,41,174.3:REM Line 3
5230 DATA 20,304,57,180.9
5240 DATA 21,269,76,186.1
5250 DATA 22,228,101,188.8
5260 DATA 23,205,122,191.3
5270 DATA 24,172,131,193.7
5280 DATA 25,147,143,196.2
5290 DATA 26,109,166,199.1
5300 DATA 27,72,185,202.4
5310 DATA 28,61,216,205.8
5320 DATA 29,45,260,212.2
5330 DATA 30,205,32,178.5
5340 DATA 31,236,41,182.3:REM Line 4
5350 DATA 32,287,101,183.6
```

```
5360 DATA 33,339,156,186.5
5370 DATA 34,133,13,178.3:REM Line 5
5380 DATA 35,155,42,182.8
5390 DATA 36,182,80,188.1
5400 DATA 37,224,136,187.2
5410 DATA 38,265,196,189.3
5420 DATA 39,288,229,200.9
5430 DATA 40,309,259,211.1
5440 DATA 41,329,277,215.7
5450 DATA 42,21,26,178.2:REM Line 6
5460 DATA 43,54,88,187.3
5470 DATA 44,76,119,193.9
5480 DATA 45,132,202,193.8
5490 DATA 46,171,256,197.4
5500 DATA 47,187,277,206.2
5510 DATA 48,96,51,182.9
5520 DATA 49,135,110,193.5
5530 DATA 50,229,238,196.2
```

3.3.7 PC-Random – computer printout of Demo file

```
PC-Random/RData File - Glenmore - Date :01-01-1987
****************************************************************

X-coordinate Left hand side        = 0
X-coordinate Right hand side       = 400
Y-coordinate at foot of plot       = 0
Y-coordinate at top  of plot       = 300
Number of data points              = 50
```

Point No.	X-coord	Y-coord	Z-coord
1	372	24	173.3
2	370	74	176.8
3	368	122	185.6
4	366	148	187.7
5	364	185	192.9
6	362	218	204.2
7	360	254	213.8
8	358	273	216.8
9	342	97	177.2
10	309	122	178.5
11	272	148	180.4
12	241	171	182.6
13	205	196	186.5
14	176	219	189.3
15	144	242	194.8
16	121	261	204.1
17	99	278	208.3
18	251	276	209.2
19	336	41	174.3
20	304	57	180.9
21	269	76	186.1
22	228	101	188.8
23	205	122	191.3
24	172	131	193.7
25	147	143	196.2
26	109	166	199.1
27	72	185	202.4
28	61	216	205.8
29	45	260	212.2
30	205	32	178.5
31	236	41	182.3

```
32          287          101          183.6
33          339          156          186.5
34          133          13           178.3
35          155          42           182.8
36          182          80           188.1
37          224          136          187.2
38          265          196          189.3
39          288          229          200.9
40          309          259          211.1
41          329          277          215.7
42          21           26           178.2
43          54           88           187.3
44          76           119          193.9
45          132          202          193.8
46          171          256          197.4
47          187          277          206.2
48          96           51           182.9
49          135          110          193.5
50          229          238          196.2
*****************************************************
```

3.3.8 PC-Random/FData

This is the third program in the PC-Random suite and is provided for those who may already have random data stored in a disc file, for example, as generated by two programs in the PC-Survey suite of programs for land surveying. Two of the programs from *BASIC Programs for Land Surveying* allow field data collected by land surveying or topographic mapping, using either tacheometry or EDM (Electromagnetic Distance Measurement) to be plotted on various plotters and stored on disc for access by PC-Random/FData.

This program follows the same format as PC-Random/RData and only the lines listed in Section 3.3.9 need to be altered. The main changes occur in lines 1010–1060, where instead of reading data statements, the data is read from a disc file.

3.3.9 PC-Random/FData – BASIC program

Changes to PC-Random/RData

```
10 REM  <RANDFDAT> Program for IBM PC and Compatibles

25 REM  Reads DATA File stored under F$+".DAT" by PC-Survey

170 PRINT "  Select application required :-"
180 PRINT
190 PRINT "  <1> Read DATA and present details"
200 PRINT
210 PRINT "  <2> Read DATA and store on disc"
220 PRINT
230 PRINT "  <3> Quit ";
260 INPUT RAND%:IF RAND%<1 OR RAND%>3 THEN 100
270 IF RAND%=3 THEN 9500
```

```
995 REM  ***** DATA File Read Routine *****
1000 CLS:GOSUB 9000
1010 INPUT "  Enter DATA file name ";F$
1020 D$=F$+".DAT"
1030 OPEN "I",#4,D$
1040 INPUT#4,XL,XR
1050 INPUT#4,YB,YT
1060 INPUT#4,N

1130 INPUT#4,P(I),X(I),Y(I),Z(I)
1140  IF Z(I) > MAX THEN MAX = Z(I)
1150  IF Z(I) < MIN THEN MIN = Z(I)
1160  NEXT I
1165 CLOSE#4

3210 LOCATE 1,2:PRINT "PC-Random/FData - ";F$

4520 PRINT#2,"PL":PRINT#2,"PC-Random/FData File - ";F$;" : Date :";DA
TE$

4620 PRINT#2,"LAPC-Random/FData File - ";F$;" : Date :";DATE$

4820 PRINT#2,"LB":PRINT#2,"PC-Random/FData File - ";F$;" : Date :";DA
TE$

6100 LPRINT "PC-Random/FData File - ";F$;" - Date :";DATE$
```

3.4 PC-CONTOUR

This program accesses disc files generated by either PC-Grid or PC-Random and can be run with the two demonstration data files supplied, that is **“GLAPWELL”** for spoil-heap contours and **“GLENMORE”** for a tacheometric survey.

On running PC-Contour, the user is asked to enter the file name for the disc data file. After reading the data file, the data information is displayed on screen with advice on the size of the plot, the numbers of rows and columns in the grid, and the maximum and minimum values for the plot. These latter values will be the limits for the contour values when plotting on screen or plotter.

3.4.1 PC-Contour – subroutine index

	Line numbers	*Function*
(a)	10–90	Initialization and control
(b)	100–1170	Data read routine
(c)	2000–2180	Data display and plot selection
(d)	2500–3990	Graphics screen display
(e)	4000–5190	Plotter routines
(f)	9000–9050	Routine to clear screen
(g)	9500–9540	Termination of program

(a) Initialization and control Line numbers 10–90

The program is initialized and the installation file checked for the availability of graphics screens and colour for subsequent display of the contours.

(b) Data read routine Line numbers 100–1170

This routine requests the data file name and prompts the user with the two demonstration data files prepared in Sections 3.2 and 3.3.

(c) Data display and plot selection Line numbers 2000–2180

The presentation of the data information is similar to that described in Sections 3.2 and 3.3. The user should note the maximum and minimum Z-values to be used in the selection of contour levels, before selecting either the graphics screen or plotter for the contour display.

(d) Graphics screen display Line numbers 2500–3990

The user is first given a choice of low-resolution (**SCREEN 1**) or high-resolution (**SCREEN 2**) for the graphics display which uses the format shown in Fig. 1.4. In addition to the normal box outline, grid nodes are plotted at the junction of every ten grid lines.

The right-hand text column is used to enter the lowest contour level, the highest contour level and the contour interval. Contours are interpolated between each set of four grid points in turn using an inverse square distance weighted interpolation. During plotting, the current contour value is displayed on screen. Each time a new contour is encountered its value will be noted alongside the contour. It is possible at the end of the plot to go back and insert more contours or to start again with different contour values.

Thus it is suggested that for the first plot, a contour interval of 10 is selected, followed by 5, for example. As the graphics plotter routines take longer than the screen plot, it is often useful to run the screen display first to check the output before using the plotter. On completion, a screen dump of the graphics display can be obtained on a dot-matrix graphics printer (Fig. 3.3) for comparison with the data (Figs 3.1 and 3.2) of the PC-Random file **"GLENMORE"**. The PC-Grid plot file **"GLAPWELL"** is shown on Fig. 3.4.

(e) Plotter routines Line numbers 4000–5190

Three plotter routines are provided for the Apple 410, Epson HI-80 and Hewlett-Packard HP-7475A colour plotters as discussed in Chapter 2. Most of the colour plotters take four pens and the default settings are 1-black, 2-red, 3-green, 4-blue. The border frame and grid nodes are plotted by pen 1 (black), the 10-unit contour is plotted by pen 3 (green), the 5-unit contour by pen 4 (blue) and the 2.5-unit contour by pen 2 (red). Each time a new contour is encountered its values will be noted alongside the contour. All

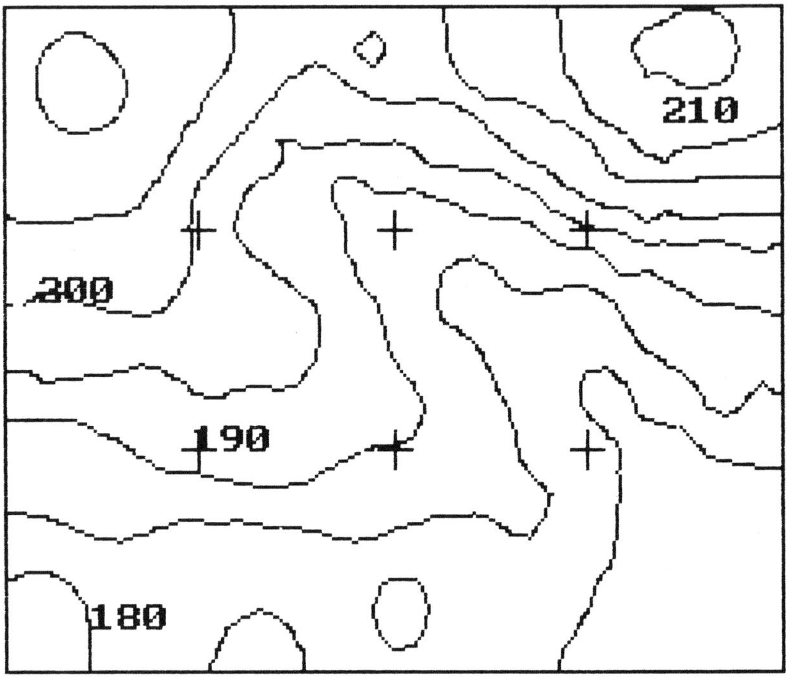

Fig. 3.3 PC-Contour plot of contours for Glenmore tacheometry survey on Screen 1 (medium-resolution).

other unit values (e.g. 2) will be plotted by pen 3 (green). The final title, location and date will be plotted by pen 2 (red). If, however, the user wishes to change these colours, the appropriate pen selection should be changed using the commands in Table 2.1.

The contour plot for the PC-Random file **"GLENMORE"** is shown in Fig. 3.5 for comparison with Fig. 3.3. The contour plot for the PC-Grid file **"GLAPWELL"** is shown in Fig. 3.6 for comparison with Fig. 3.4.

(f) Routine to clear screen Line numbers 9000–9050

This routine is provided to clear the screen before each plot and provide text at the top of the screen to identify the program title and data file presentation together with the current date.

(g) Termination of program Line numbers 9500–9540

This is the final display at the completion of the program and the user is given an option to run PC-Contour again or return to BASIC command level.

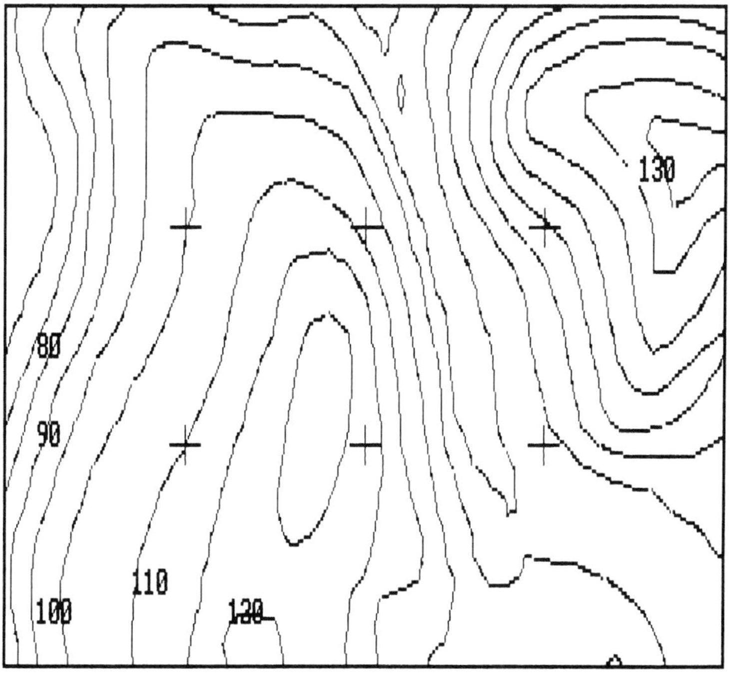

Fig. 3.4 PC-Contour plot of Glapwell spoil-heap contours of Screen 2 (high-resolution).

3.4.2 PC-Contour – numeric variables

A	= Variable used in contour interpolation
B	= Variable used in contour interpolation
C	= Variable used in contour interpolation
CC	= Current contour value
CCOL	= Column number for text display on graphics screen
CROW	= Row number for text display on graphics screen
D **DR** **DS**	} Variables used in contour interpolation
DX	= *X*-distance of plotted area
DY	= *Y*-distance of plotted area
GPXØ–2	= Variables to plot grid nodes on screen
GPYØ–2	= Variables to plot grid nodes on screen
GX	= *X*-coordinate at lower left-hand corner of plot
GY	= *Y*-coordinate at lower left-hand corner of plot
HC	= Highest contour on plot
HP%	= Integer variable to identify plotter
H(NR,NC)	= Array to read *Z*-values

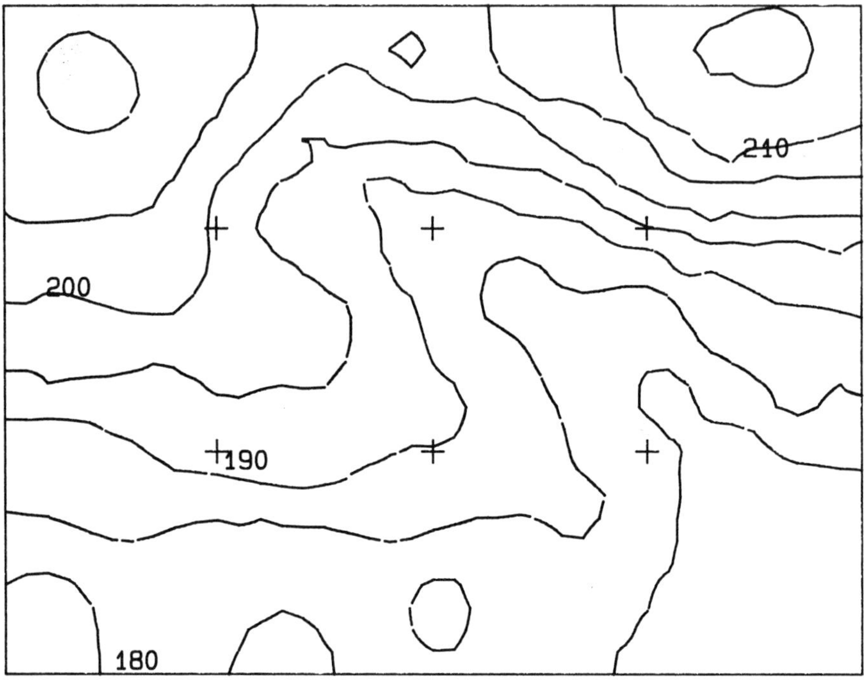

Fig. 3.5 PC-Contour plot of contours for Glenmore tacheometry survey on plotter.

I, **J**, **K**	Integer variables used in loops
LC	= Lowest contour on plot
MAX	= Maximum *Z*-value in data
MG	= Multiplication factor for plotted display
MH	= Multiplication factor for horizontal resolution
MIN	= Minimum *Z*-value in data
MV	= Multiplication factor for vertical resolution
MX	= Multiplication factor for Epson HI-80 plotter
NC	= Number of columns in grid
NP	= Number of points in contour interpolation
NR	= Number of rows in grid
PGXØ–2	= Variables to plot grid nodes on plotter
PGYØ–2	= Variables to plot grid nodes on plotter
PP	= Screen display variable
PX	= *X*-coordinate at upper right-hand corner of plot
PY	= *Y*-coordinate at upper right-hand corner of plot
PLOT%	= Integer variable for plotter selection
RGB%	= Integer variable to select line colour
SX	= Horizontal display width in user units
SY	= Vertical display height in user units

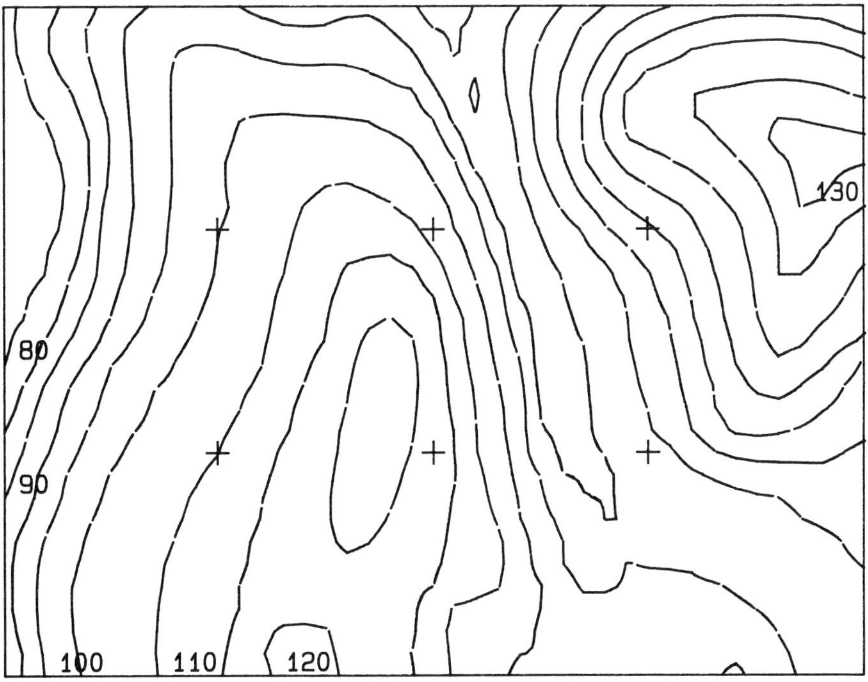

Fig. 3.6 PC-Contour plot of Glapwell spoil-heap contours on plotter.

SCD%	= Integer variable to define screen display
SCR%	= Integer variable to select screen or plotter
SR%	= Integer variable for screen resolution
TL	= Variable used in titling contour level
X,XØ,X1	= Variables used in contour interpolation
Y,YØ,Y1	= Variables used in contour interpolation
Z1–Z4	= *Z*-levels of grid points in contour interpolation
ZT	= Variable used in contour interpolation

3.4.3 PC-Contour – string variables

DATE$	Variable used to return the date of the computer's calendar/clock in US format MM–DD–YYYY.
F$	File name
GD$	Entry for graphics dump on printer
M$	= **"MORE"**, **"ERASE"** or **"END"**
P$	= **"PC-Contour"**
PCF$	= **"PCSCREEN"**, installation file for graphics
Q$	Question response (Y/N) or (y/n)
T$	= **P$** + **F$**

3.4.4 PC-Contour – BASIC program

```
10  REM  <CONTOUR> Program for IBM PC and Compatibles
12  REM  Version PC-1.0, 01-01-1987
14  REM  (C) Copyright  1987  P.H. Milne
16  REM  All rights reserved
20  REM  CONTOUR Plot on <1> Apple 410, <2> Epson HI-80, or <3> HPGL Colour Plotter
22  REM  Setting in line 4090 "COM1:1200 N,7,1,CS2000,DS15000"
25  REM  This version includes Graphics printer dump of Screen
30  CLEAR
35 PCF$="PCSCREEN"
40 OPEN "I",#3,PCF$
45 INPUT#3,SCR%
50 INPUT#3,SCD%
55 CLOSE#3
60 ON (SCR%+1) GOTO 62,64,66
62 SCREEN 0:WIDTH 40:GOTO 80
64 SCREEN 1:GOTO 68
66 SCREEN 1:GOTO 80
68 IF SCD%=1 THEN 80
70 COLOR 8,0
80 KEY OFF
90 F$=" ":RGB%=0:SR%=1
100  CLS
110  GOSUB 9000
120  PRINT
130  PRINT " THIS PROGRAM READS AN ELEVATION FILE"
140  PRINT " FROM A REGULAR SQUARE GRID TO DISPLAY"
150  PRINT " A CONTOUR MAP ON SCREEN OR PLOTTER."
160  PRINT
200 GOSUB 1000: REM Read DATA FILE
210 GOTO 2000: REM Select Output
995 REM ***** Read Data from FILE *****
1000  PRINT:PRINT " PC-Grid/Demo   file called GLAPWELL":PRINT " PC-Random/Demo file called GLENMORE"
1005  PRINT:PRINT:PRINT " Enter DATA FILE name  ";
1010  INPUT F$:IF F$="" THEN 1005
1015 PRINT:PRINT:PRINT " One moment please ..... reading data":PRINT
1020  OPEN "I",#1,F$
1030  INPUT#1, NR
1040  INPUT#1, NC
1050  INPUT#1, MAX
1060  INPUT#1, MIN
1070  INPUT#1, DX
1080  INPUT#1, DY
1090  DIM H((NR - 1),(NC - 1))
1100  FOR I = (NR-1) TO 0 STEP -1
1110  FOR J = 0 TO (NC - 1)
1120  INPUT#1, H1
1130 H(I,J) = H1
1140  NEXT J
1150  NEXT I
1160  CLOSE #1
1170 RETURN
1995 REM ***** Data Information and Plot Selection *****
2000 CLS
2002 SCREEN SCR%:IF SCR%=0 THEN 2010
2004 IF SCR%=1 THEN 2008
2006 SCREEN 1:GOTO 2010
2008 COLOR 8,0
2010 GOSUB 9000
2020  PRINT " THE DATA FILE ";F$;" HAS BEEN READ"
2030  PRINT
2040  PRINT "    No. Rows = ";NR;"  No. Cols.= ";NC
2050  PRINT "    Minimum value = ";MIN
2060  PRINT "    Maximum value = ";MAX
```

```
2070  PRINT "    Range          = ";INT((MAX - MIN)*100+.1)/100
2080  PRINT "    1 Map row      = ";(DY / (NR-1));" UNITS"
2090  PRINT "    1 Map col.     = ";(DX / (NC-1));" UNITS"
2100 BEEP
2110 PRINT:PRINT:PRINT " Select Display :-"
2120 PRINT:PRINT " <1> SCREEN, <2> PLOTTER or <3> QUIT"
2130 IF SCR%<> 0 THEN 2150
2140 PRINT:PRINT " N.B. Option <1> not available":PRINT
2150 INPUT SELECTION
2160 IF SELECTION<1 OR SELECTION>3 THEN 2100
2170 IF SCR%=0 THEN IF SELECTION=1 THEN 2000
2180 ON SELECTION GOTO 2500,4000,9500
2495 REM Choose Screen for Plotting
2500 CLS
2510 GOSUB 9000
2520 PRINT:PRINT TAB(3);"Select SCREEN";TAB(20);"<1> Colour -"
2530 PRINT TAB(20);"    Low resolution"
2540 PRINT:PRINT TAB(20);"<2> Black & White -"
2550 PRINT TAB(20);"    High resolution"
2560 PRINT TAB(3);:INPUT SR%
2570 IF SR%<1 OR SR%>2 THEN 2520
2580 IF SR%=1 THEN 3000
2600 SCREEN 2
2995 REM ***** Screen Plot *****
3000  CLS
3010 GX=5*SR%:GY=5:MG=0:MH=0:MV=0:SX=240*SR%:SY=180:PP=0
3020 MH=SX/((NC-1)*SR%):MV=SY/(NR-1)
3030 IF MH<=MV THEN MG=MH ELSE MG=MV
3040 WINDOW (0,0) - (320*SR%,200)
3050 IF SR%=1 THEN RGB%=2 ELSE RGB%=3
3060 PX=GX+(MG*(NC-1)*SR%):PY=GY+MG*(NR-1)
3070 LINE (GX,GY)-(PX,PY),RGB%,B
3080 IF SR%=1 THEN 3100
3090 LINE (GX-1,GY)-(PX-1,PY),RGB%,B
3100 FOR I=10 TO (NC-10) STEP 10
3110 FOR J=10 TO (NR-10) STEP 10
3120 GPX0=GX+(MG*I*SR%)
3130 GPX1=GPX0-5*SR%:GPX2=GPX0+5*SR%
3140 GPY0=GY+(MG*J)
3150 GPY1=GPY0-5:GPY2=GPY0+5
3160 LINE (GPX1,GPY0) - (GPX2,GPY0),RGB%
3170 LINE (GPX0,GPY1) - (GPX0,GPY2),RGB%
3180 NEXT J
3190 NEXT I
3195 IF PP=1 THEN 3880
3200 ' Start of 'MORE' contours
3220 LOCATE 1,2:PRINT "PC-Contour - ";F$
3230 LOCATE 1,30*SR%:PRINT DATE$
3240  LOCATE 3,32*SR%:PRINT "CONTOURS"
3250  LOCATE 4,32*SR%:PRINT "LOW VAL:"
3252  LOCATE 5,32*SR%:PRINT "        "
3255  LOCATE 5,32*SR%:INPUT LC
3260  LOCATE 6,32*SR%:PRINT "HIGH VAL:"
3262  LOCATE 7,32*SR%:PRINT "        "
3265  LOCATE 7,32*SR%:INPUT HC
3270  LOCATE 9,32*SR%:PRINT "CONTOUR"
3272  LOCATE 10,32*SR%:PRINT "          "
3275  LOCATE 10,32*SR%:INPUT "INT.";CI
3280  IF LC < (MIN - CI) THEN LC = LC + CI
3290 IF HC>MAX THEN HC=MAX
3300  FOR CC = LC TO HC STEP CI
3310  LOCATE 11,32*SR%:PRINT "CURRENT"
3312  LOCATE 12,32*SR%:PRINT "        "
3315  LOCATE 12,32*SR%:PRINT "=";CC
3320  FOR I = 0 TO (NR-2)
```

```
3330 Y0 = MG* I
3340  FOR J = 0 TO (NC - 2)
3350 X0 = MG* J
3360 NP = 0
3370 Z1 = H(I,J)
3380  IF CC > Z1 THEN NP = NP + 1
3390 Z2 = H(I,(J + 1))
3400  IF CC > Z2 THEN NP = NP + 1
3410 Z3 = H((I + 1),J)
3420  IF CC > Z3 THEN NP = NP + 1
3430 Z4 = H((I + 1),(J + 1))
3440  IF CC> Z4 THEN NP = NP + 1
3450  IF NP = 0 OR NP = 4 GOTO 3840
3460 A = Z1
3470 B = Z2 - A
3480 C = Z3 - A
3490 D = Z4 - A - B - C
3500 ZT = 0
3510  FOR Y1 = 0 TO 1 STEP .25
3520 DR = B + D * Y1
3530  IF DR = 0 GOTO 3680
3540 X1 = (CC - A - C * Y1) / DR
3550  IF X1 < 0 OR X1 > 1 GOTO 3680
3560 X = X0 + MG* X1
3570 Y = Y0 + MG* Y1
3580  IF X > PX OR Y > PY GOTO 3670
3590  IF ZT = 0 THEN 3620
3600  LINE - ((X*SR%+GX),(Y+GY)),1
3610 GOTO 3670
3620 PSET ((X*SR%+GX),(Y+GY)),1
3630 IF TL>0 THEN 3670
3631 IF CC/10 <> INT(CC/10) THEN 3670
3632 CCOL=(X*SR%+GX)/8:CROW=(Y+GY)/8
3633 IF CCOL<2*SR% THEN CCOL=2*SR%
3634 IF CCOL>27*SR% THEN CCOL=27*SR%
3635 IF CROW<=2 THEN CROW=2
3636 IF CROW>22 THEN CROW=22
3637 IF CC/20 <> INT(CC/20) THEN 3639
3638 LOCATE (25-CROW),CCOL:GOTO 3640
3639 LOCATE (24-CROW),CCOL
3640 PRINT CC
3650 TL=TL+1
3660 GOTO 3620
3670 ZT = ZT + 1
3680  NEXT Y1
3690 ZT = 0
3700 FOR X1 = 0 TO 1 STEP .25
3710 DS = C + D * X1
3720  IF DS = 0 GOTO 3830
3730 Y1 = (CC - A - B * X1) / DS
3740  IF Y1 < 0 OR Y1 > 1 GOTO 3830
3750 X = X0 + MG* X1
3760 Y = Y0 + MG* Y1
3770  IF X > PX OR Y > PY GOTO 3820
3780  IF ZT = 0 THEN  3810
3790  LINE - ((X*SR%+GX),(Y+GY)),1
3800 GOTO 3820
3810 PSET ((X*SR%+GX),(Y+GY)),1
3820 ZT = ZT + 1
3830  NEXT X1
3840  NEXT J
3850  NEXT I
3860 TL=0
3870  NEXT CC
3875 PP=1:GOTO 3070
3880  LOCATE 14,32*SR%:PRINT "SCREEN"
```

```
3882  LOCATE 15,32*SR%:PRINT "DUMP(y/n)"
3884  LOCATE 16,32*SR%:PRINT "         "
3885 BEEP
3886  LOCATE 16,32*SR%:INPUT GD$
3888  IF GD$="n" THEN 3910
3890  IF GD$<>"y" THEN 3880
3900 LCOPY
3910  LOCATE 18,32*SR%:PRINT "'MORE'"
3920  LOCATE 19,32*SR%:PRINT "CONTOURS"
3930  LOCATE 20,32*SR%:PRINT "'ERASE',"
3940  LOCATE 21,32*SR%:PRINT "or 'END'"
3945 BEEP
3950  LOCATE 22,32*SR%:PRINT "         "
3955  LOCATE 22,32*SR%:INPUT M$
3960  IF M$ = "MORE" GOTO 3200
3970  IF M$ = "ERASE" GOTO 3000
3980 M$=" ":SR%=1
3990  CLS:GOTO 2000
3995 REM ***** Colour Plotter Display *****
4000 IF M$="MORE" THEN 4230
4010 GX=0:GY=0:MG=0:MH=0:MV=0:SX=192:SY=120
4020 MH=SX/(NC-1):MV=SY/(NR-1):MX=2670/SX
4030 IF MH<=MV THEN MG=MH ELSE MG=MV
4040 PX=MG*(NC-1):PY=MG*(NR-1)
4050 CLS: GOSUB 9000
4051 PRINT:PRINT " Select Plotter <1> Apple 410"
4052 PRINT:PRINT "                <2> Epson HI-80":PRINT:PRINT "
        <3> Hewlett-Packard ";:INPUT PLOT%
4054 IF PLOT%<1 OR PLOT%>3 THEN 4050
4056 IF PLOT%=1 THEN HP=3:REM Apple
4058 IF PLOT%=2 THEN HP=4:REM Epson
4059 IF PLOT%=3 THEN HP=1:REM Hewlett-Packard (HPGL)
4070 PRINT:PRINT:PRINT "  Place paper on Plotter and switch on"
4080 PRINT:PRINT "  To Continue Press <Enter> ":WHILE R$="":R$=INKEY$
:WEND
4090 OPEN "COM1:1200 N,7,1,CS2000,DS15000" FOR OUTPUT AS #2
4100 ON HP GOTO 5100,5100,4105,4200
4103 REM ***** Apple 410 *****
4105 PRINT#2,"SP3;"
4110 PRINT#2,"WD -5,-5,192,125;"
4120 PRINT#2,"PS1;"
4140 PRINT#2,"MA";GX;",";GY;";"
4141 PRINT#2,"DA";PX;",";GY;";"
4142 PRINT#2,"DA";PX;",";PY;";"
4143 PRINT#2,"DA";GX;",";PY;";"
4144 PRINT#2,"DA";GX;",";GY;";"
4150 FOR I=10 TO (NC-10) STEP 10
4155 FOR J=10 TO (NR-10) STEP 10
4160 PGX0=GX+(MG*I)
4161 PGX1=PGX0-2:PGX2=PGX0+2
4162 PGY0=GY+(MG*J)
4163 PGY1=PGY0-2:PGY2=PGY0+2
4170 PRINT#2,"MA";PGX1;",";PGY0;";":PRINT#2,"DA";PGX2;",";PGY0;";"
4175 PRINT#2,"MA";PGX0;",";PGY1;";":PRINT#2,"DA";PGX0;",";PGY2;";"
4180 NEXT J
4185 NEXT I
4190 GOTO 4230
4195 REM ***** Epson HI-80 *****
4200 PRINT#2,"IW0,0,2670,1920"
4201 PRINT#2,"MA";GX;",";GY
4202 PRINT#2,"OR"
4203 PRINT#2,"SP1":PRINT#2,"LT0"
4204 PRINT#2,"MA";GX;",";GY
4205 PRINT#2,"DA";PX*MX;",";GY;",";PX*MX;",";PY*MX;",";GX;",";PY*MX;"
,";GX;",";GY
4210 FOR I=10 TO (NC-10) STEP 10
```

```
4211 FOR J=10 TO (NR-10) STEP 10
4212 PGX0=GX+(MG*I*MX)
4213 PGX1=PGX0-2*MX:PGX2=PGX0+2*MX
4214 PGY0=GY+(MG*J*MX)
4215 PGY1=PGY0-2*MX:PGY2=PGY0+2*MX
4216 PRINT#2,"MA";PGX1;",";PGY0:PRINT#2,"DA";PGX2;",";PGY0
4217 PRINT#2,"MA";PGX0;",";PGY1:PRINT#2,"DA";PGX0;",";PGY2
4218 NEXT J
4220 NEXT I
4225 REM ===== Return to SCREEN =====
4230 CLS:GOSUB 9000
4240  LOCATE 8,10:PRINT "CONTOUR LIMITS:"
4250  LOCATE 10,10:INPUT " LOW VALUE  ";LC
4260  LOCATE 12,10:INPUT " HIGH VALUE ";HC
4270  LOCATE 14,10:INPUT "CONTOUR INTERVAL =";CI
4280  IF LC < (MIN - CI) THEN LC = LC + CI
4290 IF HC>MAX THEN HC=MAX
4300  FOR CC = LC TO HC STEP CI
4301  LOCATE 16,10:PRINT "CURRENT CONTOUR  =";CC
4302 IF CC/10=INT(CC/10) THEN 4305
4303 IF (CC-5)/10=INT((CC-5)/10) THEN 4310
4304 IF (CC-2.5)/5=INT((CC-2.5)/5) THEN 4315
4305 ON HP GOTO 4306,4306,4308,4309
4306 PRINT#2,"SP3;":PRINT#2,"LT;":GOTO 4320
4308 PRINT#2,"PS3;":PRINT#2,"LT1;":GOTO 4320
4309 PRINT#2,"SP3":PRINT#2,"LT0":GOTO 4320
4310 ON HP GOTO 4311,4311,4313,4314
4311 PRINT#2,"SP4;":PRINT#2,"LT;":GOTO 4320
4313 PRINT#2,"PS4;":PRINT#2,"LT1;":GOTO 4320
4314 PRINT#2,"SP4":PRINT#2,"LT0":GOTO 4320
4315 ON HP GOTO 4316,4316,4318,4319
4316 PRINT#2,"SP2;":PRINT#2,"LT;":GOTO 4320
4318 PRINT#2,"PS2;":PRINT#2,"LT1;":GOTO 4320
4319 PRINT#2,"SP2":PRINT#2,"LT0"
4320  FOR I = 0 TO (NR-2)
4330 Y0 = MG* I
4340  FOR J = 0 TO (NC - 2)
4350 X0 = MG* J
4360 NP = 0
4370 Z1 = H(I,J)
4380  IF CC > Z1 THEN NP = NP + 1
4390 Z2 = H(I,(J + 1))
4400  IF CC > Z2 THEN NP = NP + 1
4410 Z3 = H((I + 1),J)
4420  IF CC > Z3 THEN NP = NP + 1
4430 Z4 = H((I + 1),(J + 1))
4440  IF CC> Z4 THEN NP = NP + 1
4450  IF NP = 0 OR NP = 4 GOTO 4840
4460 A = Z1
4470 B = Z2 - A
4480 C = Z3 - A
4490 D = Z4 - A - B - C
4500 ZT = 0
4510  FOR Y1 = 0 TO 1 STEP .25
4520 DR = B + D * Y1
4530  IF DR = 0 GOTO 4680
4540 X1 = (CC - A - C * Y1) / DR
4550  IF X1 < 0 OR X1 > 1 GOTO 4680
4560 X = X0 + MG* X1
4570 Y = Y0 + MG* Y1
4580  IF X > PX OR Y > PY GOTO 4670
4590  IF ZT = 0 THEN 4615
4600 ON HP GOTO 4601,4601,4603,4605
4601 PRINT#2,"PD;PA";X+GX;",";Y+GY;";":GOTO 4670
4603 PRINT#2,"DA";X+GX;",";Y+GY;";":GOTO 4670
4605 PRINT#2,"DA";(X+GX)*MX;",";(Y+GY)*MX
```

```
461Ø GOTO 467Ø
4615 ON HP GOTO 4616,4616,4618,4619
4616 PRINT#2,"PU;PA";X+GX;",";Y+GY;";":GOTO 462Ø
4618 PRINT#2,"MA";X+GX;",";Y+GY;";":GOTO 462Ø
4619 PRINT#2,"MA";(X+GX)*MX;",";(Y+GY)*MX
462Ø IF TL>Ø THEN 467Ø
4621 IF CC/1Ø=INT(CC/1Ø) THEN 4625
4622 IF (CC-5)/1Ø=INT((CC-5)/1Ø) THEN 463Ø
4623 IF (CC-2.5)/5=INT((CC-2.5)/5) THEN 4635
4624 GOTO 467Ø
4625 ON HP GOTO 4626,4626,4628,4629
4626 PRINT#2,"SIØ.18,Ø.22;PU;PRØ,1;":GOTO 464Ø
4628 PRINT#2,"MR -1,1;":GOTO 464Ø
4629 PRINT#2,"MR";(-1)*MX;",";MX:GOTO 464Ø
463Ø ON HP GOTO 4631,4631,4633,4634
4631 GOTO 467Ø
4633 PRINT#2,"MR -1,3;":GOTO 464Ø
4634 PRINT#2,"MR";(-1)*MX;",";3*MX:GOTO 464Ø
4635 ON HP GOTO 4636,4636,4638,4639
4636 GOTO 467Ø
4638 PRINT#2,"MR -1,5;":GOTO 464Ø
4639 PRINT#2,"MR";(-1)*MX;",";5*MX
464Ø ON HP GOTO 4641,4641,4643,4644
4641 PRINT#2,"LB";CC:PRINT#2,CHR$(3):GOTO 465Ø
4643 PRINT#2,"LS1;":PRINT#2,"PL":PRINT#2, CC:PRINT#2,CHR$(13) CHR$(1Ø
) CHR$(3):GOTO 465Ø
4644 PRINT#2,"SI25,2Ø":PRINT#2,"LA";CC
465Ø TL=TL+1
466Ø GOTO 4615
467Ø ZT = ZT + 1
468Ø  NEXT Y1
469Ø ZT = Ø
47ØØ FOR X1 = Ø TO 1 STEP .25
471Ø DS = C + D * X1
472Ø  IF DS = Ø GOTO 483Ø
473Ø Y1 = (CC - A - B * X1) / DS
474Ø  IF Y1 < Ø OR Y1 > 1 GOTO 483Ø
475Ø X = XØ + MG* X1
476Ø Y = YØ + MG* Y1
477Ø  IF X > PX OR Y > PY GOTO 482Ø
478Ø  IF ZT = Ø THEN  481Ø
479Ø ON HP GOTO 4791,4791,4793,4794
4791 PRINT#2,"PD;PA";X+GX;",";Y+GY;";":GOTO 482Ø
4793 PRINT#2,"DA";X+GX;",";Y+GY;";":GOTO 482Ø
4794 PRINT#2,"DA";(X+GX)*MX;",";(Y+GY)*MX
48ØØ GOTO 482Ø
481Ø ON HP GOTO 4811,4811,4813,4814
4811 PRINT#2,"PU;PA";X+GX;",";Y+GY;";":GOTO 482Ø
4813 PRINT#2,"MA";X+GX;",";Y+GY;";":GOTO 482Ø
4814 PRINT#2,"MA";(X+GX)*MX;",";(Y+GY)*MX
482Ø ZT = ZT + 1
483Ø  NEXT X1
484Ø  NEXT J
485Ø  NEXT I
486Ø TL=Ø
487Ø  NEXT CC
4875 ON HP GOTO 488Ø,488Ø,4885,489Ø
488Ø  PRINT#2,"SP2;SIØ.2,Ø.3;":PRINT#2,"PU;PA 1Ø,-3;"
4881  PRINT#2,"LB":PRINT#2,"PC-Contour Data File - ";F$;" - Date : ";
DATE$
4882  PRINT#2, CHR$(3)
4883 GOTO 49ØØ
4885  PRINT#2,"PS2;":PRINT#2,"LS2;":PRINT#2,"MA 1Ø,-3;"
4886  PRINT#2,"PL":PRINT#2,"PC-Contour Data File - ";F$;" - Date : ";
DATE$
4887  PRINT#2,CHR$(13) CHR$(1Ø) CHR$(3)
```

```
4888 GOTO 4900
4890  PRINT#2,"SP2":PRINT#2,"SI40,30":PRINT#2,"MA";10*MX;",";(-3)*MX
4891  PRINT#2,"LAPC-Contour Data File - ";F$;" - Date : ";DATE$
4900 CLS:GOSUB 9000
4910  LOCATE 10,10:PRINT "INPUT 'MORE' for"
4920  LOCATE 12,10:PRINT "CONTOURS added,"
4930  LOCATE 14,10:PRINT "'NEW' for fresh plot,"
4940  LOCATE 16,10:PRINT "                              "
4950  LOCATE 16,10:INPUT "or 'END'"; M$
4960  IF M$ = "MORE" GOTO 4230
4965 CLOSE#2
4970  IF M$ = "NEW" GOTO 4000
4980 M$=" "
4990  CLS:GOTO 2000
4995 REM  ***** Hewlett-Packard (HP-GL) *****
5100 PRINT#2,"PS3;"
5105 PRINT#2,"IP170,602,15370,10602;"
5110 PRINT#2,"SC -5,192,-5,135;"
5120 PRINT#2,"SP1;"
5140 PRINT#2,"PU;PA";GX;",";GY;";"
5141 PRINT#2,"PD;PA";PX;",";GY;";"
5142 PRINT#2,"PA";PX;",";PY;";"
5143 PRINT#2,"PA";GX;",";PY;";"
5144 PRINT#2,"PA";GX;",";GY;";"
5150 FOR I=10 TO (NC-10) STEP 10
5155 FOR J=10 TO (NR-10) STEP 10
5160 PGX0=GX+(MG*I)
5161 PGX1=PGX0-2:PGX2=PGX0+2
5162 PGY0=GY+(MG*J)
5163 PGY1=PGY0-2:PGY2=PGY0+2
5170 PRINT#2,"PU;PA";PGX1;",";PGY0;";":PRINT#2,"PD;PA";PGX2;",";PGY0;
";"
5175 PRINT#2,"PU;PA";PGX0;",";PGY1;";":PRINT#2,"PD;PA";PGX0;",";PGY2;
";"
5180 NEXT J
5185 NEXT I
5190 GOTO 4230
8995 REM ***** PC-Contour Heading *****
9000 P$="PC-Contour : ":T$=P$+F$
9010 LOCATE 1,2:PRINT T$
9020 LOCATE 1,30*SR%:PRINT DATE$
9030 PRINT " ====================================="
9040 PRINT
9050 RETURN
9500 CLS
9510 GOSUB 9000
9520 PRINT:PRINT " Program complete"
9530 PRINT:PRINT
9540 END
```

3.5 PC-SURFPLOT

As mentioned at the start of this chapter, a plan or contour drawing presents information in 2-dimensions. There are often occasions when there are advantages in surface modelling, that is, using an isometric projection to display the data. This program can use the same data files, either prepared by PC-Grid or PC-Random, and previously used by PC-Contour, to display four different views of the area at 90° to one another. In all cases there will be some vertical exaggeration to enhance the presentation, and this is automatically adjusted to present the display in the centre of the screen. The user has the option of zooming in to enlarge the grid details if necessary. As before, screen dumps and plots are available.

3.5.1 PC-Surfplot – subroutine index

	Line numbers	*Function*
(a)	10–90	Initialization and control
(b)	100–280	Data file selection
(c)	1500–1700	Data file read from disc
(d)	2000–2640	Data file information and screen selection
(e)	4000–4330	Graphics screen display
(f)	5000–5390	Selection of zoom or rotate for graphics
(g)	6000–6420	Data preparation for plotting
(h)	7000–8500	Plotter routines
(i)	9000–9110	Clear screen routine
(j)	9500–9560	Termination of program

(a) Initialization and control Line numbers 10–90

The program is initialized and the installation file checked for the availability of graphics screens and colour for subsequent display of the isometric projections. This program requires a Color/Graphics Adapter to be able to zoom and rotate the displays.

(b) Data file selection Line numbers 100–280

This program gives the user an opportunity to display either Glapwell or Glenmore, stored earlier from PC-Grid and PC-Random, or their own data file.

(c) Data file read from disc Line numbers 1500–1700

The user's data file is entered when requested, the data read from the disc and the level of the grid points stored in **ZD(NR,NC)**.

(d) Data file information and screen selection Line numbers 2000–2640

The data file information display is similar to that of the previous programs in this chapter. The user is then given a choice of displaying the isometric projection in colour (medium-resolution) or monochrome (high-resolution) on the graphics screen. This program requires a Color/Graphics Adapter card to enable the display to be rotated, etc., before plotting.

(e) Graphics screen display Line numbers 4000–4330

This display uses the alternative screen presentation of Fig. 1.5, allowing a wider plotting area, with two lines of text above and below the graphics window.

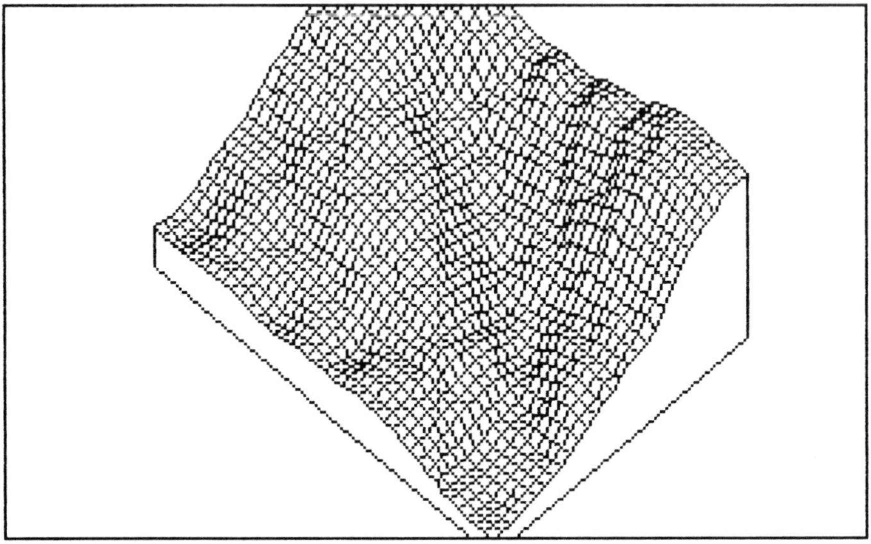

Fig. 3.7 PC-Surfplot screen dump of isometric projection (View 1) of Glenmore on Screen 1 (medium-resolution).

The program presented in this section takes each block of grid points in turn and calculates their levels **Y1–Y4** with *X*-coordinates **X1–X4**, and this grid square is plotted on the screen. The routine shown uses a hidden line algorithm so that old plotting underneath the new grid square is removed using the **PAINT** command. On completion of the mesh diagram, a block base is drawn showing the vertical height differences at the three visible corners of the plot, Fig. 3.7.

(f) Selection of zoom or rotate for graphics Line numbers 5000–5390

The first plot of any new data file will be View 1, i.e. a standard 0° isometric projection, Fig. 3.8. The user is now given a choice of zooming in or out, or of rotating the surface model, Figs 3.9, 3.10 and 3.11. In the zoom mode, changes are set to 20% of the original, but this factor can be changed by the user if required.

If the screen presentation is satisfactory and **e** for **END** selected, a further list of plotting options is presented:

Do you wish to save Screen to
⟨1⟩ Disc ⟨2⟩ Printer ⟨3⟩ Plotter ⟨4⟩ End

thus allowing a screen version to be saved as a picture to disc for future recall, as a screen dump to a printer, or as a hardcopy on a plotter.

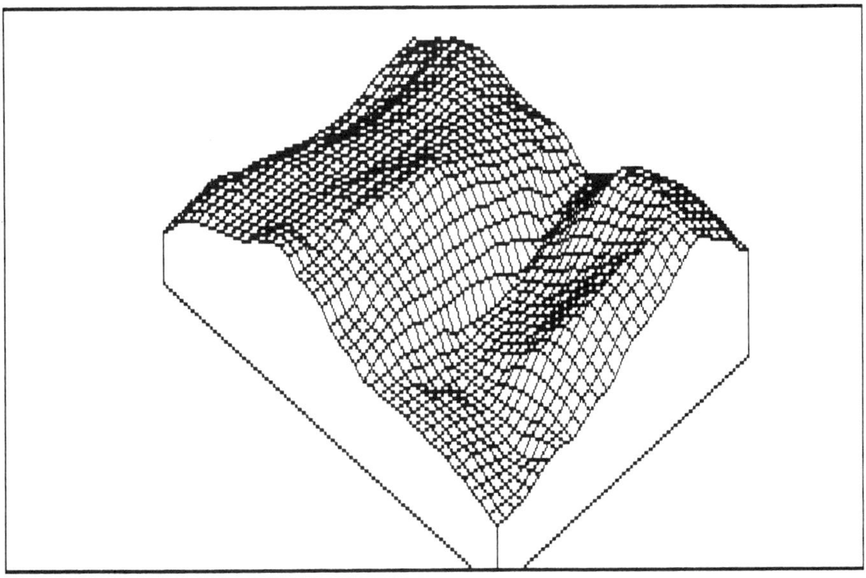

Fig. 3.8 PC-Surfplot screen dump of isometric projection (View 1) of Glapwell on Screen 2 (high-resolution).

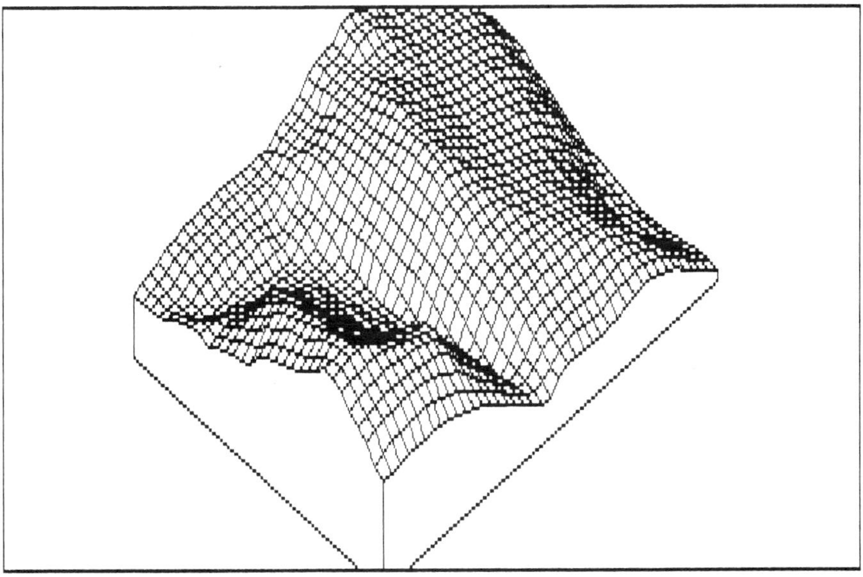

Fig. 3.9 PC-Surfplot screen dump of isometric projection (View 2) of Glapwell on Screen 2 (high-resolution).

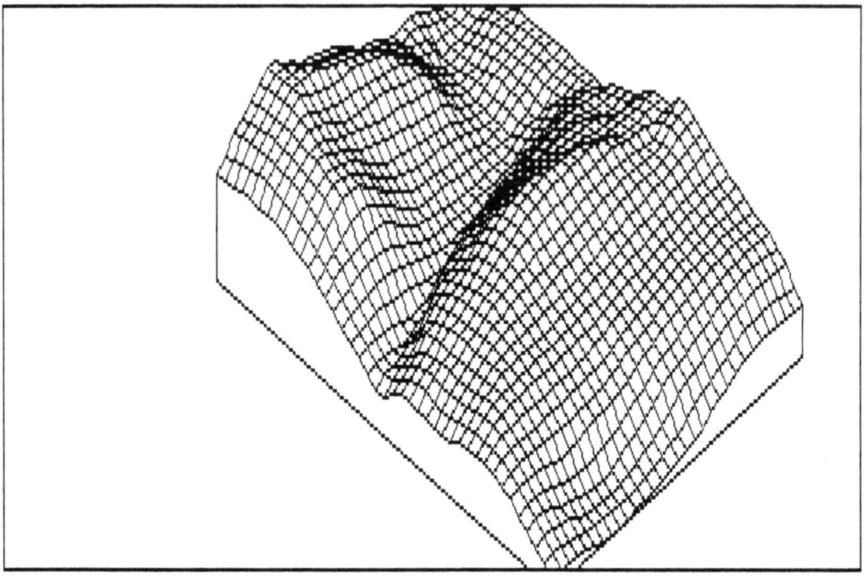

Fig. 3.10 PC-Surfplot screen dump of isometric projection (View 3) of Glapwell on Screen 2 (high-resolution).

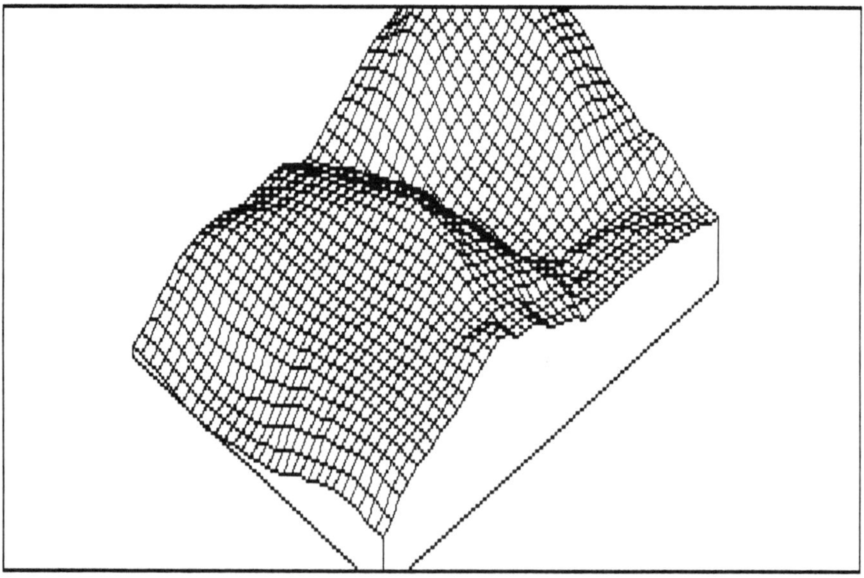

Fig. 3.11 PC-Surfplot screen dump of isometric projection (View 4) of Glapwell on Screen 2 (high-resolution).

(g) Data preparation for plotting Line numbers 6000–6420

The data needs to be rotated before plotting and this routine prepares the data depending on the value of **ROT%**. This routine is used by both the screen display and the plotter.

(h) Plotter routines Line numbers 7000–8500

Plotter routines are provided for the three plotters discussed earlier, Apple 410, Epson HI-80 and Hewlett-Packard HP-7475A. The frame is drawn by pen 1 (black), the grid mesh by pen 3 (green), the block base by pen 4 (blue) and the plot titled by pen 2 (red), as shown in Fig. 3.12.

(i) Clear screen routine Line numbers 9000–9110

This routine is used to clear the graphics screen and identify the program and isometric projection viewed.

(j) Termination of program Line numbers 9500–9560

This is the final display screen and returns the user to BASIC command level.

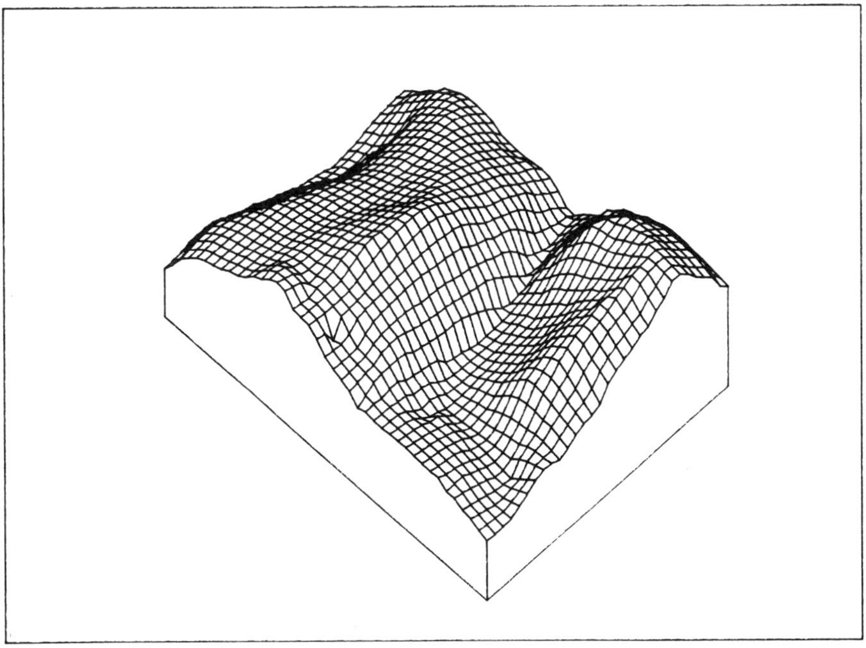

Fig. 3.12 PC-Surfplot isometric projection (View 1) of Glapwell spoil-heap on plotter for comparison with Fig. 3.8.

3.5.2 PC-Surfplot – numeric variables

B	= Number of pixels in isometric grid (horizontal)
BBY1–3	= Variables used to define block base
BTY1–3	= Variables used to define block base
BX1–3	= Variables used to define block base
BY1–3	= Variables used to define block base
DF%	= Data file section
DISP%	= Selected screen save location
GCOL	= Graphics screen column variable
GROW	= Graphics screen row variable
GX	= *X*-coordinate at lower left-hand corner of plot
GY	= *Y*-coordinate at lower left-hand corner of plot
I **J** **K** **L**	Integer variables used in loops
MAX	= Maximum *Z*-value in data
MIN	= Minimum *Z*-value in data
MX	= Multiplication factor used by Epson HI-80 plotter
NC	= Number of columns in grid
NCOL	= Column variable in isometric routine
NR	= Number of rows in grid
NROW	= Row variable in isometric routine
PRTX	= *X*-location of title on plot
PRTY	= *Y*-location of title on plot
PX	= *X*-coordinate at upper right-hand corner of plot
PY	= *Y*-coordinate at upper right-hand corner of plot
ROT%	= Integer variable to select angle of rotation
S	= Number of pixels in isometric grid (vertical)
SCD%	= Integer variable to select screen display
SCR%	= Integer variable for screen selection
SEL%	= Integer variable to select program
SR%	= Integer variable for graphics screen resolution
SX	= Horizontal display width in user units
SY	= Vertical display height in user units
VD%	= Video display variable
VEX	= Vertical exaggeration
XØ–X4	= *X*-coordinates of grid points
XCEN	= *X*-coordinate of centre of plot
YØ–Y4	= *Y*-coordinate of grid points
YCEN	= *Y*-coordinate of centre of plot
YSMIN	= Minimum value of *Z* on screen
ZH	= Read *Z* value of grid points
Z(5Ø,5Ø)	= Array to store *Z*-values of grid points
ZD(5Ø,5Ø)	= Array to store *Z*-values of grid points

3.5.3 PC-Surfplot – string variables

A$	= **INKEY$**
DATE$	Variable used to return the date of the computer's calendar/clock, in US format MM–DD–YYYY
F$	File name for data file
Q$	Question response (Y/N) or (y/n)
SCR$	File name to save picture of screen on disc
SCRPIC$	= **SCR$** + **".PIC"**
TIME$	Variable used to return the time of the built-in clock in HH:MM:SS format
V$	Variable to store view shown in isometric projection

3.5.4 PC-Surfplot – BASIC program

```
10 REM  <SURFPLOT> Program for IBM PC and Compatibles
12 REM  Version 01-01-1987
14 REM  (C)  Copyright  1987  P.H.Milne
16 REM  All rights reserved
20 REM  Output available to <1> Apple 410, <2> Epson HI-80, <3> HPGL
Colour Plotter
22 REM RS-232C Setting in Line 4400 - OPEN "COM1:9600,E,7,1,CS2000,DS
15000" FOR OUTPUT AS #2
24 REM  Can access data files on disc ZOOM and ROTATE display
25 REM  This version includes Graphics printer dump of Screen
30 CLEAR
35 PCF$="PCSCREEN"
40 OPEN "I",#3,PCF$
45 INPUT#3,SCR%
50 INPUT#3,SCD%
55 CLOSE#3
60 ON (SCR%+1) GOTO 62,64,66
62 SCREEN 0:WIDTH 40:GOTO 80
64 SCREEN 1:GOTO 68
66 SCREEN 1:GOTO 80
68 IF SCD%=1 THEN 80
70 COLOR 8,0
80 KEY OFF
90 DIM Z(50,50), ZD(50,50)
95 SR%=1:DF%=0:DISP%=1:ROT%=1:VD%=0:F$="   "
100 CLS:GOSUB 9000
120 PRINT " =======================================":PRINT
130 PRINT " THIS PROGRAM READS INFORMATION FROM "
140 PRINT " REGULAR SQUARE GRID POINTS INTO A "
150 PRINT " DATA FILE FOR PLOTTING BY PC-SurfPlot"
160 PRINT
170 PRINT " To Continue, Press <Return> ":WHILE R$="":R$=INKEY$:WEND
180 PRINT:PRINT
200 PRINT " Select Data File for Display"
210 PRINT:PRINT " <1> Glapwell, <2> Glenmore or":PRINT " <3> Other di
sc file";:INPUT DF%
220 IF DF%<1 OR DF%>3 THEN 200
230 PRINT:PRINT
240 ON DF% GOTO 250,270,1500
250 F$="Glapwell"
260 GOTO 1520
270 F$="Glenmore"
```

```
280 GOTO 1520
1495 REM ***** Data File Disc Routine *****
1500 PRINT:PRINT " Enter DATA File Name ";
1510 INPUT F$
1520 PRINT:PRINT " One moment please .... reading data"
1530 OPEN "I",#1,F$
1540 INPUT#1, NR
1550 INPUT#1, NC
1560 INPUT#1, MAX
1570 INPUT#1, MIN
1580 INPUT#1, DX
1590 INPUT#1, DY
1600 FOR I=0 TO (NR-1)
1610 FOR J=0 TO (NC-1)
1620 INPUT#1, ZH
1630 ZD(I,J)=ZH
1640 NEXT J
1650 NEXT I
1660 CLOSE#1
1670 PRINT
1680 PRINT " Do you wish to check Data File (y/n)";:INPUT Q$
1690 IF Q$="n" THEN 2120
1700 IF Q$<>"y" THEN 1670
1995 REM ***** Data File Information *****
2000 CLS:SCREEN 1
2010 GOSUB 9000
2020 PRINT " ======================================="
2030 PRINT:PRINT:PRINT " THE DATA FILE - ";F$:PRINT:PRINT " HAS NOW B
EEN READ, GIVING :-"
2040 PRINT
2050 PRINT " No.  Rows = ";NR;"  No.  Cols. = ";NC
2060 PRINT " Minimum value  = ";MIN
2070 PRINT " Maximum value  = ";MAX
2080 PRINT " Range          = ";(MAX-MIN)
2090 PRINT " 1 Map row      = ";DY/(NR-1);" Units"
2100 PRINT " 1 Map column   = ";DX/(NC-1);" Units"
2110 IF Q$="y" THEN 2140
2120 CLS:GOSUB 9000
2130 PRINT " ======================================="
2140 PRINT:PRINT
2150 PRINT " Do you wish to";TAB(18);"<1> Plot Surface":PRINT TAB(18)
;"<2> Run PC-Contour "
2160 PRINT TAB(18);"<3> Quit ";:INPUT SEL%
2170 IF SEL%<1 OR SEL%>3 THEN 2140
2180 IF SEL%=3 THEN 9500
2190 IF SEL%=1 THEN 2500
2200 LOAD "A:CONTOUR",R
2495 REM ***** Screen Selection *****
2500 CLS
2510 GOSUB 9000
2520 PRINT " ======================================="
2530 PRINT:PRINT TAB(3);"Select SCREEN";TAB(20);"<1> Colour -"
2540 PRINT TAB(20);"    Low resolution"
2550 PRINT:PRINT TAB(20);"<2> Black & White -"
2560 PRINT TAB(20);"    High resolution"
2570 PRINT TAB(20);:INPUT SR%
2580 IF SR%<1 OR SR%>2 THEN 2530
2590 IF ROT%=1 THEN NCOL=NC-1:NROW=NR-1
2600 PRINT:PRINT:PRINT " One moment please .... preparing data"
2610 IF SR%=1 THEN 6000
2620 SCREEN 2
2630 YCEN=0:XCEN=0:SX=160*SR%:SY=100
2640 GOTO 6000
3995 REM ***** Screen Plotting Routine *****
4000 DISP%=1:YCEN=0:XCEN=0:SX=160*SR%:SY=100
4005 IF SR%=1 THEN RGB%=2 ELSE RGB%=3
```

```
4010 VIEW SCREEN (8,18)-(310*SR%,180),,RGB%
4020 WINDOW (XCEN-SX,YCEN-SY)-(XCEN+SX,YCEN+SY)
4030 VEX=50/(MAX-MIN):YSMIN=INT((MIN*VEX)/10)*10+5
4040 X0=-10*SR%:Y0=(70-YSMIN)
4050 CLS
4060 GOSUB 9000
4070 S=2.25:B=3*SR%
4080 IF ROT%=2 OR ROT%=4 THEN X0=10*SR%
4090 FOR GCOL=1 TO NCOL
4100 FOR GROW=1 TO NROW
4110 X1=X0+(GCOL-GROW)*B
4120 Y1=Y0+Z(GROW,GCOL)*VEX-(GROW+GCOL)*S
4130 X2=X1-B
4140 Y2=Y0+Z(GROW,GCOL-1)*VEX-(GROW+GCOL-1)*S
4150 X3=X1+B
4160 Y3=Y0+Z(GROW-1,GCOL)*VEX-(GROW-1+GCOL)*S
4170 X4=X1
4180 Y4=Y0+Z(GROW-1,GCOL-1)*VEX-(GROW-1+GCOL-1)*S
4190 LINE (X1,Y1)-(X2,Y2),2
4200 LINE -(X4,Y4),2:LINE -(X3,Y3),2
4210 LINE -(X1,Y1),2
4220 Y=(Y1+Y4)/2
4230 PAINT (X4,Y),0,2
4240 LINE (X1,Y1)-(X2,Y2),1
4250 LINE -(X4,Y4),1:LINE -(X3,Y3),1
4260 LINE -(X1,Y1),1
4270 NEXT GROW,GCOL
4280 LINE (X0+(-NROW)*B,Y0-(NROW)*S+Z(NROW,0)*VEX)-(X0+(-NROW)*B,Y0-(
NROW)*S+YSMIN)
4290 LINE (X0+(NCOL-NROW)*B,Y0-(NCOL+NROW)*S+Z(NROW,NCOL)*VEX)-(X0+(N
COL-NROW)*B,Y0-(NCOL+NROW)*S+YSMIN)
4300 LINE (X0+(NCOL)*B,Y0-(NCOL)*S+Z(0,NCOL)*VEX)-(X0+(NCOL)*B,Y0-(NC
OL)*S+YSMIN)
4310 LINE (X0+(-NROW)*B,Y0-(NROW)*S+YSMIN)-(X0+(NCOL-NROW)*B,Y0-(NCOL
+NROW)*S+YSMIN)
4320 LINE -(X0+(NCOL)*B,Y0-(NCOL)*S+YSMIN)
4330 GOTO 5000
4395 REM ***** Plotter Routine *****
4400 OPEN "COM1:9600,E,7,1,CS2000,DS15000" FOR OUTPUT AS #2
4405 DISP%=2
4410 IF SR%=2 THEN SX=SX/2
4420 IF SR%=2 THEN B=B/2
4430 IF ROT%=2 OR ROT%=4 THEN X0=10
4440 VEX=50/(MAX-MIN):YSMIN=INT((MIN*VEX)/10)*10+5
4450 GX=XCEN-SX:GY=YCEN-SY:PX=XCEN+SX:PY=YCEN+SY:MX=2670/320
4460 PRTX=GX+2:PRTY=GY+2
4470 IF ROT%=1 THEN NCOL=NC-1:NROW=NR-1
4490 ON HP% GOTO 4500,4500,4550,4575
4499 REM ***** HP=1, (HPGL) *****
4500 PRINT#2,"PS3;"
4502 PRINT#2,"IP 170,602,15370,10602;"
4505 PRINT#2,"SC";GX;",";PX;",";GY;",";PY;";"
4510 PRINT#2,"SP1;"
4515 PRINT#2,"PU;PA";GX;",";GY;";"
4516 PRINT#2,"PD;PA";PX;",";GY;";"
4517 PRINT#2,"PA";PX;",";PY;";"
4518 PRINT#2,"PA";GX;",";PY;";"
4519 PRINT#2,"PA";GX;",";GY;";"
4520 PRINT#2,"SP3;"
4521 GOTO 6000
4549 REM ***** HP=3, Apple 410 *****
4550 PRINT#2,"SP3;"
4555 PRINT#2,"WD";GX;",";GY;",";PX;",";PY;";"
4560 PRINT#2,"PS1;"
```

```
4565 PRINT#2,"MA";GX;",";GY;";":PRINT#2,"DA";PX;",";GY;";":PRINT#2,"D
A";PX;",";PY;";":PRINT#2,"DA";GX;",";PY;";":PRINT#2,"DA";GX;",";GY;";
"
4570 PRINT#2,"PS3;":GOTO 6ØØØ
4574 REM ***** HP=4, Epson *****
4575 PRINT#2,"IWØ,Ø,267Ø,192Ø"
458Ø PRINT#2,"MA1335,96Ø":PRINT#2,"OR"
4581 PRINT#2,"SP1":PRINT#2,"LTØ"
4582 PRINT#2,"MA";GX*MX;",";GY*MX
4583 PRINT#2,"DA";PX*MX;",";GY*MX
4584 PRINT#2,"DA";PX*MX;",";PY*MX
4585 PRINT#2,"DA";GX*MX;",";PY*MX
4586 PRINT#2,"DA";GX*MX;",";GY*MX
459Ø PRINT#2,"SP3"
4595 REM ***** Plotter Output *****
46ØØ FOR GCOL = Ø TO NCOL
461Ø FOR GROW = Ø TO (NROW-1)
462Ø X1=XØ+(GCOL-GROW)*B
4625 Y1=YØ+Z(GROW,GCOL)*VEX-(GROW+GCOL)*S
463Ø X2=X1-B:Y2=YØ+Z(GROW+1,GCOL)*VEX-(GROW+1+GCOL)*S
464Ø IF GROW<>Ø THEN 466Ø
465Ø ON HP% GOTO 4651,4651,4653,4654
4651 PRINT#2,"PU;PA";X1;",";Y1;";":GOTO 466Ø
4653 PRINT#2,"MA";X1;",";Y1;";":GOTO 466Ø
4654 PRINT#2,"MA";X1*MX;",";Y1*MX
466Ø ON HP% GOTO 4661,4661,4663,4664
4661 PRINT#2,"PD;PA";X2;",";Y2;";":GOTO 467Ø
4663 PRINT#2,"DA";X2;",";Y2;";":GOTO 467Ø
4664 PRINT#2,"DA";X2*MX;",";Y2*MX
467Ø NEXT GROW,GCOL
468Ø FOR GROW = Ø TO NROW
469Ø FOR GCOL = Ø TO (NCOL-1)
47ØØ X1=XØ+(GCOL-GROW)*B
471Ø Y1=YØ+Z(GROW,GCOL)*VEX-(GROW+GCOL)*S
472Ø X2=XØ+(GCOL+1-GROW)*B
473Ø Y2=YØ+Z(GROW,GCOL+1)*VEX-(GROW+GCOL+1)*S
474Ø IF GCOL<>Ø THEN 476Ø
475Ø ON HP% GOTO 4751,4751,4753,4754
4751 PRINT#2,"PU;PA";X1;",";Y1;";":GOTO 476Ø
4753 PRINT#2,"MA";X1;",";Y1;";":GOTO 476Ø
4754 PRINT#2,"MA";X1*MX;",";Y1*MX
476Ø ON HP% GOTO 4761,4761,4763,4764
4761 PRINT#2,"PD;PA";X2;",";Y2;";":GOTO 477Ø
4763 PRINT#2,"DA";X2;",";Y2;";":GOTO 477Ø
4764 PRINT#2,"DA";X2*MX;",";Y2*MX
477Ø NEXT GCOL,GROW
4775 BX1=XØ+(-NROW)*B:BTY1=YØ-(NROW)*S+Z(NROW,Ø)*VEX:BBY1=YØ-(NROW)*S
+YSMIN
478Ø BX2=XØ+(NCOL-NROW)*B:BTY2=YØ-(NCOL+NROW)*S+Z(NROW,NCOL)*VEX:BBY2
=YØ-(NCOL+NROW)*S+YSMIN
4785 BX3=XØ+(NCOL)*B:BTY3=YØ-(NCOL)*S+Z(Ø,NCOL)*VEX:BBY3=YØ-(NCOL)*S+
YSMIN
479Ø ON HP% GOTO 48ØØ,48ØØ,486Ø,492Ø
4799 ***** HP=1, (HP-GL) *****
48ØØ PRINT#2,"SP4;"
4801 PRINT#2,"PU;PA";BX1;",";BTY1;";"
4805 PRINT#2,"PD;PA";BX1;",";BBY1;";"
481Ø PRINT#2,"PA";BX2;",";BBY2;";"
4815 PRINT#2,"PA";BX3;",";BBY3;";"
482Ø PRINT#2,"PA";BX3;",";BTY3;";PU;"
4825 PRINT#2,"PA";BX2;",";BTY2;";"
483Ø PRINT#2,"PD;PA";BX2;",";BBY2;";PU;"
4835 PRINT#2,"SP2;":PRINT#2,"SIØ.2,Ø.3;"
484Ø PRINT#2,"PU;PA ";PRTX;",";PRTY;";"
4845 PRINT#2,"LB";"PC-SurfPlot - ";F$;" - View ";V$;" - Date : ";DATE
$
```

```
4850 PRINT#2, CHR$(3)
4855 GOTO 4980
4859 REM ***** HP=3, Apple 410 *****
4860 PRINT#2,"MA";BX1;",";BTY1;";"
4865 PRINT#2,"DA";BX1;",";BBY1;";"
4870 PRINT#2,"DA";BX2;",";BBY2;";"
4875 PRINT#2,"DA";BX3;",";BBY3;";"
4880 PRINT#2,"DA";BX3;",";BTY3;";"
4885 PRINT#2,"MA";BX2;",";BTY2;";"
4890 PRINT#2,"DA";BX2;",";BBY2;";"
4895 PRINT#2,"PS2;":PRINT#2,"LS3;"
4900 PRINT#2,"MA";PRTX;",";PRTY;";"
4905 PRINT#2,"PL":PRINT#2,"PC-SurfPlot - ";F$;" - View ";V$;" - Date
: ";DATE$
4910 PRINT#2, CHR$(13) CHR$(10) CHR$(3)
4915 GOTO 4980
4919 REM ***** HP=4, Epson *****
4920 PRINT#2,"MA";BX1*MX;",";BTY1*MX
4925 PRINT#2,"DA";BX1*MX;",";BBY1*MX
4930 PRINT#2,"DA";BX2*MX;",";BBY2*MX
4935 PRINT#2,"DA";BX3*MX;",";BBY3*MX
4940 PRINT#2,"DA";BX3*MX;",";BTY3*MX
4945 PRINT#2,"MA";BX2*MX;",";BTY2*MX
4950 PRINT#2,"DA";BX2*MX;",";BBY2*MX
4955 PRINT#2,"SP2":PRINT#2,"SI40,30"
4960 PRINT#2,"MA";(GX*MX+10);",";(GY*MX+5)
4965 PRINT#2,"LA":PRINT#2,"PC-SurfPlot - ";F$;" - View ";V$;" - Date
: ";DATE$
4980 CLOSE#2
4990 IF SR%=2 THEN SX=SX*2:B=B*2
4995 REM ***** Selection of ZOOM or ROTATE *****
5000 BEEP:DISP%=1:LOCATE 24,2:PRINT "ZOOM = z, ROTATE = r, END = e ";
5010 A$=INKEY$
5020 IF A$="" THEN GOTO 5010
5030 IF INSTR("zre",A$)=0 THEN GOTO 5000
5040 IF A$="z" THEN 5130
5050 IF A$="e" THEN 5270
5060 LOCATE 24,2:PRINT "ROTATE - Select angle of rotation ";:LOCATE 2
5,2:PRINT"<1> 0,<2> 90,<3> 180,<4> 270,<5> END";
5070 A$=INKEY$:IF A$="" THEN 5070
5080 IF INSTR("12345",A$)=0 THEN 5070
5090 ROT%=VAL(A$)
5100 LOCATE 24,2:PRINT "                                        ";:LOCATE 2
5,2:PRINT"                                        ";
5110 IF ROT%=5 THEN 9500
5120 GOTO 6000
5130 LOCATE 24,2:PRINT "ZOOM IN=i, OUT=o,                      ";:LOCATE 2
5,2:PRINT"UP=u, DOWN=d, RIGHT=r, LEFT=l, END=e";
5140 A$=INKEY$
5150 IF A$="" THEN 5140
5160 IF INSTR("ioudelr",A$)=0 THEN GOTO 5130
5170 LOCATE 24,2:PRINT "                                        ";:LOCATE 2
5,2:PRINT"                                        ";
5180 IF A$="e" THEN 5270
5190 IF A$="o" THEN SX=SX/.8:SY=SY/.8
5200 IF A$="i" THEN SX=SX*.8:SY=SY*.8
5210 IF A$="r" THEN XCEN=XCEN-SX*.2
5220 IF A$="l" THEN XCEN=XCEN+SX*.2
5230 IF A$="u" THEN YCEN=YCEN-SY*.2
5240 IF A$="d" THEN YCEN=YCEN+SY*.2
5250 WINDOW (XCEN-SX,YCEN-SY)-(XCEN+SX,YCEN+SY)
5260 GOTO 4050
5270 LOCATE 24,2:PRINT "                                        ";:LOCATE
 25,2:PRINT"                                        ";
```

```
5280 LOCATE 24,2:PRINT "Do you wish to save Screen to";:LOCATE 25,2:P
     RINT"<1>Disc<2>Plotter<3>Printer<4>End";
5290 INPUT PLOT%:IF PLOT%<1 OR PLOT%>4 THEN 5270
5300 LOCATE 24,2:PRINT "                                        ";:LOCATE
 25,2:PRINT"                                        ";
5310 ON PLOT% GOTO 5320,5400,5500,9500
5320 LOCATE 24,2:PRINT "                                       ";:LOCATE 2
5,2:PRINT"                                        ";
5330 LOCATE 24,2:PRINT "Enter File Name for saved Screen ";:LOCATE 25
,2:PRINT"(8 letters max.)";
5340 INPUT SCR$:IF SCR$="" THEN 5320
5350 DEF SEG = &HB800
5360 SCRPIC$=SCR$+".PIC"
5370 BSAVE SCRPIC$,0,&H4000
5380 LOCATE 24,2:PRINT "                                       ";:LOCATE 2
5,2:PRINT"                                        ";
5390 GOTO 5000
5400 LOCATE 24,2:PRINT "                                        ";:LOCATE
 25,2:PRINT"                                        ";
5410 LOCATE 24,2:PRINT "Select Plotter -<1> Apple, <2> Epson";:LOCATE
 25,2:PRINT "or <3> Hewlett-Packard (HP-GL) ";
5420 INPUT PX%:IF PX%<1 OR PX%>3 THEN 5400
5430 LOCATE 24,2:PRINT "                                        ";:LOCATE
 25,2:PRINT"                                        ";
5440 IF PX%=1 THEN HP%=3
5450 IF PX%=2 THEN HP%=4
5460 IF PX%=3 THEN HP%=1
5470 GOTO 4400
5495 REM ***** Graphics Dump *****
5500 LCOPY
5510 GOTO 5000
5995 REM ***** Data preparation for Plotting *****
6000 VD%=1
6010 ON ROT% GOTO 6020,6100,6210,6330,9500
6020 NCOL=NC-1:NROW=NR-1
6030  FOR I = 0 TO NROW
6040  FOR J = 0 TO NCOL
6050  Z(I,J)=ZD(I,J)
6060  NEXT J
6070  NEXT I
6080 IF DISP%=1 THEN 4000
6090 IF DISP%=2 THEN 4600
6100 NCOL=NR-1:NROW=NC-1
6110 K=NCOL
6120  FOR I = 0 TO NROW
6130  FOR J = 0 TO NCOL
6140  Z(I,J)=ZD(K,I)
6150 K=K-1
6160  NEXT J
6170 K=NCOL
6180  NEXT I
6190 IF DISP%=1 THEN 4050
6200 IF DISP%=2 THEN 4600
6210 NCOL=NC-1:NROW=NR-1
6220 K=NROW:L=NCOL
6230 FOR I=0 TO NROW
6240 FOR J=0 TO NCOL
6250 Z(I,J)=ZD(K,L)
6260 L=L-1
6270 NEXT J
6280 K=K-1
6290 L=NCOL
6300 NEXT I
6310 IF DISP%=1 THEN 4050
6320 IF DISP%=2 THEN 4600
6330 NCOL=NR-1:NROW=NC-1
```

```
6340 K=NROW
6350  FOR I = 0 TO NROW
6360  FOR J = 0 TO NCOL
6370  Z(I,J)=ZD(J,K)
6380  NEXT J
6390 K=K-1
6400  NEXT I
6410 IF DISP%=1 THEN 4050
6420 IF DISP%=2 THEN 4600
8995 REM ***** PC-SurfPlot Heading *****
9000 LOCATE 1,2:PRINT "PC-SurfPlot"
9010 LOCATE 1,31*SR%:PRINT DATE$
9020 LOCATE 2,2:PRINT F$
9030 IF VD%=0 THEN 9100
9040 IF ROT%=1 THEN V$="1:  0 Degs"
9050 IF ROT%=2 THEN V$="2: 90 Degs"
9060 IF ROT%=3 THEN V$="3:180 Degs"
9070 IF ROT%=4 THEN V$="4:270 Degs"
9080 IF ROT%=5 THEN V$="          "
9090 LOCATE 2,12*SR%:PRINT "View ";V$
9100 LOCATE 2,31*SR%:PRINT TIME$
9110 RETURN
9495 REM ***** Program End *****
9500 CLS:VD%=0
9510 LOCATE 2,2:PRINT "                                        "
9520 GOSUB 9000
9530 VIEW SCREEN (8,18)-(310*SR%,180),,0
9540 LOCATE 24,2:PRINT "                                        ";:LOCATE 2
5,2:PRINT"                                        ";
9550 LOCATE 8,5:PRINT " Program Complete":PRINT:PRINT:PRINT:PRINT
9560 END
```

4

Highway alignment design

4.1 INTRODUCTION

The geometric design and setting out of curves is an important part of an engineer's or surveyor's task in establishing the horizontal and vertical alignment of transportation facilities. Alignment design is, of course, not confined to highways but also covers railways, canals, pipelines, transmission lines and rapid transit systems. However, the methods developed for highways are equally applicable to the other means of transport, and consequently this chapter refers principally to highway alignment.

In Chapter 2, dealing with earthworks, mention was made of the curvature corrections required when a road curve was encountered (see Section 2.3.1(m)). In designing the horizontal alignment, the type of curve selected will depend on the design requirements, for example a circular curve (PC-HCurve) would be used for low speed in built-up areas, whereas transition curves (PC-TCurve) would be required on high-speed motorways. After the horizontal alignment is chosen, the vertical alignment (PC-VCurve) will be considered from an analysis of the earthworks and mass-haul diagrams (discussed in Chapter 2) in relation to the topography, (discussed in Chapter 3). If it is found that there is an overlap of horizontal and vertical curve designs, these are normally phased together to ensure compatibility, as described in Section 4.5 (PC-RODEOS).

4.2 HORIZONTAL CIRCULAR CURVES

Road curve design is one of the essential tasks involved in highway engineering and was discussed in Chapter 5 of *BASIC Programs for Land Surveying*. Once the road curve is designed, it is the duty of the engineering surveyor to set out the curve on the ground. This may be an easy task on a green-field site, but it may be more complex on construction sites where buildings have to be demolished or excavation work is in progress. In congested areas, for example, where clear lines of sight are not available for setting out by deflection angles, recourse has normally to be made to linear methods, either offsets from the tangents or long chord.

There are therefore advantages in being able to display the horizontal curve setting-out features, either on the graphics screen or plotter to check intervisibility, etc. The user is given a choice of selecting three setting-out methods with hardcopies available from both the graphics screen and the plotter.

4.2.1 PC-HCurve – subroutine index

	Line numbers	*Function*
(a)	10–90	Initialization and control
(b)	100–110	Screen header display
(c)	200–240	Selection of angle format
(d)	500–680	Entry of curve location, etc.
(e)	700–880	Entry of curve design data
(f)	900–990	Menu selection for setting out
(g)	1000–1340	⟨1⟩ Offsets from tangent
(h)	2000–2380	⟨2⟩ Offsets from long chord
(i)	3000–3590	⟨3⟩ Tangential deflection angles
(j)	4000–4440	Graphics display and screen dump
(k)	5000–5190	Plotter initialization and selection
(l)	5200–5940	Plotter routine for HPGL
(m)	6000–6080	Routine to print curve location, etc.
(n)	6100–6390	Printout for ⟨1⟩
(o)	6400–6580	Printout for ⟨2⟩
(p)	6600–6940	Printout for ⟨3⟩ – right-hand curve
(q)	7000–7330	Printout for ⟨3⟩ – left-hand curve
(r)	8000–8040	Routine to convert DD.MMSS to decimals
(s)	8050–8140	Routine to convert decimals to DD.MMSS
(t)	8200–8940	Plotter routine for Apple 410
(u)	9000–9060	Routine to clear screen
(v)	9100–9130	Routine to display error message
(w)	9500–9620	Termination of program
(x)	10200–10940	Plotter routine for Epson HI-80

(a) Initialization and control Line numbers 10–90

The program is initialized and the installation file checked for the availability of graphics screens and colour for subsequent display. The required numeric and string variables are initialized and two **DEF FN** functions used to define arccosine and arcsine. Several arrays for chainage, etc. are dimensioned with a maximum of 100 chords or offsets. If a larger number is required, the arrays should be redimensioned.

(b) Screen header display Line numbers 100–110

At the start of each program selection the screen is cleared using **GOSUB 9000** to identify on screen the program suite and title.

(c) Selection of angle format Line numbers 200–240

The operator has a choice in line 200 of entering angles in decimal or DD.MMSS format, where the **D9** flag is set to 1 for decimals and 2 for DD.MMSS in line 210.

(d) Entry of curve location, etc. Line numbers 500–680

This routine is provided to keep a record of site location, operator, date, etc., with an opportunity to obtain a printout if required.

(e) Entry of curve design data Line numbers 700–880

Only three data entries, the deviation angle (D_F), the radius (R), and the chainage of the intersection point (F_3), are required to calculate the curve design parameters as detailed in lines 820–880. The design parameters are shown in Fig. 4.1.

(f) Menu selection for setting out Line numbers 900–990

The required method for setting out the curve is now chosen from the menu:

⟨1⟩ Offsets from tangent
⟨2⟩ Offsets from long chord
⟨3⟩ Tangential deflection angles
⟨4⟩ Graphical output
⟨5⟩ Plotter output

The choice is stored in the variable **P1** and the program branches accordingly from line 990.

(g) ⟨1⟩ Offsets from tangent Line numbers 1000–1340

This is the simpler of the two linear setting-out methods, since all the measurements are either along, or perpendicular to, the straight joining the tangent point A to the intersection point I (Fig. 4.1). The initial calculations determine the maximum tangential distance to the mid-ordinate M and the user is asked to enter the interval spacing of the tangential offsets (maximum 100). After displaying the setting-out details a computer printout is available together with a plot of the curve.

(h) ⟨2⟩ Offsets from long chord Line numbers 2000–2380

In this case, the long chord (L_C) between the tangent points A and C must be capable of being marked out on the ground, and hence the advantage of a graphical plot to check chainage distances from the tangent points and the

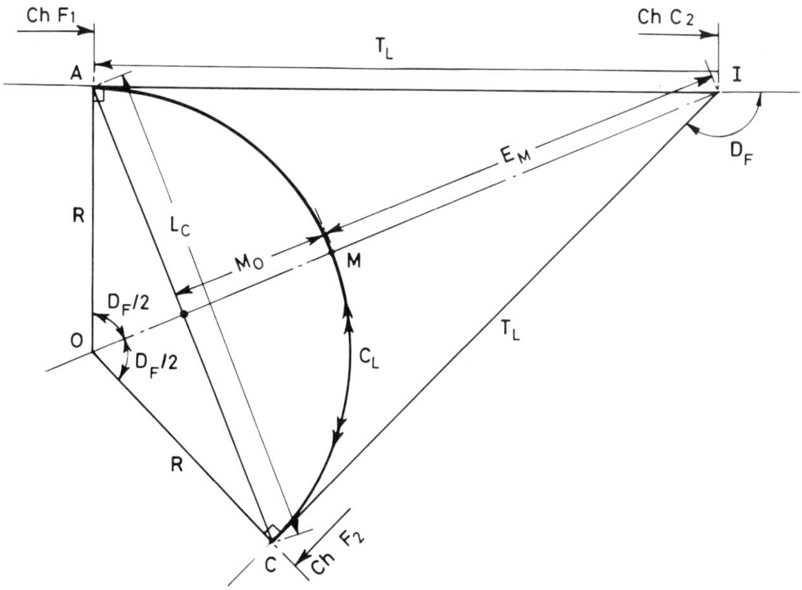

Fig. 4.1 Nomenclature for standard circular curve design.

intersection point I (Fig. 4.1). The initial calculations determine the maximum chord distance, i.e. half the long chord ($L_C/2$), and the user is asked to enter the interval spacing of the chord offsets (maximum 100). After displaying the setting-out details a computer printout is available, together with a plot of the curve. This routine is also used for the graphics displays (j) and (k).

(i) ⟨3⟩ Tangential deflection angles Line numbers 3000–3590

The tangential deflection angle method of setting out, although requiring the use of a theodolite at a tangent point for angle measurement, is more accurate than the linear measurement only techniques described in (g) and (h) above. This method assumes that the distance along the arc of a circle is equal to the chord. This assumption will only be valid for short chords, and depends on the accuracy required in the setting out. The user is advised at the outset of the length of the curve, and asked to select a suitable chord length. Advice is presented for three regularly used orders of accuracy where the equivalent chord lengths are:

$$1/2500 \text{ accuracy}, \ C < R/10$$
$$1/5000 \text{ accuracy}, \ C < R/14$$
$$1/10000 \text{ accuracy}, \ C < R/20,$$

and the user asked to choose a chord length (C_H).

As it is seldom that the first tangent point is at zero chainage, it means that if pegs are to be set out round the curve, a short chord will be required at the outset in order to establish the pegs at standard chainages, which are multiple intervals of the chord length. The chainage of the first point on the curve is therefore calculated as a multiple of the chord length C_H, in line 3090, and the length of the first chord C_1, computed in line 3100. The chainage of the last point on the curve is then calculated using a similar process in line 3110, and a short chord at the end of the curve (C_2) used to bring the curve to the final tangent point. The number (N) of chords C_H long required to set out the curve can now be computed in line 3130.

Before computing the chainages and calculating the relevant setting-out angles, the user is asked to specify whether it is a left-hand or right-hand curve. The latter is set out clockwise from zero, whilst the former is set out anti-clockwise from 360°. The screen display then gives all the relevant details and setting-out angles with the option of a hardcopy printout, together with a plot of the curve. A check is also made that the final angle set out is equal to half the deviation angle.

(j) Graphics display and screen dump Line numbers 4000–4440

The display screen chosen for the presentation of the layout of the curve is shown in Fig. 1.5 with text at the top and bottom of the graphics window. The intersection point I is centred on the screen and the curve plotted symmetrically on either side on the centre line. There is no need to specify a scale for the plot as the display will automatically present the maximum plot, either for the long chord or the central line. The chainages of the first and last tangent points together with the chainage of the intersection point are displayed. The deviation angle and radius are also noted in the top left-hand corner of the screen for reference. The user can obtain a screen dump to a dot-matrix graphics printer (Fig. 4.2) before returning to the main menu and the plotter selection.

(k) Plotter initialization and selection Line numbers 5000–5190

After initializing the plotter variables to give a maximum plot width or height, a choice of three plotters is given, Apple 410, Epson HI-80 and Hewlett-Packard. As discussed earlier in (j), the curve is plotted centrally on the plotter and the setting-out design details listed in the upper left-hand corner. The chainages of the first and last tangent points together with the chainage of the intersection point are displayed on the plot. The curve design lengths (Fig. 4.1) are also labelled on the plot for reference, Fig. 4.3.

(l) Plotter routine for HPGL Line numbers 5200–5940

This routine is suitable for Hewlett-Packard plotters and their equivalents using HPGL as discussed in routine (k) of Section 2.2. After scaling the

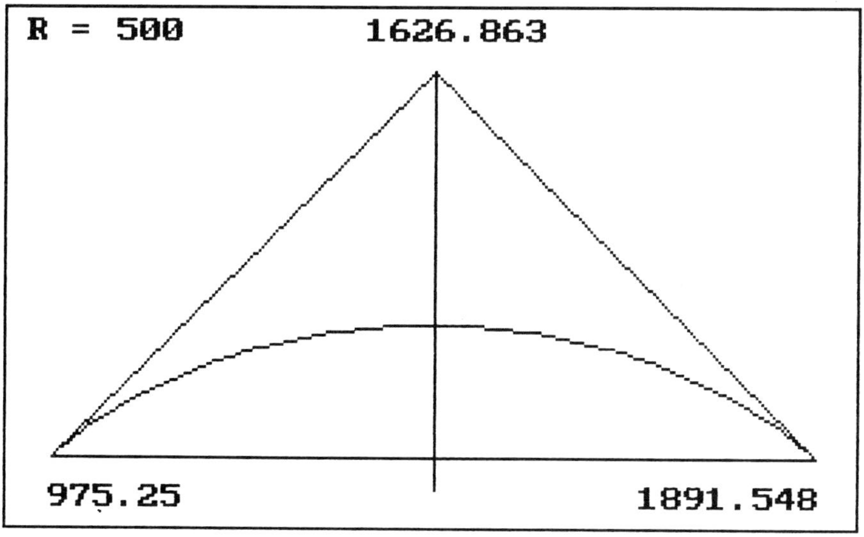

Fig. 4.2 PC-HCurve – screen dump of horizontal circular curve whose details are given in the computer printout in Section 4.2.5.

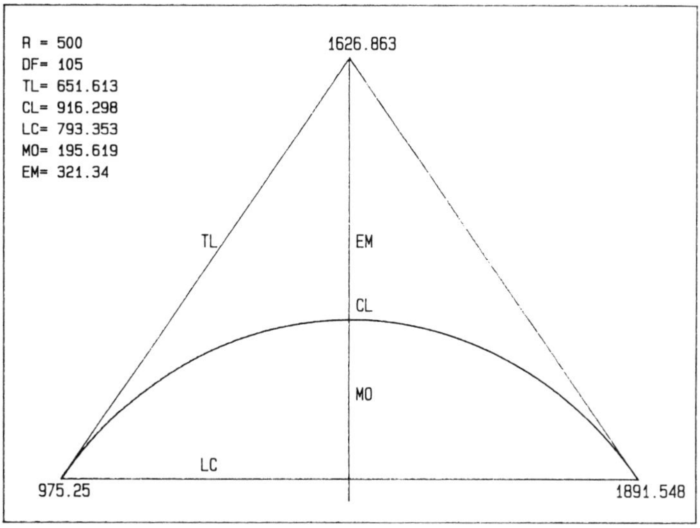

Fig. 4.3 PC-HCurve – plot of horizontal circular curve shown in Fig. 4.2.

plotter into user units, a frame is drawn round the plotting area by pen 1 (black). Pen 4 (blue) is chosen to draw the centre line, tangents and long chord, and then the curve plotted by pen 3 (green). Finally the chainages, labels and design details are printed by pen 2 (red).

(m) Routine to print curve location, etc. Line numbers 6000–6080

This routine is provided for record purposes to annotate the printout, assuming a standard IBM PC type printer. It may require alteration for different microcomputers and printers.

(n) Printout for ⟨1⟩ Line numbers 6100–6390

The first part of this printout, lines 6120–6220, are accessed by each of the printout routines, to list the basic curve design parameters and relevant chainages. The latter part of the printout details the point numbers, chainages, tangential distances and tangential offsets for setting-out the complete curve.

(o) Printout for ⟨2⟩ Line numbers 6400–6580

After listing the curve design parameters in (n) above, the printout gives the point numbers, chainages, chord distances and chord offsets for setting out the complete curve.

(p) Printout for ⟨3⟩ right-hand curve Line numbers 6600–6940

After listing the curve design parameters in (o) above, the point numbers, chainages, chords and cumulative setting-out angles from zero, for a right-hand curve, are printed.

(q) Printout for ⟨3⟩ left-hand curve Line numbers 7000–7330

This routine is similar to (p) above, except that the cumulative setting-out angles are calculated from 360° for a left-hand curve.

(r) Routine to convert DD.MMSS to decimals Line numbers 8000–8040

Earlier in the program the user was given the option of entering angles in either the decimals or degrees, minutes, seconds (DD.MMSS) format. If the angles are entered in DD.MMSS then they must be converted to decimal degrees prior to computation using this routine.

(s) Routine to convert decimals to DD.MMSS Line numbers 8050–8140

Where angles have been entered in DD.MMSS and converted for computation using (r), they are converted back to DD.MMSS using this routine before display and printout.

(t) Plotter routine for Apple 410 Line numbers 8200–8940

This routine is similar to (l) where the plotter commands are altered as shown in Table 2.1.

(u) Routine to clear screen Line numbers 9000–9060

This routine is used to clear the screen and provide two lines of text at the head of each screen giving the title of the selected software suite and the selected program, together with the date and time of calculation.

(v) Routine to display error message Line numbers 9100–9130

If a wrong keyboard entry has been made, the error is trapped and this error message displayed and re-entry requested.

(w) Termination of program Line numbers 9500–9620

When the **QUIT** selection is made in the program, an end-of-program message is displayed to advise the user that computations are complete.

(x) Plotter routine for Epson HI-80 Line numbers 10200–10940

This routine is similar to (l) where the plotter commands are altered as shown in Table 2.1.

4.2.2 PC-HCurve – numeric variables

AØ–A2	= Chord setting-out angles
A9	= Angle solution for arcsine
A(N + 1)	= Array containing setting-out angles
C	= Counter in error trace routine
C1	= First chord on curve
C2	= Last chord on curve
C9	= Chainage of last point on curve
CH	= Chord length
CH(N)	= Array for chainage of points on curve
CL	= Curve length
CL(N)	= Array for distance along curve
D,DØ,D1,D2,D5	= Degrees of angles
D9	= Angles in decimals ⟨1⟩ or DD.MMSS ⟨2⟩
DA(N)	= Array of angles used in offsets from long chord solution
DF	= Deviation angle
DX	= Half deviation angle
E	= Error number
EM	= External distance at mid-ordinate
F1,F2	= Chainages of tangent points
F3	= Chainage of IP
F4	= Chainage of *M*, mid-ordinate
GX	= *X*-coordinate at lower left-hand corner of plot
GY	= *Y*-coordinate at lower left-hand corner of plot
HP%	= Integer variable to identify plotter
I	= Loop integer
IPX	= *X*-coordinate of intersection point
IPY	= *Y*-coordinate of intersection point
J	= Loop integer
K	= Loop integer

LC	= Long chord
M,MØ,M1	= Minutes of angle
MO	= Mid-ordinate
MX	= Multiplication factor for Epson HI-80 plotter
N	= Number of chords or offsets
P1	= Program selection number
P7	= Counter in printout routine
PLOT%	= Integer variable for plotter selection
PX	= *X*-coordinate at upper right-hand corner of plot
PY	= *Y*-coordinate at upper right-hand corner of plot
R	= Radius of curve
R9	= 57.2957795
RGB%	= Integer variable to select screen colour
S,SØ	= Seconds of angle
SCD%	= Monitor display, monochrome/colour
SCR%	= Screen resolution (installation file)
SCRM	= Scale magnification to fit screen/plotter
SR%	= Screen resolution
SX	= Horizontal display width in user units
SY	= Vertical display height in user units
TL	= Tangent length
TPLX	= *X*-coordinate of tangent point (left)
TPLY	= *Y*-coordinate of tangent point (left)
TPRX	= *X*-coordinate of tangent point (right)
TPRY	= *Y*-coordinate of tangent point (right)
UM	= Maximum tangent distance at mid-ordinate
U(N)	= Array of chord or tangent distances
VM	= Tangent offset at mid-ordinate
V(N)	= Array of chord or tangent offset distances
W	= Offset interval

4.2.3 PC-HCurve – string variables

A$		Storage for location of curve
B$		Storage for operator's name
C$		Storage for date of calculations
CC$	=	**"PC-Circular Curves"**
DATE$		Variable used to return the date of the computer's calendar/clock in US format MM–DD–YYYY
EO$	=	**"SORRY, DATA ERROR . . . PLEASE RE-ENTER"**
H$	=	**"PC-Circular Curve Setting Out Program**
K$		Storage for instrument used
L$	=	Either **"L"** left-handed or **"R"** for right-handed curve
P$	=	**"DO YOU WISH PRINTOUT OF ABOVE DATA (Y/N)"**
PCF$	=	**"PCSCREEN"**
P1$	=	**"⟨1⟩ OFFSETS FROM TANGENT"**

P2$	= **"⟨2⟩ OFFSETS FROM LONG CHORD"**
P3$	= **"⟨3⟩ TANGENTIAL DEFLECTION ANGLES"**
Q$	= Answer response to question (Y/N)
T$	= **"PC-Graphics"**
TIME$	Variable used to return the time of the built-in clock in HH:MM:SS format

4.2.4 PC-HCurve – BASIC program

```
10  REM  <PC-HCURV> Program for IBM PC or Compatible
12  REM  BASIC Programs for Land Surveying, Chapter 5
13  REM  + Computer Graphics for Surveying, Chapter 4
14  REM  (C)  COPYRIGHT  1984-87  P.H.Milne
16  REM  ALL RIGHTS RESERVED
18  REM  Version PC-1.0, 01-01-87
20 CLEAR:KEY OFF
25 DEFDBL A,C-F,L,M,R-V
30 A = 0:B = 0:E = 0:G = 0:J = 0:K = 0:N = 0:S = 0:U = 0:V = 0:W = 0
32 CK=0:SR%=1
35  DIM A(100),CH(100),CL(100),DA(100),DF(100),U(100),V(100)
40 E0$ = "SORRY, DATA ERROR ... PLEASE RE-ENTER."
42 H$ = "PC-Circular Curve Setting Out Program"
45 P$ = "DO YOU WISH PRINTOUT OF ABOVE DATA (Y/N)"
50 PCF$="PCSCREEN"
52 OPEN "I",#3,PCF$
54 INPUT#3,SCR%
56 INPUT#3,SCD%
58 CLOSE#3
60 ON (SCR%+1) GOTO 62,64,66
62 SCREEN 0:WIDTH 40:GOTO 80
64 SCREEN 1:GOTO 68
66 SCREEN 1:GOTO 80
68 IF SCD%=1 THEN 80
70 COLOR 8,0
80  DEF  FN C9(C9) = (1.5707964# -  ATN (C9 /  SQR (1 - C9 * C9))) *
R9: REM  ARCCOSINE
90  DEF  FN S9(S9) =  ATN (S9 /  SQR (1 - S9 * S9)) * R9: REM  ARCSIN
E
95  REM  ***************************************************
100  GOSUB 9000
110  PRINT
200  PRINT "ARE ANGLES IN DECIMALS <1>, DD.MMSS <2> OR DO YOU WISH TO
 QUIT <3>."
210  INPUT D9
220  IF D9 < 1 OR D9 > 3 THEN 200
230  ON D9 GOTO 240,240,9500
240 R9 = 57.2957795#:P1 = 0
500  GOSUB 9000
540  PRINT
550  PRINT "ENTER LOCATION OF WORK, ETC. WHEN       REQUESTED, AND PR
ESS <RETURN>"
560  PRINT "ENTER LOCATION OF CURVE :-"
570  INPUT A$
580  PRINT "ENTER OPERATOR'S NAME :-"
590  INPUT B$
600  PRINT "ENTER DATE OF CALCULATION AS DD/MM/YY :-"
610  INPUT C$
620  PRINT "ENTER INSTRUMENT USED :-"
630  INPUT K$
640  PRINT
650  PRINT "DO YOU WISH PRINTOUT OF LOCATION ETC.,  (Y/N) ";
```

```
660  INPUT Q$
670  IF Q$ = "Y" THEN 6000
680  IF Q$ <  > "N" THEN 650
700  GOSUB 9000
710  PRINT:PRINT H$
720  PRINT : PRINT "*****************************************"
730  PRINT : PRINT "ENTER DEFLECTION ANGLE ";: INPUT DF
740  IF DF < 0 OR DF > 180 THEN 730
750  PRINT : PRINT "ENTER RADIUS OF CURVE ";: INPUT R
760  IF R <  = 0 THEN 750
770  PRINT : PRINT "ENTER CHAINAGE OF 1ST TANGENT POINT"
780  INPUT F1: IF F1 < 0 THEN 770
790  PRINT : PRINT "CALCULATING DESIGN CONSTANTS ...."
800  IF D9 = 1 THEN 820
805  REM  CONVERT DF IN DD.MMSS TO DECIMALS
810 D5 = DF: GOSUB 8000: IF CK = 1 THEN 730
815 DF = D5
820 TL =  INT ((R *  TAN (DF / (2 * R9))) * 1000 + .5) / 1000
830 LC =  INT ((2 * R *  SIN (DF / (2 * R9))) * 1000 + .5) / 1000
840 CL =  INT ((R * DF / R9) * 1000 + .5) / 1000
850 MO =  INT ((R * (1 -  COS (DF / (2 * R9)))) * 1000 + .5) / 1000
860 EM =  INT ((TL *  TAN (DF / (4 * R9))) * 1000 + .5) / 1000
870 F2=INT((F1+CL)*1000+.5)/1000
880 F3=INT((F1+TL)*1000+.5)/1000
900  GOSUB 9000:PRINT: PRINT H$
910  PRINT : PRINT "SELECT SETTING OUT METHOD FROM MENU :-"
920  PRINT
930  PRINT "    <1> OFFSETS FROM TANGENT"
940  PRINT "    <2> OFFSETS FROM LONG CHORD"
950  PRINT "    <3> TANGENTIAL DEFLECTION ANGLES"
960  PRINT "    <4> GRAPHICAL OUTPUT"
965  PRINT "    <5> PLOTTER OUTPUT"
970  PRINT "    <6> QUIT             ";
975  INPUT P1
980  IF P1 < 1 OR P1 > 6 THEN 900
985 PRINT:PRINT " One moment ... preparing data ..."
990  ON P1 GOTO 1000,2000,3000,4000,5000,9500
993  REM  ***********************************************
995  REM  <1> LINEAR MEASUREMENTS FROM TANGENT
1000  GOSUB 9000:P1$ = "<1> OFFSETS FROM TANGENT"
1010  PRINT H$: PRINT P1$
1020  PRINT "*****************************************"
1025  REM  KNOW DF,R,F1,F2,TL,LC,CL,MO,EM
1030 UM =  INT ((TL * R / (EM + R)) * 1000 + .5) / 1000
1040 VM =  INT ((EM * R / (EM + R)) * 1000 + .5) / 1000
1050  PRINT : PRINT "MAX. OFFSET AT MID-ORDINATE = ";UM
1060  PRINT : PRINT "SELECT OFFSET INTERVAL ";: INPUT W
1065  IF W <  = 0 THEN 1060
1070 N =  INT (UM / W)
1072  IF N < 100 THEN 1080
1074  GOSUB 9100
1076  PRINT "TOO MANY OFFSETS !"
1078  GOTO 1060
1080  IF D9 = 1 THEN 1100
1090 D = DF: GOSUB 8050:DF = D
1100  GOSUB 9000
1110  PRINT H$: PRINT P1$
1120  PRINT "*****************************************"
1130  PRINT "PT.  CHAINAGE      TAN.DIST  TAN.OFFSET"
1140  PRINT "========================================"
1145 F1=INT(F1*1000+.5)/1000
1150 PRINT "TP1";TAB(6);F1;TAB(20);" 0";TAB(30);" 0"
1160  FOR I = 0 TO N - 1
1170 U(I) = (I + 1) * W
1180 V(I) =  INT ((R -
```

```
1190 A9 =  SQR (U(I) * U(I) + V(I) * V(I)) / (2 * R)
1200 DF(I) = 2 *  FN S9(A9)
1210 CL(I) =  INT ((R * DF(I) / R9) * 1000 + .5) / 1000
1215 U(I)=INT(U(I)*1000+.5)/1000
1220  PRINT (I + 1); TAB( 6);(F1 + CL(I)); TAB( 20);U(I); TAB( 30);V(
I)
1230  NEXT I
1250  PRINT "MO"; TAB( 6); INT ((F1 + CL / 2) * 1000 + .5) / 1000; TA
B( 20);UM; TAB( 30);VM
1260  FOR J = N - 1 TO 0 STEP  - 1
1270  PRINT (2 * N - J); TAB( 6);F2 - CL(J); TAB( 20);U(J); TAB( 30);
V(J)
1280  NEXT J
1290 PRINT "TP2";TAB(6);F2;TAB(20);" 0";TAB(30);" 0"
1300  PRINT "****************************************"
1310  PRINT P$;: INPUT A$
1320  IF A$ = "Y" THEN 6100
1330  IF A$ <  > "N" THEN 1310
1340  GOTO 100
1993  REM  ************************************************
1995  REM  <2> LINEAR MEASUREMENT FROM LONG CHORD
2000  GOSUB 9000:P2$ = "<2> OFFSETS FOM LONG CHORD"
2010  PRINT H$: PRINT P2$
2020  PRINT "****************************************"
2025  REM  KNOW DF,R,F1,F2,F3,TL,LC,CL,MO,EM
2030  PRINT : PRINT "MAX. OFFSET AT MID-ORDINATE = ";LC / 2
2040  PRINT : PRINT "SELECT OFFSET INTERVAL ";: INPUT W
2045  IF W <  = 0 THEN 2040
2050 N =  INT (LC / (2 * W))
2052  IF N < 100 THEN 2060
2054  GOSUB 9100
2056  PRINT "TOO MANY OFFSETS !"
2058  GOTO 2040
2059  REM  CALCULATE CHAINAGE OF M = F4
2060 F4 =  INT ((F1 + CL / 2) * 1000 + .5) / 1000
2070  FOR I = 0 TO N - 1
2080 U(I) = (I + 1) * W
2090 V(I) =  INT (( SQR (R * R - U(I) * U(I)) - (R - MO)) * 1000 + .5
) / 1000
2100 A9 = U(I) / R
2110 DA(I) =  FN S9(A9)
2120 CL(I) =  INT ((R * DA(I) / R9) * 1000 + .5) / 1000
2130  NEXT I
2160  IF D9 = 1 THEN 2180
2170 D = DF: GOSUB 8050:DF = D
2180 IF P1=4 THEN 4200
2190 IF P1=5 THEN 5100
2200  GOSUB 9000
2210  PRINT H$: PRINT P2$
2220  PRINT "****************************************"
2230  PRINT "PT.  CHAINAGE      CH.DIST   CH.OFFSET"
2240  PRINT "========================================"
2250 PRINT "TP1";TAB(6);F1;TAB(20);LC/2;TAB(30);" 0"
2260  FOR J = N - 1 TO 0 STEP  - 1
2265 U(J)=INT(U(J)*1000+.5)/1000
2270  PRINT (N - J); TAB( 6);(F4 - CL(J)); TAB( 20);U(J); TAB( 30);V(
J)
2280  NEXT J
2290 PRINT "MO";TAB(6);F4;TAB(20);" 0";TAB(30);MO
2300  FOR K = 0 TO N - 1
2310  PRINT (N + 1 + K); TAB( 6);(F4 + CL(K)); TAB( 20);U(K); TAB( 30
);V(K)
2320  NEXT K
2330 PRINT "TP2";TAB(6);F2;TAB(20);LC/2;TAB(30);" 0"
2340  PRINT "****************************************"
```

```
2350  PRINT : PRINT P$;: INPUT A$
2360  IF A$ = "Y" THEN 6400
2370  IF A$ <  > "N" THEN 2350
2380  GOTO 100
2993  REM  **************************************************
2995  REM  <3> ANGULAR MEASUREMENTS FROM TANGENT POINT
3000  GOSUB 9000:P3$ = "<3> TANGENTIAL DEFLECTION ANGLES"
3010  PRINT H$: PRINT P3$
3020  PRINT "****************************************"
3025  REM  KNOW DF,R,F1,F2,F3,TL,LC,CL,MO,EM
3030  PRINT : PRINT "TOTAL CURVE LENGTH = ";CL
3040  PRINT : PRINT "SELECT CHORD LENGTH C FROM :-"
3050  PRINT "  1/2500  ACCURACY, C < "; INT (R / 10)
3060  PRINT "  1/5000  ACCURACY, C < "; INT (R / 14)
3070  PRINT "  1/10000 ACCURACY, C < "; INT (R / 20)
3080 INPUT CH:CH=INT(CH*1000+.5)/1000
3082  IF CL / CH < 100 THEN 3090
3084  GOSUB 9100
3086  PRINT "TOO MANY CHORDS !"
3088  GOTO 3040
3089  REM  CALCULATE CHAINAGE OF 1ST POINT ON CURVE
3090 CH(0) = ( INT (F1 / CH + .01) + 1) * CH
3095  REM  CALCULATE LENGTH OF 1ST CHORD
3100 C1 =  INT ((CH(0) - F1) * 1000 + .5) / 1000
3105  REM  CALCULATE CHAINAGE OF LAST POINT ON CURVE
3110 C9 = ( INT (F2 / CH)) * CH
3115  REM  CALCULATE LENGTH OF LAST CHORD
3120 C2 =  INT ((F2 - C9) * 1000 + .5) / 1000
3125  REM  CALCULATE NO.(N) OF CHORDS (CH)
3130 N =  INT ((CL - C1 - C2) / CH + .5)
3135  REM  CALCULATE TANGENTIAL ANGLES FOR CHORDS IN DECIMAL DEGREES
3140 A0 = 1718.8 * CH / (R * 60)
3150 A1 = 1718.8 * C1 / (R * 60):A(0) = A1
3160 A2 = 1718.8 * C2 / (R * 60)
3170  PRINT : PRINT "IS CURVE LEFT OR RIGHT HANDED (L/R)";: INPUT L$
3180  IF L$ = "L" THEN 3400
3190  IF L$ <  > "R" THEN 3170
3195  REM  SET OUT CURVE RIGHT HANDED FROM ZERO DEGREES
3200  GOSUB 9000
3210  PRINT H$: PRINT P3$
3220  PRINT "****************************************"
3230  PRINT "PT.  CHAINAGE      CHORD     ANGLE(DEC)"
3240  PRINT "========================================"
3250 PRINT "TP1";TAB(6);F1;TAB(30);" .000"
3260 PRINT " 1";TAB(6);CH(0);TAB(20);C1;TAB(30);INT(A1*1000+.5)/1000
3270  FOR I = 1 TO N
3280 CH(I)=INT((CH(0)+I*CH)*1000+.5)/1000
3290 A(I) = A(0) + I * A0
3300  PRINT (I + 1); TAB( 6);CH(I); TAB( 20);CH; TAB( 30); INT (A(I)
* 1000 + .5) / 1000
3310  NEXT I
3320  PRINT "TP2"; TAB( 6);F2; TAB( 20);C2; TAB( 30); INT ((A(N) + A2
) * 1000 + .5) / 1000
3330  PRINT "========================================"
3340  PRINT "FINAL ANGLE AT TP2 SHOULD BE "; INT ((DF / 2) * 1000 + .
5) / 1000
3350  PRINT "****************************************"
3360  PRINT : PRINT P$;: INPUT A$
3370  IF A$ = "Y" THEN 6600
3380  IF A$ <  > "N" THEN 3360
3390  GOTO 100
3395  REM  SET OUT CURVE LEFT HANDED FROM 360 DEGREES
3400  GOSUB 9000
3410  PRINT H$: PRINT P3$
```

```
3420  PRINT "***************************************"
3430  PRINT "PT.  CHAINAGE      CHORD     ANGLE(DEC)"
3440  PRINT "======================================="
3450 PRINT "TP1";TAB(6);F1;TAB(30);" 360.000"
3460 PRINT " 1";TAB(6);CH(0);TAB(20);C1;TAB(30);INT((360-A1)*1000+.5)
/1000
3470  FOR I = 1 TO N
3480 CH(I)=INT((CH(0)+I*CH)*1000+.5)/1000
3490 A(I) = (360 - A(0)) - (I * A0)
3500  PRINT (I + 1); TAB( 6);CH(I); TAB( 20);CH; TAB( 30); INT (A(I)
* 1000 + .5) / 1000
3510  NEXT I
3520  PRINT "TP2"; TAB( 6);F2; TAB( 20);C2; TAB( 30); INT ((A(N) - A2
) * 1000 + .5) / 1000
3530  PRINT "======================================="
3540  PRINT "FINAL ANGLE AT TP2 SHOULD BE "; INT ((360 - DF / 2) * 10
00 + .5) / 1000
3550  PRINT "***************************************"
3560  PRINT : PRINT P$;: INPUT A$
3570  IF A$ = "Y" THEN 7000
3580  IF A$ <  > "N" THEN 3560
3590  GOTO 100
3995 REM ***** Graphics Screen Display *****
4000  CLS:GOSUB 9000
4010  IF SR%=1 THEN RGB%=2 ELSE RGB%=3
4020 WINDOW (0,0)-(320*SR%,200)
4030 LINE (8,18)-(310*SR%,180),RGB%,B
4100 TPLY=40:TPRY=40
4110 SCRM=(275*SR%)/(LC*1.6)
4120 IF SCRM*(MO+EM) > 110 THEN SCRM=110/(MO+EM)
4130 IPX=160*SR%
4140 IPY=50+SCRM*(MO+EM)
4150 TPLX=160*SR%-(SCRM*1.6*LC/2)
4160 TPRX=160*SR%+(SCRM*1.6*LC/2)
4170 W=15
4180 GOTO 2050
4200 LINE (IPX,TPLY-10)-(IPX,IPY),3
4210 LINE -(TPLX,TPLY),3
4220 LINE -(TPRX,TPRY),3
4230 LINE -(IPX,IPY),3
4232 CCOL=(TPLX*SR%-32)/8
4234 IF CCOL<3 THEN CCOL=3
4240 LOCATE 22,CCOL:PRINT F1
4242 CCOL=(TPRX*SR%-32)/8
4244 IF CCOL>29 THEN CCOL=29
4250 LOCATE 22,CCOL:PRINT F2
4260 LOCATE 4,17*SR%:PRINT F3
4270 LOCATE 4,3*SR%:PRINT "R =";INT(R*1000+.5)/1000
4280 LOCATE 1,2:PRINT CC$
4290 LOCATE 2,2:PRINT "Defln Angle = ";DF
4300 PSET (TPLX,TPLY),1
4310 FOR J=N-1 TO 0 STEP -1
4320 LINE -((IPX-U(J)*SCRM*1.6*SR%),(TPLY+V(J)*SCRM)),1
4330 NEXT J
4340 LINE -(IPX,TPLY+MO*SCRM),1
4350 FOR K=0 TO N-1
4360 LINE -((IPX+U(K)*SCRM*1.6*SR%),(TPLY+V(K)*SCRM)),1
4370 NEXT K
4380 LINE -(TPRX,TPRY),1
4390 P1=0
4400 LOCATE 24,2:PRINT "Screen Dump (y/n) ";
4410 Q$=INKEY$:IF Q$ = "" THEN 4410
4420 IF INSTR("yn",Q$)=0 THEN 4400
4430 IF Q$ = "n" THEN 900
```

```
4440 LCOPY
4450 GOTO 900
4995 REM ***** Plotter Output *****
5000 GX=0:GY=0:SX=180:SY=120:MX=2670/SX:PX=GX+SX:PY=GY+SY
5010 SCRM=150/LC
5020 IF SCRM*(MO+EM) > 100 THEN SCRM = 100/(MO+EM)
5030 IPX=GX+SX/2
5040 IPY=GY+10+SCRM*(MO+EM)
5050 TPLX=IPX-(SCRM*LC/2)
5060 TPRX=IPX+(SCRM*LC/2)
5070 TPLY=GY+10:TPRY=GY+10
5080 W=10
5090 GOTO 2050
5100 GOSUB 9000
5110 PRINT:PRINT " Select Plotter -";TAB(18);"<1> Apple 410":PRINT:PR
INT TAB(18);"<2> Epson HI-80"
5120 PRINT:PRINT TAB(18);"<3> Hewlett-Packard":PRINT:PRINT TAB(18);"<
4> QUIT          ";:INPUT PLOT%
5130 IF PLOT% <1 OR PLOT% >4 THEN 5100
5140 IF PLOT%=4 THEN 900
5145 IF PLOT%=1 THEN HP%=3
5150 IF PLOT%=2 THEN HP%=4
5155 IF PLOT%=3 THEN HP%=1
5160 PRINT:PRINT " Place paper on Plotter and switch on"
5170 PRINT:PRINT " To Continue Press <Enter> ":WHILE R$="":R$=INKEY$:
WEND
5180 GOSUB 50000
5190 ON HP% GOTO 5200,5200,8200,10200
5195 REM ***** HP-GL Plotter (HP-7475A) *****
5200 GOSUB 51000
5210 PRINT#2,"SP4;"
5220 PRINT#2,"PU;PA";IPX;",";(TPLY-5);";"
5230 PRINT#2,"PD;PA";IPX;",";IPY;";"
5240 PRINT#2,"PA";TPLX;",";TPLY;";"
5250 PRINT#2,"PA";TPRX;",";TPRY;";"
5260 PRINT#2,"PA";IPX;",";IPY;";"
5290 PRINT#2,"SP3;"
5300 PRINT#2,"PU;PA";TPLX;",";TPLY";"
5310 FOR J=N-1 TO 0 STEP -1
5320 PRINT#2,"PD;PA";(IPX-U(J)*SCRM);",";(TPLY+V(J)*SCRM);";"
5330 NEXT J
5340 PRINT#2,"PA";IPX;",";(TPLY+MO*SCRM);";"
5350 FOR K=0 TO N-1
5360 PRINT#2,"PA";(IPX+U(K)*SCRM);",";(TPLY+V(K)*SCRM);";"
5370 NEXT K
5380 PRINT#2,"PA";TPRX;",";TPRY;";"
5390 PRINT#2,"SP2;":PRINT#2,"SI0.2,0.3;"
5400 PRINT#2,"PU;PA";(TPLX-8);",";(TPLY-3);";"
5410 PRINT#2,"LB";F1
5420 PRINT#2,CHR$(3)
5430 PRINT#2,"PU;PA";(TPRX-8);",";(TPLY-3);";"
5440 PRINT#2,"LB";F2
5450 PRINT#2,CHR$(3)
5460 PRINT#2,"PU;PA";(IPX-8);",";(IPY+2);";"
5470 PRINT#2,"LB";F3
5480 PRINT#2,CHR$(3)
5500 PRINT#2,"PU;PA";(IPX-1-SCRM*LC/4);",";(TPLY+2);";"
5510 PRINT#2,"LBLC"
5520 PRINT#2,CHR$(3)
5530 PRINT#2,"PU;PA";(IPX-1-SCRM*LC/4);",";((TPLY+IPY)/2+5);";"
5540 PRINT#2,"LBTL"
5550 PRINT#2,CHR$(3)
5560 PRINT#2,"PU;PA";(IPX+2);",";((TPLY+IPY)/2+5);";"
5570 PRINT#2,"LBEM"
5580 PRINT#2,CHR$(3)
```

```
5600 PRINT#2,"PU;PA";(IPX+2);",";(TPLY+MO*SCRM+2);";"
5610 PRINT#2,"LBCL"
5620 PRINT#2,CHR$(3)
5630 PRINT#2,"PU;PA";(IPX+2);",";(TPLY+MO*SCRM/2);";"
5640 PRINT#2,"LBMO"
5650 PRINT#2,CHR$(3)
5660 PRINT#2,"PU;PA";(GX+5);",";(GY+SY-8);";"
5670 PRINT#2,"LBR =";R
5680 PRINT#2,CHR$(3)
5700 PRINT#2,"PU;PA";(GX+5);",";(GY+SY-12);";"
5710 PRINT#2,"LBDF=";DF
5720 PRINT#2,CHR$(3)
5730 PRINT#2,"PU;PA";(GX+5);",";(GY+SY-14);";"
5740 PRINT#2,"LBTL=";TL
5750 PRINT#2,CHR$(3)
5760 PRINT#2,"PU;PA";(GX+5);",";(GY+SY-16);";"
5770 PRINT#2,"LBCL=";CL
5780 PRINT#2,CHR$(3)
5800 PRINT#2,"PU;PA";(GX+5);",";(GY+SY-20);";"
5810 PRINT#2,"LBLC=";LC
5820 PRINT#2,CHR$(3)
5830 PRINT#2,"PU;PA";(GX+5);",";(GY+SY-24);";"
5840 PRINT#2,"LBMO=";MO
5850 PRINT#2,CHR$(3)
5860 PRINT#2,"PU;PA";(GX+5);",";(GY+SY-28);";"
5870 PRINT#2,"LBEM=";EM
5880 PRINT#2,CHR$(3)
5900 PRINT#2,"PU,PA5,-3;"
5910 PRINT#2,"LBPC-Curves - ";A$;" : Date : ";DATE$
5920 PRINT#2,CHR$(3)
5930 CLOSE#2
5940 GOTO 900
5985  REM  ************************************************
5990  REM  PRINTOUT OF LOCATION, ETC
6000 LPRINT "Location of HCurve :- ";A$
6010 LPRINT "************************************************"
6020 LPRINT "OPERATOR'S NAME :- ";B$
6030 LPRINT "DATE OF CALCS.  :- ";C$
6040 LPRINT "INSTRUMENT USED :- ";K$
6060 LPRINT "************************************************"
6070 LPRINT
6080  GOTO 700
6093  REM  ************************************************
6095  REM  PRINTOUT FOR P1=1
6100 LPRINT H$
6110 LPRINT P1$:P7=0
6120 LPRINT "****************************************"
6130 LPRINT "RADIUS            = ";INT(R*1000+.5)/1000
6140 LPRINT "DEFLECTION ANGLE = ";DF
6150 LPRINT "CHAINAGE I.P.     = ";F3
6160 LPRINT "TANGENT LENGTH    = ";TL
6170 LPRINT "CHAINAGE TP1      = ";F1
6180 LPRINT "CURVE LENGTH      = ";CL
6190 LPRINT "CHAINAGE TP2      = ";F2
6200 LPRINT "LONG CHORD        = ";LC
6210 LPRINT "MID-ORDINATE      = ";MO
6220 LPRINT "EXTERNAL MO-DIST = ";EM
6230  IF P7 = 1 THEN  RETURN
6240 LPRINT "========================================"
6250 LPRINT "PT.  CHAINAGE      TAN.DIST  TAN.OFFSET"
6260 LPRINT "========================================"
6270 LPRINT "TP1";TAB(6);F1;TAB(20);" 0";TAB(30);" 0"
6280  FOR I = 0 TO N - 1
6290 LPRINT (I+1);TAB(6);(F1+CL(I));TAB(20);U(I);TAB(30);V(I)
6300  NEXT I
```

```
6310 LPRINT "MO";TAB(6);INT((F1+CL/2)*1000+.5)/1000;TAB(20);UM;TAB(30
);VM
6320  FOR J = N - 1 TO 0 STEP  - 1
6330 LPRINT (2*N-J);TAB(6);(F2-CL(J));TAB(20);U(J);TAB(30);V(J)
6340  NEXT J
6350 LPRINT "TP2";TAB(6);F2;TAB(20);" 0";TAB(30);" 0"
6360 LPRINT "****************************************"
6370 LPRINT
6380 GOTO 900
6393  REM  ************************************************
6395  REM  PRINTOUT FOR P1=2
6400 LPRINT H$
6410 LPRINT P2$
6420 P7 = 1: GOSUB 6120
6430 LPRINT "========================================"
6440 LPRINT "PT.  CHAINAGE      CH.DIST   CH.OFFSET"
6450 LPRINT "========================================"
6460 LPRINT "TP1";TAB(6);F1;TAB(20);LC/2;TAB(30);" 0"
6470  FOR J = N - 1 TO 0 STEP  - 1
6480 LPRINT (N-J);TAB(6);(F4-CL(J));TAB(20);U(J);TAB(30);V(J)
6490  NEXT J
6500 LPRINT "MO";TAB(6);F4;TAB(20);" 0";TAB(30);MO
6510  FOR K = 0 TO N - 1
6520 LPRINT (N+1+K);TAB(6);(F4+CL(K));TAB(20);U(K);TAB(30);V(K)
6530  NEXT K
6540 LPRINT "TP2";TAB(6);F2;TAB(20);LC/2;TAB(30);" 0"
6550 LPRINT "****************************************"
6560 LPRINT
6570 GOTO 900
6580 :
6593  REM  ************************************************
6595  REM  <3> RIGHT HANDED CURVE PRINTOUT
6600 LPRINT H$:LPRINT P3$
6610 P7=1:IF D9=1 THEN 6640
6620 D = DF / 2: GOSUB 8050:DX = D
6630 D = DF: GOSUB 8050:DF = D
6640  GOSUB 6120
6650 LPRINT "========================================"
6660 LPRINT "PT.  CHAINAGE       CHORD     ANGLE"
6670 LPRINT "========================================"
6680 LPRINT "TP1";TAB(6);F1;TAB(30);" .000"
6690  IF D9 = 2 THEN 6800
6695  REM  D9=1, ANGLES IN DECIMALS
6700 LPRINT " 1";TAB(6);CH(0);TAB(20);C1;TAB(30);INT(A1*1000+.5)/1000
6710  FOR J = 1 TO N
6720 LPRINT (J+1);TAB(6);CH(J);TAB(20);CH;TAB(30);INT(A(J)*1000+.5)/1
000
6730  NEXT J
6740 LPRINT "TP2";TAB(6);F2;TAB(20);C2;TAB(30);INT((A(N)+A2)*1000+.5)
/1000
6750 LPRINT "========================================"
6760 LPRINT "FINAL ANGLE AT TP2 SHOULD BE  ";INT((DF/2)*1000+.5)/1000
6770 LPRINT "****************************************"
6780 LPRINT
6790 GOTO 900
6795  REM  D9=2, ANGLES REQUIRED IN DD.MMSS
6800 D = (A(N) + A2): GOSUB 8050:A(N + 1) = D
6810  FOR J = 0 TO N
6820 D = A(J): GOSUB 8050:A(J) = D
6830  NEXT J
6840 LPRINT " 1";TAB(6);CH(0);TAB(20);C1;TAB(30);A(0)
6850  FOR K = 1 TO N
6860 LPRINT (K+1);TAB(6);CH(K);TAB(20);CH;TAB(30);A(K)
6870  NEXT K
6880 LPRINT "TP2";TAB(6);F2;TAB(20);C2;TAB(30);A(N+1)
```

```
6890 LPRINT "========================================="
6900 LPRINT "FINAL ANGLE AT TP2 SHOULD BE  ";DX
6910 LPRINT "*****************************************"
6920 LPRINT
6930 GOTO 9ØØ
6993  REM  ************************************************
6995  REM  <3> LEFT HANDED CURVE PRINTOUT
7ØØØ LPRINT H$:LPRINT P3$
7Ø1Ø P7=1:IF D9=1 THEN 7Ø4Ø
7Ø2Ø D = (36Ø - DF / 2): GOSUB 8Ø5Ø:DX = D
7Ø3Ø D = DF: GOSUB 8Ø5Ø:DF = D
7Ø4Ø  GOSUB 612Ø
7Ø5Ø LPRINT "========================================="
7Ø6Ø LPRINT "PT.  CHAINAGE      CHORD     ANGLE"
7Ø7Ø LPRINT "========================================="
7Ø8Ø LPRINT "TP1";TAB(6);F1;TAB(3Ø);" 36Ø.ØØØ"
7Ø9Ø  IF D9 = 2 THEN 72ØØ
7Ø95  REM  D9=1, ANGLES IN DECIMALS
71ØØ LPRINT " 1";TAB(6);CH(Ø);TAB(2Ø);C1;TAB(3Ø);INT((36Ø-A1)*1ØØØ+.5
)/1ØØØ
711Ø  FOR J = 1 TO N
712Ø LPRINT (J+1);TAB(6);CH(J);TAB(2Ø);CH;TAB(3Ø);INT(A(J)*1ØØØ+.5)/1
ØØØ
713Ø  NEXT J
714Ø LPRINT "TP2";TAB(6);F2;TAB(2Ø);C2;TAB(3Ø);INT((A(N)-A2)*1ØØØ+.5)
/1ØØØ
715Ø LPRINT "========================================="
716Ø LPRINT "FINAL ANGLE AT TP2 SHOULD BE  ";INT((36Ø-DF/2)*1ØØØ+.5)/
1ØØØ
717Ø LPRINT "*****************************************"
718Ø LPRINT
719Ø GOTO 9ØØ
7195  REM  D9=2, ANGLES REQUIRED IN DD.MSS
72ØØ D = (A(N) - A2): GOSUB 8Ø5Ø:A(N + 1) = D
72Ø5 A(Ø) = 36Ø - A1
721Ø  FOR J = Ø TO N
722Ø D = A(J): GOSUB 8Ø5Ø:A(J) = D
723Ø  NEXT J
724Ø LPRINT " 1";TAB(6);CH(Ø);TAB(2Ø);C1;TAB(3Ø);A(Ø)
725Ø  FOR K = 1 TO N
726Ø LPRINT (K+1);TAB(6);CH(K);TAB(2Ø);CH;TAB(3Ø);A(K)
727Ø  NEXT K
728Ø LPRINT "TP2";TAB(6);F2;TAB(2Ø);C2;TAB(3Ø);A(N+1)
729Ø LPRINT "========================================="
73ØØ LPRINT "FINAL ANGLE AT TP2 SHOULD BE  ";DX
731Ø LPRINT "*****************************************"
732Ø LPRINT
733Ø GOTO 9ØØ
7993  REM  ************************************************
7995  REM  ROUTINE DD.MMSS TO DECIMALS
8ØØØ DØ =  INT (D5)
8ØØ5  REM  FIND NO. OF MINUTES,MØ
8Ø1Ø M1 = (D5 - DØ) * 1ØØ:MØ =
8Ø15  REM  FIND NO. OF SECONDS,SØ
8Ø2Ø SØ =  INT (M1 * 1ØØ - MØ * 1ØØ + .5)
8Ø22  IF MØ < Ø OR MØ > 59 THEN 8Ø3Ø
8Ø24  IF SØ < Ø OR SØ > 59.99 THEN 8Ø3Ø
8Ø25  REM  COLLECT DEGS,MINS,SECS
8Ø26 D5 = DØ + MØ / 6Ø + SØ / 36ØØ
8Ø28 CK = Ø: RETURN
8Ø3Ø  GOSUB 91ØØ: REM  DATA ERROR
8Ø4Ø CK = 1: RETURN
8Ø43  REM  ************************************************
8Ø45  REM  ROUTINE DECIMAL TO DD.MMSS
8Ø5Ø D1 =  INT (D)
```

```
8055  REM  FIND TOTAL NO. OF SECONDS
8060 D2 = (D - D1) * 3600
8065  REM  FIND NO. OF MINUTES
8070 M =  INT (D2 / 60)
8075  REM  FIND NO. OF SECONDS
8080 S = D2 - 60 * M + .5
8090  IF S < 60 THEN 8130
8100 M = M + 1:S = 0
8110  IF M < 60 THEN 8130
8120 D1 = D1 + 1:M = 0
8130 D = D1 * 10000 + M * 100 + S
8140 D =  INT (D) / 10000: RETURN
8195 REM ***** Apple 410 Plotter *****
8200 GOSUB 53000
8210 PRINT#2,"PS4;"
8220 PRINT#2,"MA";IPX;",";(TPLY-5);";"
8230 PRINT#2,"DA";IPX;",";IPY;";"
8240 PRINT#2,"DA";TPLX;",";TPLY;";"
8250 PRINT#2,"DA";TPRX;",";TPRY;";"
8260 PRINT#2,"DA";IPX;",";IPY;";"
8290 PRINT#2,"PS3;"
8300 PRINT#2,"MA";TPLX;",";TPLY;";"
8310 FOR J=N-1 TO 0 STEP -1
8320 PRINT#2,"DA";(IPX-U(J)*SCRM);",";(TPLY+V(J)*SCRM);";"
8330 NEXT J
8340 PRINT#2,"DA";IPX;",";(TPLY+MO*SCRM);";"
8350 FOR K=0 TO N-1
8360 PRINT#2,"DA";(IPX+U(K)*SCRM);",";(TPLY+V(K)*SCRM);";"
8370 NEXT K
8380 PRINT#2,"DA";TPRX;",";TPRY;";"
8390 PRINT#2,"PS2;":PRINT#2,"LS2;"
8400 PRINT#2,"MA";(TPLX-8);",";(TPLY-4);";"
8410 PRINT#2,"PL";F1
8420 PRINT#2,CHR$(13) CHR$(10) CHR$(3)
8430 PRINT#2,"MA";(TPRX-8);",";(TPLY-4);";"
8440 PRINT#2,"PL";F2
8450 PRINT#2,CHR$(13) CHR$(10) CHR$(3)
8460 PRINT#2,"MA";(IPX-8);",";(IPY+2);";"
8470 PRINT#2,"PL";F3
8480 PRINT#2,CHR$(13) CHR$(10) CHR$(3)
8500 PRINT#2,"MA";(IPX-SCRM*LC/4);",";(TPLY+2);";"
8510 PRINT#2,"PLLC"
8520 PRINT#2,CHR$(13) CHR$(10) CHR$(3)
8530 PRINT#2,"MA";(IPX-1-SCRM*LC/4);",";((TPLY+IPY)/2+5);";"
8540 PRINT#2,"PLTL"
8550 PRINT#2,CHR$(13) CHR$(10) CHR$(3)
8560 PRINT#2,"MA";(IPX+2);",";((TPLY+IPY)/2+5);";"
8570 PRINT#2,"PLEM"
8580 PRINT#2,CHR$(13) CHR$(10) CHR$(3)
8600 PRINT#2,"MA";(IPX+2);",";(TPLY+MO*SCRM+2);";"
8610 PRINT#2,"PLCL"
8620 PRINT#2,CHR$(13) CHR$(10) CHR$(3)
8630 PRINT#2,"MA";(IPX+2);",";(TPLY+MO*SCRM/2);";"
8640 PRINT#2,"PLMO"
8650 PRINT#2,CHR$(13) CHR$(10) CHR$(3)
8660 PRINT#2,"MA";(GX+5);",";(GY+SY-8);";"
8670 PRINT#2,"PLR =";R
8680 PRINT#2,CHR$(13) CHR$(10) CHR$(3)
8700 PRINT#2,"MA";(GX+5);",";(GY+SY-12);";"
8710 PRINT#2,"PLDF=";DF
8720 PRINT#2,CHR$(13) CHR$(10) CHR$(3)
8730 PRINT#2,"MA";(GX+5);",";(GY+SY-16);";"
8740 PRINT#2,"PLTL=";TL
8750 PRINT#2,CHR$(13) CHR$(10) CHR$(3)
8760 PRINT#2,"MA";(GX+5);",";(GY+SY-20);";"
```

```
8770 PRINT#2,"PLCL=";CL
8780 PRINT#2,CHR$(13) CHR$(10) CHR$(3)
8800 PRINT#2,"MA";(GX+5);",";(GY+SY-24);";"
8810 PRINT#2,"PLLC=";LC
8820 PRINT#2,CHR$(13) CHR$(10) CHR$(3)
8830 PRINT#2,"MA";(GX+5);",";(GY+SY-28);";"
8840 PRINT#2,"PLMO=";MO
8850 PRINT#2,CHR$(13) CHR$(10) CHR$(3)
8860 PRINT#2,"MA";(GX+5);",";(GY+SY-32);";"
8870 PRINT#2,"PLEM=";EM
8880 PRINT#2,CHR$(13) CHR$(10) CHR$(3)
8900 PRINT#2,"MA5,-4;"
8910 PRINT#2,"PLPC-Curves - ";A$;" : Date : ";DATE$
8920 PRINT#2,CHR$(13) CHR$(10) CHR$(3)
8930 CLOSE#2
8940 GOTO 900
8985  REM ****************************************
8990  REM  GOSUB ROUTINE TO CLEAR SCREEN
8995  REM ***** PC-Circular Curves Heading *****
9000  CLS:T$="PC-Curves":CC$="PC-HCurve"
9010 LOCATE 1,2:PRINT T$
9020 LOCATE 1,30*SR%:PRINT DATE$
9030 LOCATE 2,2:PRINT CC$
9040 LOCATE 2,30*SR%:PRINT TIME$
9045 IF P1=4 THEN 9060
9050 PRINT "======================================"
9060 RETURN
9093  REM **************************************************
9095  REM  GOSUB ROUTINE FOR ERRORS
9100 BEEP
9110  PRINT : PRINT E0$
9120  PRINT
9130  RETURN
9493  REM **************************************************
9495  REM  END OF PROGRAM
9500  GOSUB 9000
9510  PRINT
9520  PRINT "      END CURVE SETTING OUT PROGRAM"
9530  PRINT
9540  PRINT "****************************************"
9542  FOR I = 1 TO 500
9544  NEXT I
9550  REM  PROGRAM PREPARED BY
9560  REM  DR. P.H. MILNE
9610  REM  ***************************
9620  END
10195 REM ***** Epson HI-80 Plotter *****
10200 GOSUB 54000
10210 PRINT#2,"SP4"
10220 PRINT#2,"MA";IPX*MX;",";(TPLY-5)*MX
10230 PRINT#2,"DA";IPX*MX;",";IPY*MX
10240 PRINT#2,"DA";TPLX*MX;",";TPLY*MX
10250 PRINT#2,"DA";TPRX*MX;",";TPRY*MX
10260 PRINT#2,"DA";IPX*MX;",";IPY*MX
10290 PRINT#2,"SP3"
10300 PRINT#2,"MA";TPLX*MX;",";TPLY*MX
10310 FOR J=N-1 TO 0 STEP -1
10320 PRINT#2,"DA";(IPX-U(J)*SCRM)*MX;",";(TPLY+V(J)*SCRM)*MX
10330 NEXT J
10340 PRINT#2,"DA";IPX*MX;",";(TPLY+MO*SCRM)*MX
10350 FOR K=0 TO N-1
10360 PRINT#2,"DA";(IPX+U(K)*SCRM)*MX;",";(TPLY+V(K)*SCRM)*MX
10370 NEXT K
10380 PRINT#2,"DA";TPRX*MX;",";TPRY*MX
10390 PRINT#2,"SP2":PRINT#2,"SI25,20"
```

```
10400 PRINT#2,"MA";(TPLX-8)*MX;",";(TPLY-3)*MX
10410 PRINT#2,"LA";F1
10430 PRINT#2,"MA";(TPRX-8)*MX;",";(TPLY-3)*MX
10440 PRINT#2,"LA";F2
10460 PRINT#2,"MA";(IPX-8)*MX;",";(IPY+2)*MX
10470 PRINT#2,"LA";F3
10500 PRINT#2,"MA";(IPX-1-SCRM*LC/4)*MX;",";(TPLY+2)*MX
10510 PRINT#2,"LALC"
10530 PRINT#2,"MA";(IPX-1-SCRM*LC/4)*MX;",";((TPLY+IPY)/2+5)*MX
10540 PRINT#2,"LATL"
10560 PRINT#2,"MA";(IPX+2)*MX;",";((TPLY+IPY)/2+5)*MX
10570 PRINT#2,"LAEM"
10600 PRINT#2,"MA";(IPX+2)*MX;",";(TPLY+MO*SCRM+2)*MX
10610 PRINT#2,"LACL"
10630 PRINT#2,"MA";(IPX+2)*MX;",";(TPLY+MO*SCRM/2)*MX
10640 PRINT#2,"LAMO"
10660 PRINT#2,"MA";(GX+5)*MX;",";(GY+SY-8)*MX
10670 PRINT#2,"LAR =";R
10700 PRINT#2,"MA";(GX+5)*MX;",";(GY+SY-12)*MX
10710 PRINT#2,"LADF=";DF
10730 PRINT#2,"MA";(GX+5)*MX;",";(GY+SY-16)*MX
10740 PRINT#2,"LATL=";TL
10760 PRINT#2,"MA";(GX+5)*MX;",";(GY+SY-20)*MX
10770 PRINT#2,"LACL=";CL
10800 PRINT#2,"MA";(GX+5)*MX;",";(GY+SY-24)*MX
10810 PRINT#2,"LALC=";LC
10830 PRINT#2,"MA";(GX+5)*MX;",";(GY+SY-28)*MX
10840 PRINT#2,"LAMO=";MO
10860 PRINT#2,"MA";(GX+5)*MX;",";(GY+SY-32)*MX
10870 PRINT#2,"LAEM=";EM
10890 PRINT#2,"SI30,25"
10900 PRINT#2,"MA";5*MX;",";+2*MX
10910 PRINT#2,"LAPC-Curves - ";A$;" : Date : ";DATE$
10930 CLOSE#2
10940 GOTO 900
```

4.2.5 PC-HCurve – computer printout

Horizontal curve details for Figs 4.2 and 4.3

```
Location of HCurve :- Strathclyde
**************************************************
OPERATOR'S NAME :- John Smith
DATE OF CALCS.  :- 01/01/87
INSTRUMENT USED :- Sokkisha DT20E
**************************************************

PC-Circular Curve Setting Out Program
<3> TANGENTIAL DEFLECTION ANGLES
****************************************
RADIUS           =  500
DEFLECTION ANGLE =  105
CHAINAGE I.P.    =  1626.863
TANGENT LENGTH   =  651.613
CHAINAGE TP1     =  975.25
CURVE LENGTH     =  916.298
CHAINAGE TP2     =  1891.548
LONG CHORD       =  793.353
MID-ORDINATE     =  195.619
EXTERNAL MO-DIST =  321.34
```

PT.	CHAINAGE	CHORD	ANGLE
TP1	975.25		.000
1	990	14.75	.5042
2	1020	30	2.335
3	1050	30	4.1658
4	1080	30	6.0005
5	1110	30	7.4313
6	1140	30	9.2621
7	1170	30	11.0928
8	1200	30	12.5236
9	1230	30	14.3544
10	1260	30	16.1851
11	1290	30	18.0159
12	1320	30	19.4507
13	1350	30	21.2814
14	1380	30	23.1122
15	1410	30	24.543
16	1440	30	26.3737
17	1470	30	28.2045
18	1500	30	30.0353
19	1530	30	31.4701
20	1560	30	33.3008
21	1590	30	35.1316
22	1620	30	36.5624
23	1650	30	38.3931
24	1680	30	40.2239
25	1710	30	42.0547
26	1740	30	43.4854
27	1770	30	45.3202
28	1800	30	47.151
29	1830	30	48.5817
30	1860	30	50.4125
31	1890	30	52.2433
TP2	1891.548	1.548	52.2952

FINAL ANGLE AT TP2 SHOULD BE 52.3

4.3 TRANSITION CURVES

Mention was made in Section 4.1 that circular curves were normally selected for built-up urban areas whereas transition curves were essential for high-speed motorways, highways or freeways. The reason why circular curves cannot be used at high-speed is because of the effect of centrifugal force on the vehicle and the occupants. If there was no transition at the end of a straight before entering a high-speed curve, there would be a sideways force which could cause skidding or sideslipping. To prevent such an occurrence, a transition curve, either a clothoid or cubic parabola, is inserted between the straight and the circular curve. Most countries publish criteria for road design; for example, in the United Kingdom the information is available from the Department of the Environment.

By introducing a transition curve at either end of a circular curve, the centrifugal force will gradually increase up to the maximum when the circular curve is reached, and then decrease again until back on the straight.

This considerably reduces any tendency of the vehicle to skid and also minimizes any discomfort experienced by the passengers. Sometimes, there is insufficient space to insert a circular curve to meet the specified design conditions, and the two transition curves therefore meet at the centre of the new curve.

In designing a transition curve the setting-out points can be determined either in rectangular coordinates or by polar coordinates, as shown in Fig. 4.4. If the deviation angle ϕ_m between the tangent and the circular curve is less than 12° there is no difference between the clothoid and the cubic parabola.

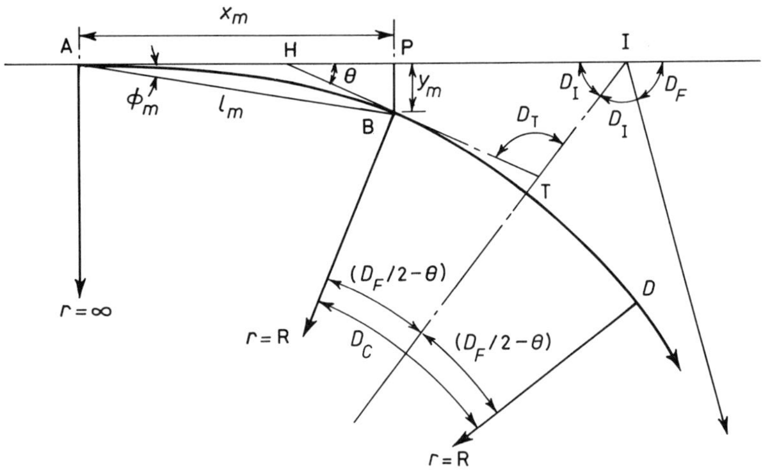

Fig. 4.4 Nomenclature for designing a transition curve where $\theta < D_F/2$.

To set out the curve on the ground, the value of K, the multiplying constant, has to be evaluated from the design criteria. Ideally, transition curves are defined for:

(i) A suitable speed (V),
(ii) A permissible centripetal ratio (F), and
(iii) A permissible rate of change of radial acceleration (C).

If each of these is specified K can be determined from the equations presented in Chapter 5 of *BASIC Programs for Land Surveying*.

If from the above it is found that the angle θ, shown in Fig. 4.4 is greater than $D_F/2$ then the curve cannot be set out to the design constants and an alternative solution must be found. Either the constants have to be reconsidered or, alternatively, two transition curves are designed to meet in the middle with no circular curve portion, Fig. 4.5. Thus, a suitable tangent length has to be chosen on site or a new minimum radius entered. Both

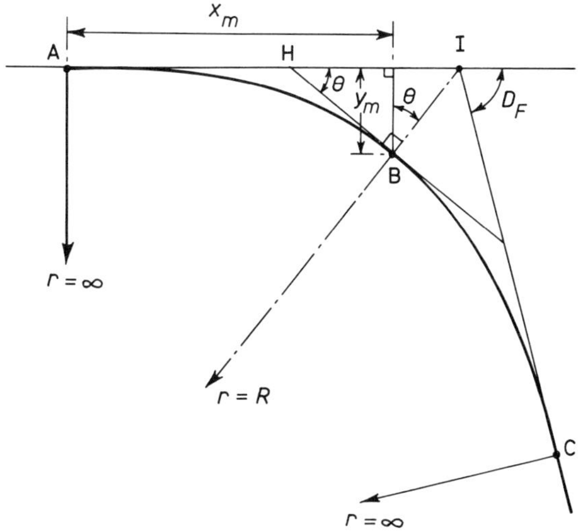

Fig. 4.5 Nomenclature for designing a transition curve where $\theta = D_F/2$.

methods of setting out curves on site are considered in the following transition curve program PC-TCurve.

4.3.1 PC-TCurve – subroutine index

	Line numbers	*Function*
(a)	10–90	Initialization and control
(b)	100–180	Screen header display
(c)	200–240	Selection of angle format
(d)	500–680	Entry of curve location, etc.
(e)	1000–1090	Selection of curve design criteria
(f)	1100–1190	Entry of curve design criteria
(g)	1200–1290	Calculation of curve parameters
(h)	1300–1490	Alternative solutions for $\theta > D_F/2$
(i)	1500–1830	Screen display of curve design
(j)	1900–1990	Curve setting out selection
(k)	2000–2740	First transition calculations
(l)	2750–3330	Second transition calculations
(m)	4000–4470	Graphics display and screen dump
(n)	5000–5190	Plotter initialization and selection
(o)	5200–5950	Plotter routine for HPGL
(p)	6000–6080	Printout of curve location, etc.
(q)	6100–6150	Printout of curve design criteria
(r)	6200–6390	Printout of first transition curve

	Line numbers	Function
(s)	6500–6630	Printout of second transition curve
(t)	8000–8040	Routine to convert DD.MMSS to decimals
(u)	8050–8140	Routine to convert decimals to DD.MMSS
(v)	8200–8950	Plotter routine for Apple 410
(w)	9000–9070	Routine to clear screen
(x)	9100–9130	Routine to display error message
(y)	9500–9620	Termination of program
(z)	10200–10950	Plotter routine for Epson HI-80

(a) Initialization and control Line numbers 10–90

The program is initialized and the installation file checked for the availability of graphics screens and colour for subsequent display. The required numeric and string variables are initialized and two **DEF FN** functions used to define arccosine and arcsine. Several arrays for chainage, etc. are dimensioned with a maximum of 100 chords or offsets. If a larger number is required, the arrays should be redimensioned.

(b) Screen header display Line numbers 100–110

At the start of each program selection the screen is cleared using **GOSUB 9000** to identify on screen the program suite and title.

(c) Selection of angle format Line numbers 200–240

The operator has a choice in line 200 of entering angles in decimal or DD.MMSS format, where the **D9** flag is set to 1 for decimals and 2 for DD.MMSS in line 210.

(d) Entry of curve location, etc. Line numbers 500–680

This routine is provided to keep a record of site location, operator, date, etc., with an opportunity to obtain a printout if required.

(e) Selection of curve design criteria Line numbers 1000–1090

The operator is given a choice of design criteria where three out of four parameters are required from velocity (**V**), permissible centripetal ratio (**F**), permissible rate of change of radial acceleration (**C**) and minimum radius (**R**). The deflection angle (**DF**) between the straights is also entered here.

(f) Entry of curve design criteria Line numbers 1100–1190

The three specified design parameters are entered from the keyboard and, depending on which selection (**P1**) was made in (e) above, the program branches accordingly from line 1190.

(g) Calculation of curve parameters Line numbers 1200–1290

The missing parameters are computed and then the multiplying constant **K** evaluated together with the total length of the transition curve (**SF**) and the angle of θ(**TH**). If the angle $\theta > D_F/2$ then the curve cannot be set out to the design parameters given, and the program branches to line 1300. If $\theta < D_F/2$ then there is a circular curve portion with deflection angle **DC** and circular curve length **CL**. The program continues at line 1500.

(h) Alternative solution for $\theta > D_F/2$ Line numbers 1300–1490

The program only branches to this routine if the curve cannot be set out to given design constants. A choice is given in line 1310 to either return to the beginning and enter new design criteria, or to force a solution either with $\theta = D_F/2$ or a new specified minimum radius.

(i) Screen display of curve design Line numbers 1500–1830

All the curve design parameters are then displayed on the screen as listed in lines 1620–1780, with the option of a printout for record purposes.

(j) Curve setting out selection Line numbers 1900–1990

Having designed the curve, an option is given to prepare a setting-out table for the transition curve portions using either rectangular or polar coordinates. If it is desired to set out the curve, the chainage of the *IP*, the length of the setting-out chords and whether the curve is left- or right-handed are requested.

(k) First transition calculations Line numbers 2000–2740

If an alternative solution was obtained in section (h) above (that is, $\theta = D_F/2$) then the tangent length is known and the program jumps to line 2100. In the standard solution, the tangent length AI in Fig. 4.4 is computed by splitting AI into three sections AP, HP and HI where AP is the maximum tangential distance (x_m) and BP is the maximum tangential offset (y_m).

(l) Second transition calculations Line numbers 2750–3330

The procedure for the second transition is the same as for (k) above except the deflection angles will be measured in the opposite direction, since a transition curve has to be set out from both ends, whereas a circular curve can be set out from one point. If there is a centre circular portion, the design parameters should be entered in the program PC-HCurve discussed previously in Section 4.2.

(m) Graphics display and screen dump Line numbers 4000–4470

The display screen chosen for the presentation of the layout of the curve is shown in Fig. 1.5 with text at the top and bottom of the graphics window.

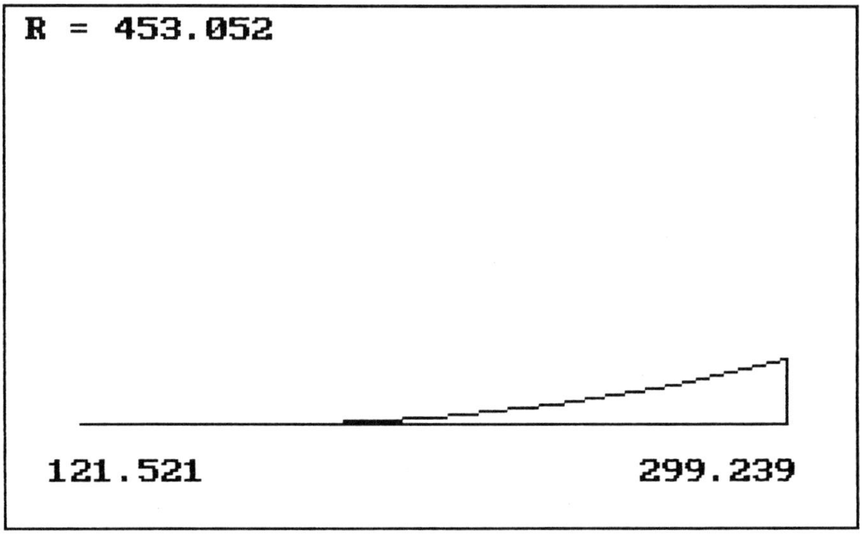

Fig. 4.6 PC-TCurve – screen dump of first transition for a left-handed transition curve.

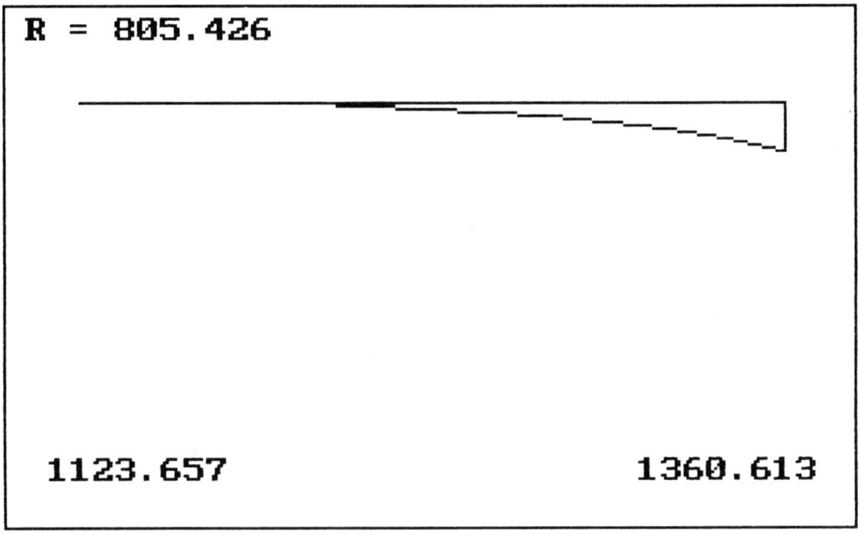

Fig. 4.7 PC-TCurve – screen dump of first transition for a right-handed transition curve.

The curve chainage is considered from left to right across the screen, with the tangent point for the first transition appearing at the left-hand side together with the chainage. If the curve is left-handed, it is plotted at the foot of the screen with the spiral curve above (i.e. to the left of) a horizontal

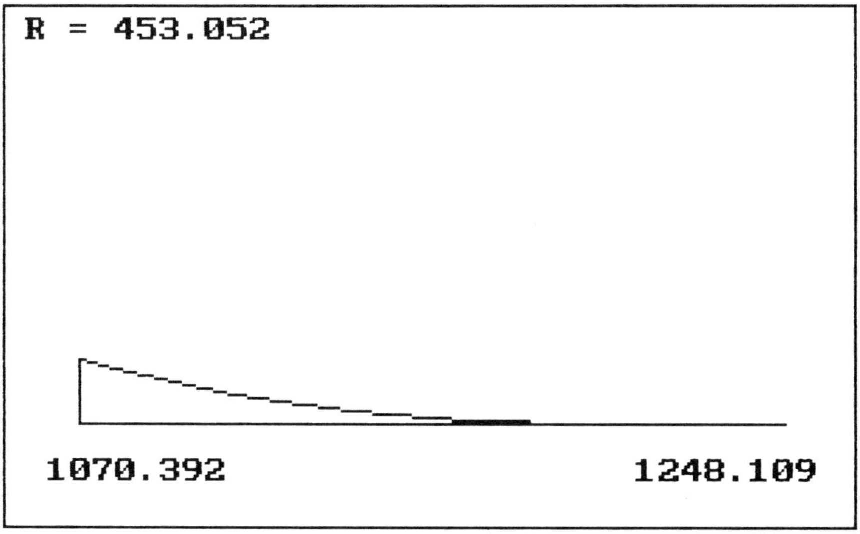

Fig. 4.8 PC-TCurve – screen dump of second transition for a left-handed transition curve.

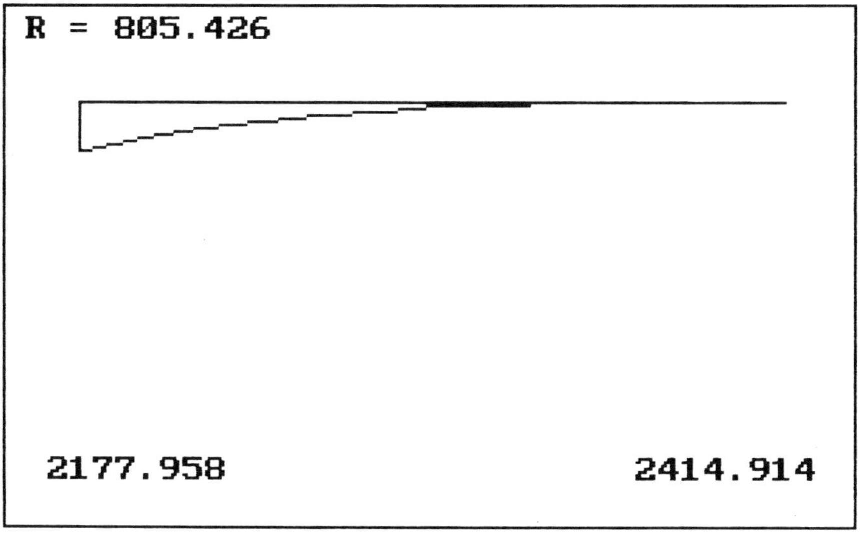

Fig. 4.9 PC-TCurve – screen dump of second transition for a right-handed transition curve.

tangent line to the intersection point, Fig. 4.6. If the curve is right-handed, it is plotted in the middle of the screen with the spiral below (i.e. to the right of) a horizontal tangent line to the intersection point, Fig. 4.7. There is no need to specify a scale for the plot as the display will automatically present the transition at maximum width on the screen. An offset exaggeration (× 2) is used to show the offset distances more clearly.

The second transition is also similarly plotted with the chainages presented on screen with the final tangent point at the right-hand side of the screen, Figs 4.8 and 4.9. After both the first and second transition screen displays the user can obtain a screen dump to a dot-matrix printer for record purposes. The deflection angle and minimum radius are also noted on screen for reference. The program then returns to the graphics display menu to allow for plotter output if required.

(n) Plotter initialization and selection Line numbers 5000–5190

After initializing the plotter variables to give a maximum plot width, a choice of three plotters is given, Apple 410, Epson HI-80 and Hewlett-Packard. Unlike the graphics screen output described in (m) where only one transition curve is displayed, this routine allows both transition curves to be plotted. The first transition curve is plotted from left to right in the middle of the sheet with the chainages of the start and finish of the transition noted. Fig. 4.10. The second transition is plotted on the lower half of the sheet, again

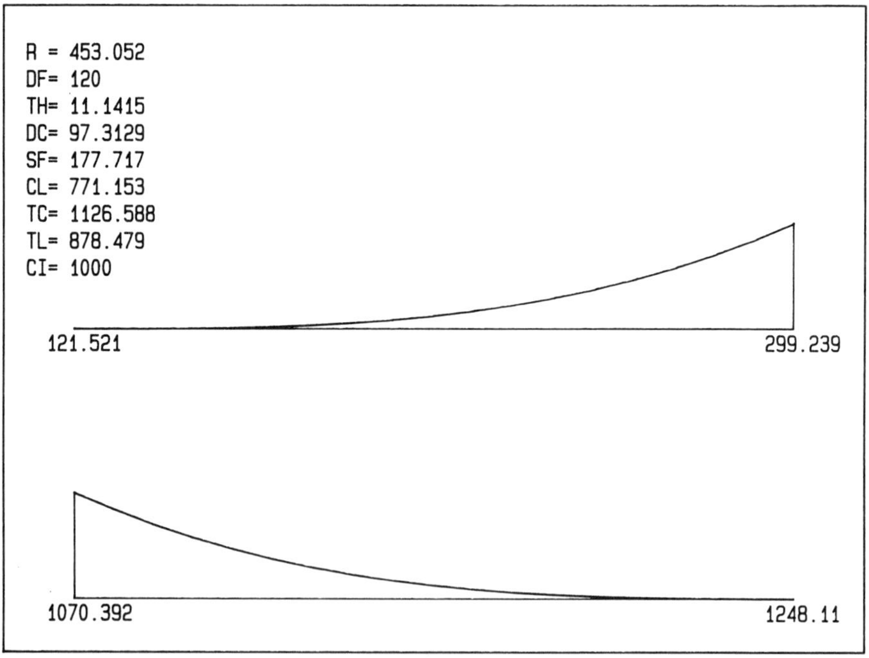

Fig. 4.10 PC-TCurve – plot of both spirals for a left-handed transition curve for comparison with Figs. 4.6 and 4.8.

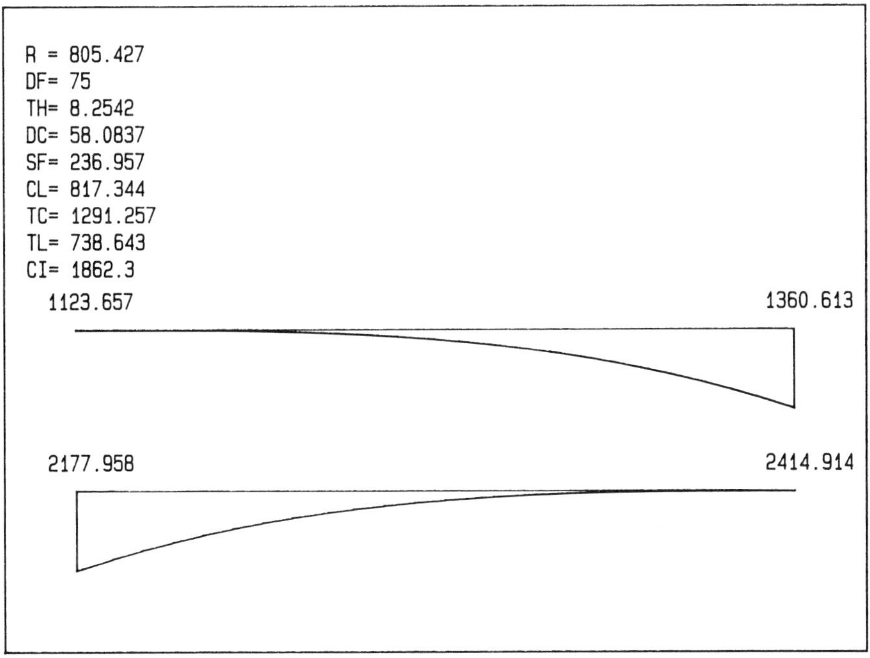

Fig. 4.11 PC-TCurve – plot of both spirals for a right-handed transition curve for comparison with Figs. 4.7 and 4.9.

from left to right with the final tangent point in the lower right-hand corner. For record purposes the curve design parameters are noted in the top left-hand corner of the plot, and a title, location and date added at the foot. Fig. 4.10 shows the two transitions for a left-hand curve and Fig. 4.11, that of a right-hand curve.

(o) Plotter routine for HPGL Line numbers 5200–5950

This routine is suitable for Hewlett-Packard plotters and their equivalents using HPGL as discussed in routine (k) of Section 2.2. After scaling the plotter into user units, a frame is drawn round the plotting area by pen 1 (black). Pen 4 (blue) is chosen to draw the tangent and the maximum offset, and then the curve plotted by pen 3 (green). Finally the chainages, labels and design details are printed by pen 2 (red).

(p) Routine to print curve locations, etc. Line numbers 6000–6080

This routine is provided for record purposes to annotate the printout, assuming a standard IBM PC type printer. It may require alteration for different microcomputers and printers.

(q) Printout of curve design criteria Line numbers 6100–6150

This printout is similar to the screen display from line numbers 1620–1790.

(r) Printout of first transition curve Line numbers 6200–6390

This printout combines the rectangular and polar coordinate computations and also lists the chainage of the IP and the tangent length for reference.

(s) Printout of second transition curve Line numbers 6500–6630

This is similar to (r) above.

(t) Routine to convert DD.MMSS to decimals Line numbers 8000–8040

Earlier in the program the user was given the option of entering angles in either the decimals or degrees, minutes, seconds (DD.MMSS) format. If the angles are entered in DD.MMSS then they must be converted to decimal degrees prior to computation using this routine.

(u) Routine to convert decimals to DD.MMSS Line numbers 8050–8140

Where angles have been entered in DD.MMSS and converted to decimals for computation using routine (t), they are converted back to DD.MMSS using this routine before display and printout.

(v) Plotter routine for Apple 410 Line numbers 8200–8950

This routine is similar to (o) where the plotter commands are altered as shown in Table 2.1.

(w) Routine to clear screen Line numbers 9000–9070

This routine is used to clear the screen and provide two lines of text at the head of each screen giving the title of the selected software suite and the selected program, together with the date and time of calculation.

(x) Routine to display error message Line numbers 9100–9130

If a wrong keyboard entry has been made in entering data, the error is trapped and this error message displayed and re-entry requested.

(y) Termination of program Line numbers 9500–9620

When the **QUIT** selection is made in the program, an end-of-program message is displayed to advise the operator that computations are complete.

(z) Plotter routine for Epson HI-80 Line numbers 10200–10950

This routine is similar to (o) where the plotter commands are altered as shown in Table 2.1.

4.3.2 PC-TCurve – numeric variables

A	= 0 or 360 degrees
BT	= Distance *BT* in Fig. 4.4
C	= Rate of change of radial acceleration
C1	= First chord in transition
CCOL	= Column number for text display on screen
CH	= Standard chord length
CI	= Chainage of IP
CK	= Check on correct angle entry
CL	= Circular curve length
D–D5	= Degrees of angle
D9	= Flag for decimals ⟨1⟩ or DD.MMSS ⟨2⟩
DC	= Angular arc for circular curve
DF	= Deflection angle at IP
DI	= Internal angle at IP
DISP%	= Choice of display, screen or plotter
DT	= Angle in triangle *HIT*
F	= Centripetal force ratio
F1–F4	= Chainages of transition curve points
G	= Acceleration due to gravity
GX	= *X*-coordinate at lower left-hand corner of plot
GY	= *Y*-coordinate at lower left-hand corner of plot
HB,HI,HT,HP	= Distances in Fig. 4.4.
HP%	= Integer variable to identify plotter
I,IN	= Integer loop variables
J,JN	= Integer loop variables
K	= Multiplying constant
LF	= Polar distance to end point
LF(I)	= Polar distance array
M–M1	= Minutes of angle
MG	= Switch for left-hand or right-hand curve
MX	= Multiplication factor for Epson HI-80 plotter
N,N1	= Number of chords in first and second transitions
P1,P2	= Program selection numbers
P7	= Counter in GOSUB routine
PH	= Total angle ϕ at end point
PH(I)	= Polar angle array
PLOT%	= Integer variable for plotter selection
PTXT	= Flag to indicate previous text printout
PX	= *X*-coordinate at upper right-hand corner of plot
PY	= *Y*-coordinate at upper right-hand corner of plot
R	= Minimum radius
R9	= 57.2957795
RGB%	= Integer variable to select screen colour
S,SØ	= Seconds of angle
SCD%	= Monitor display, monochrome/colour

SCR%	=	Screen resolution (installation file)
SCRM	=	Scale magnification to fit screen/plotter
SF	=	Full scale transition curve length
SM	=	Model scale value of **SF**
SPIR	=	Flag for first or second spiral
SR%	=	Screen resolution
SX	=	Horizontal display width in user units
SY	=	Vertical display height in user units
TC	=	Total transition curve length
TD	=	Tangent distance in model
TH	=	Angle θ
TL	=	Tangent distance at full scale
TOLY	=	*Y*-coordinate of tangent offset on left
TORY	=	*Y*-coordinate of tangent offset on right
TPLL	=	Variable to hold tangent point chainage on left
TPLX	=	*X*-coordinate of tangent point on left
TPLY	=	*Y*-coordinate of tangent point on left
TPRL	=	Variable to hold tangent point chainage on right
TPRX	=	*X*-coordinate of tangent point on right
TPRY	=	*Y*-coordinate of tangent point on right
V,VØ	=	Design of velocity of curve
XCEN	=	*X*-coordinate of centre of screen or plotter
XF,YF	=	Full scale values of *X* and *Y* at end of transition
XM,YM	=	Model scale values of *X* and *Y* at end of transition
XF(I)	=	Array of *X*-measurements
YF(I)	=	Array of *Y*-measurements

4.3.3 PC-TCurve – string variables

A$		Storage for location of curve
B$		Storage for operator's name
C$		Storage for date of calculations
CC$	=	**"PC-TCurve"**
DATE$		Variable used to return the date of the computer's calendar/clock in US format MM–DD–YYYY
EO$	=	**"SORRY, DATA ERROR . . . PLEASE RE-ENTER"**
H$	=	**"PC-Transition Curve Design"**
K$		Storage for instrument used
L$	=	**"L"** for left and **"R"** for right-handed curve
P$	=	**"DO YOU WISH PRINTOUT OF ABOVE DATA (Y/N)"**
PCF$	=	**"PCSCREEN"**
Q$	=	Answer response to question (Y/N)
R$	=	**"R"** for rectangular and **"P"** for polar coordinates
T$	=	**"PC-Curves"**
TIME$		Variable used to return the time of the built-in clock in HH:MM:SS format

4.3.4 PC-TCurve – BASIC program

```
10  REM  <PC-TCURV> Transition Curves Program for IBM PC or Compatibl
e
12  REM  BASIC Programs for Land Surveying, Chapter 5
13  REM  + Computer Graphics for Surveying, Chapter 4
14  REM  (C)  COPYRIGHT  1984-87  P.H.Milne
16  REM  ALL RIGHTS RESERVED
18  REM  Version PC-1.0, 01-01-87
20 CLEAR:KEY OFF
25 DEFDBL C,D,F,H,K,M,P,R-T,V,X,Y
30 A = 0:C0 = 0:CL = 0:G = 9.810001:P1 = 0:P2 = 0:TL = 0:V0 = 0
32 CK=0:SR%=1:PTXT=0:DISP%=0
35  DIM CH(100),CL(100),LF(100),PH(100),SM(100),XF(100),XM(100),YF(10
0),YM(100), XP(100),YP(100)
40 E0$ = "SORRY, DATA ERROR ... PLEASE RE-ENTER."
42 H$ = " PC-Transition Curve Design"
45 P$ = "DO YOU WISH PRINTOUT OF ABOVE DATA (Y/N)"
50 PCF$="PCSCREEN"
52 OPEN "I",#3,PCF$
54 INPUT#3,SCR%
56 INPUT#3,SCD%
58 CLOSE#3
60 ON (SCR%+1) GOTO 62,64,66
62 SCREEN 0:WIDTH 40:GOTO 80
64 SCREEN 1:GOTO 68
66 SCREEN 1:GOTO 80
68 IF SCD%=1 THEN 80
70 COLOR 8,0
80  DEF  FN C9(C9) = (1.5707964# -  ATN (C9 /  SQR (1 - C9 * C9))) *
R9: REM  ARCCOSINE
90  DEF  FN S9(S9) =  ATN (S9 /  SQR (1 - S9 * S9)) * R9: REM  ARCSIN
E
95  REM  **************************************************
100  GOSUB 9000:SPIR=1
110  PRINT
200  PRINT "ARE ANGLES IN DECIMALS <1>, DD.MMSS <2> OR DO YOU WISH TO
 QUIT <3>. ";
210  INPUT D9
220  IF D9 < 1 OR D9 > 3 THEN 200
230  ON D9 GOTO 240,240,9500
240 R9 = 57.29577950000001#:P1 = 0:TL = 0
500  GOSUB 9000
540  PRINT
550  PRINT "ENTER LOCATION OF WORK, ETC. WHEN       REQUESTED, AND PR
ESS <Enter>"
560  PRINT:PRINT "ENTER LOCATION OF CURVE :-"
570  INPUT A$
580  PRINT "ENTER OPERATOR'S NAME :-"
590  INPUT B$
600  PRINT "ENTER DATE OF CALCULATION AS DD/MM/YY -"
610  INPUT C$
620  PRINT "ENTER INSTRUMENT USED :-"
630  INPUT K$
640  PRINT
650  PRINT "DO YOU WISH PRINTOUT OF LOCATION ETC.,  (Y/N) ";
660  INPUT Q$
670  IF Q$ = "Y" THEN 6000
680  IF Q$ <  > "N" THEN 650
993  REM  **************************************************
995  REM  KEYBOARD ENTRY OF DESIGN CRITERIA
1000  GOSUB 9000: PRINT H$: PRINT "**********************************
******"
1010  PRINT : PRINT "SELECT CURVE DESIGN CRITERIA :-"
1020  PRINT "VELOCITY (V),"
```

```
1Ø3Ø  PRINT "RATE OF CHANGE OF RADIAL ACC. (C),"
1Ø4Ø  PRINT "CENTRIPETAL RATIO (F), MIN. RADIUS (R)"
1Ø5Ø  PRINT : PRINT "<1> V,C,F : <2> V,C,R : <3> F,C,R ";
1Ø6Ø  INPUT P1: IF P1 < 1 OR P1 > 3 THEN 1ØØØ
1Ø7Ø  PRINT : PRINT "ENTER DEFLECTION ANGLE ";: INPUT DF
1Ø75  IF DF <  = Ø OR DF > 179.999 THEN 1Ø7Ø
1Ø8Ø  IF D9 = 1 THEN 11ØØ
1Ø9Ø D5 = DF: GOSUB 8ØØØ: IF CK = 1 THEN 1Ø7Ø
1Ø95 DF = D5
11ØØ  GOSUB 9ØØØ: PRINT H$
111Ø  PRINT "****************************************"
112Ø  IF P1 = 3 THEN 114Ø
113Ø  PRINT : PRINT "ENTER DESIGN VELOCITY (KM/HR) ";: INPUT VØ
1135  IF VØ <  = Ø THEN 113Ø
114Ø  PRINT : PRINT "ENTER RATE OF CHANGE OF RADIAL          ACCELERA
TION (M/S3) ";: INPUT C
1145  IF C <  = Ø THEN 114Ø
115Ø  IF P1 = 2 THEN 118Ø
116Ø  PRINT : PRINT "ENTER CENTRIPETAL RATIO ";: INPUT F
1165  IF F <  = Ø THEN 116Ø
117Ø  IF P1 = 1 THEN 119Ø
118Ø  PRINT : PRINT "ENTER MINIMUM RADIUS (M) ";: INPUT R
1185  IF R <  = Ø THEN 118Ø
119Ø  ON P1 GOTO 12ØØ,122Ø,124Ø
12ØØ V = VØ * 1Ø / 36:R = V * V / (F * G)
121Ø  GOTO 125Ø
122Ø V = VØ * 1Ø / 36:F = V * V / (R * G)
123Ø  GOTO 125Ø
124Ø V =  SQR (F * R * G)
125Ø K =  SQR (V ^ 3 / C):SF = F * V * G / C
126Ø TH = (SF * SF / (2 * K * K)) * R9
127Ø  IF TH > (DF / 2) THEN 13ØØ
128Ø DC = DF - 2 * TH:CL = R * DC / R9
129Ø TC = CL + 2 * SF: GOTO 15ØØ
1295  REM  FAULTY DESIGN CRITERIA
13ØØ  GOSUB 91ØØ
131Ø  PRINT "DO YOU WISH TO ENTER NEW CURVE DESIGN   CRITERIA (Y/N):
IF (N) TRANSITIONS WILL MEET AT CENTRE ";: INPUT Q$
132Ø  IF Q$ = "Y" THEN 1ØØØ
133Ø  IF Q$ <  > "N" THEN 131Ø
134Ø TH = DF / 2:SM =  SQR (DF / R9)
135Ø XM = SM - (SM ^ 5 / 4Ø):YM = SM ^ 3 / 6 - (SM ^ 7 / 336)
136Ø TD = XM + YM *  TAN (TH / R9)
137Ø  PRINT : PRINT "IS TANGENT DISTANCE FROM I.P. KNOWN     (Y/N) ";
: INPUT Q$
138Ø  IF Q$ = "N" THEN 145Ø
139Ø  IF Q$ <  > "Y" THEN 137Ø
14ØØ  PRINT : PRINT "ENTER TANGENT DISTANCE (M) ";: INPUT TL
14Ø5  IF TL <  = Ø THEN 14ØØ
141Ø P2 = 1:K = TL / TD
1415  REM  ASSUME C AND F AS BEFORE
142Ø V = (K * K * C) ^ (1 / 3):R = V * V / (F * G)
143Ø SF = SM * K:XF = K * XM:YF = K * YM
144Ø TC = 2 * SF: GOTO 15ØØ
145Ø  PRINT : PRINT "DO YOU WISH TO SPECIFY MIN. RADIUS (Y/N) IF (N)
PROGRAM WILL REVERT TO START";
146Ø  INPUT Q$: IF Q$ = "N" THEN 1ØØ
147Ø  IF Q$ <  > "Y" THEN 145Ø
148Ø  PRINT : PRINT "ENTER MINIMUM RADIUS (M) ";: INPUT R
1482  IF R <  = Ø THEN 148Ø
1485  REM  PROGRAM REVERTS TO P1=3 CRITERIA
149Ø P1 = 3: GOTO 124Ø
1495  REM  PREPARE ANGLES FOR DISPLAY
15ØØ  IF D9 = 1 THEN 155Ø
15Ø5  REM  CONVERT DECIMAL TO DD.MMSS
151Ø D = DF: GOSUB 8Ø5Ø:DF = D
```

```
1520 D = TH: GOSUB 8050:TH = D
1530 D = DC: GOSUB 8050:DC = D
1540  GOTO 1600
1550 TH =  INT (TH * 1000 + .5) / 1000
1560 DC =  INT (DC * 1000 + .5) / 1000
1595  REM  SCREEN DISPLAY OF DESIGN DATA
1600  GOSUB 9000: PRINT H$:P7 = 0
1610  PRINT "****************************************"
1620  PRINT "DESIGN CRITERIA FOR CURVE :-"
1630  IF V0 = 0 THEN 1650
1635 V0=INT(V0*1000+.5)/1000
1640  PRINT "VELOCITY (KPH)                 = ";V0
1650  PRINT "VELOCITY (MPS)                 = ";INT(V * 1000 + .5)/1000
1655 C=INT(C*1000+.5)/1000
1660  PRINT "RADIAL ACCELERATION (MPS3)     = ";C
1670  PRINT "CENTRIPETAL RATIO              = ";INT(F * 1000 + .5)/1000
1680  PRINT "MINIMUM RADIUS (M)             = ";INT(R * 1000 + .5)/1000
1690  PRINT
1700  PRINT "MULTIPLYING CONSTANT           = ";INT (K * 1000 + .5)/1000
1710  PRINT "DEFLECTION ANGLE AT I.P.       = ";DF
1720  PRINT "TRANSITION CURVE ANGLE         = ";TH
1730  PRINT "CIRCULAR CURVE ANGLE           = ";DC
1740  PRINT "SPIRAL TRANSITION LENGTH       = ";INT (SF * 1000 + .5)/1000
1750  PRINT "CIRCULAR CURVE LENGTH          = ";INT(CL * 1000 + .5)/1000
1760  PRINT "TOTAL CURVE LENGTH             = ";INT(TC * 1000 + .5)/1000
1770  IF TL = 0 THEN 1790
1775 TL=INT(TL*1000+.5)/1000
1780  PRINT "MEASURED TANGENT LENGTH        = ";TL
1790  PRINT "****************************************"
1800  IF P7 = 1 THEN  RETURN
1810  PRINT : PRINT P$;: INPUT Q$
1820  IF Q$ = "Y" THEN 6100
1830  IF Q$ <  > "N" THEN 1810
1840 :
1895  REM  SETTING OUT ?
1900  GOSUB 9000: PRINT H$
1910  PRINT "****************************************"
1920  PRINT : PRINT "DO YOU WISH TO SET OUT CURVE (Y/N) ";: INPUT Q$
1930  IF Q$ = "N" THEN 100
1940  IF Q$ <  > "Y" THEN 1900
1950  PRINT : PRINT "SELECT POLAR (P) OR RECTANGULAR (R)     CO-ORDIN
ATES (P/R) ";: INPUT R$
1960  IF R$ <  > "P" AND R$ <  > "R" THEN 1950
1970  PRINT : PRINT "ENTER CHAINAGE OF I.P. ";: INPUT CI
1975  IF CI < 0 THEN 1970
1980  PRINT : PRINT "ENTER MULTIPLE CHORD INTERVALS (M) ";: INPUT CH
1982  IF CH <  = 0 THEN 1980
1984  IF SF / CH < 100 THEN 1990
1986  GOSUB 9100: PRINT "TOO SMALL A CHORD INTERVAL !"
1988  GOTO 1980
1990  PRINT : PRINT "IS CURVE LEFT OR RIGHT HANDED (L/R) ";: INPUT L$
: IF L$ <  > "L" AND L$ <  > "R" THEN 1990
1995  REM  FIND CHAINAGE OF 1ST TANGENT POINT
2000  IF P2 = 1 THEN 2100
2010 XF = SF - SF ^ 5 / (40 * K ^ 4):YF = SF ^ 3 / (6 * K ^ 2) - SF ^
 7 / (336 * K ^ 6)
2020  IF D9 = 1 THEN 2050
2025  REM  D9=2, CONVERT ANGLES DECIMALS
2030 D5 = DF: GOSUB 8000:DF = D5
2040 D5 = TH: GOSUB 8000:TH = D5
2045  REM  SOLVE TRIANGLE HPB
2050 HP = YF / ( TAN (TH / R9)):HB = YF / ( SIN (TH / R9))
2055  REM  SOLVE TRIANGLE OBT
2060 BT = R *  TAN ((DF / 2 - TH) / R9):HT = HB + BT
2065  REM  SOLVE TRIANGLE HIT USING SINE RULE
2070 DI = (180 - DF) / 2:DT = (180 - TH - DI)
```

```
2080 HI = (HT *  SIN (DT / R9)) /  SIN (DI / R9)
2090 TL = HI + XF - HP
2095  REM  FIND CHAINAGE OF A, START TRANSITION
2100 F1 = CI - TL
2105  REM  FIND CHAINAGE AT END 1ST TRANSITION
2110 F2 = F1 + SF
2115  REM  FIND CHAINAGE AT START 2ND TRANSITION
2120 F3 = F2 + CL
2125  REM  FIND CHAINAGE AT NEW STRAIGHT
2130 F4 = F1 + TC
2135  REM  FIND 1ST POINT ON CURVE
2140 CH(Ø) = ( INT (F1 / CH) + 1) * CH
2145  REM  FIND LENGTH FIRST CHORD
2150 C1 =  INT ((CH(Ø) - F1) * 1000 + .5) / 1000
2155  REM  FIND LAST POINT ON CURVE
2160 C9 = ( INT (F2 / CH)) * CH
2165  REM  FIND NUMBER OF CHORDS
2170 N =  INT ((SF - C1) / CH)
2180 CL(Ø) = C1
2190 F1 =  INT (F1 * 1000 + .5) / 1000:F2 =  INT (F2 * 1000 + .5) / 1
000:F3 =  INT (F3 * 1000 + .5) / 1000:F4 =  INT (F4 * 1000 + .5) / 10
00
2195  REM  SCREEN DISPLAY OF SETTING OUT DETAILS
2200 A = Ø:P7 = Ø
2202 IF D9=1 THEN 2208
2204 D=TH:GOSUB 8050:TH=D
2206 GOTO 2210
2208 TH=INT(TH*1000+.5)/1000
2210  GOSUB 9000: PRINT H$: PRINT "***********************************
******"
2220  IF R$ = "P" THEN 2240
2230  PRINT "CHAINAGE  LENGTH   TANG.DIST  TANG.OFF": GOTO 2250
2240  PRINT "CHAINAGE  LENGTH   POLAR DIST  ANGLE"
2250  PRINT "=======================================": IF P7 = 1 THE
N  RETURN
2260  IF R$ = "P" THEN 2280
2270 PRINT F1;TAB(11);" Ø";TAB(20);" Ø";TAB(31);" Ø":GOTO 2300
2280  IF L$ = "L" THEN A = 360
2290 PRINT F1;TAB(11);" Ø";TAB(20);" Ø";TAB(31);A
2295  REM  START LOOP FOR CURVE SETTING OUT POINTS
2300  FOR I = Ø TO N
2310 CH(I)=INT((CH(Ø)+CH*I)*1000+.5)/1000
2320 CL(I)=INT((CL(Ø)+CH*I)*1000+.5)/1000
2340 XF(I) = CL(I) - (CL(I)) ^ 5 / (40 * K ^ 4)
2350 YF(I) = (CL(I)) ^ 3 / (6 * K ^ 2) - (CL(I)) ^ 7 / (336 * K ^ 6)
2355  REM  FIND POLAR CO-ORDINATES
2360 LF(I) = INT((SQR((XF(I))^2 + (YF(I))^2)) * 1000 + .5)/1000
2370 PH(I) =  ATN (YF(I) / XF(I)) * R9
2380  IF L$ = "L" THEN 2430
2390  IF D9 = 1 THEN 2410
2400 D = PH(I): GOSUB 8050:PH(I) = D: GOTO 2470
2410 PH(I) =  INT ((PH(I)) * 1000 + .5) / 1000
2420  GOTO 2470
2430 PH(I) = 360 - PH(I)
2440  IF D9 = 1 THEN 2460
2450 D = PH(I): GOSUB 8050:PH(I) = D: GOTO 2470
2460 PH(I) =  INT ((PH(I)) * 1000 + .5) / 1000
2470 XF(I) =  INT ((XF(I)) * 1000 + .5) / 1000
2480 YF(I) =  INT ((YF(I)) * 1000 + .5) / 1000
2485  REM  ARRIVE HERE WITH BOTH "R" AND "P" RESULTS
2490  IF R$ = "P" THEN 2520
2495  REM  RECTANGULAR CO-ORDINATES DISPLAY
2500  PRINT CH(I); TAB( 11);CL(I); TAB( 20);XF(I); TAB( 31);YF(I)
2510  GOTO 2530
2515  REM  POLAR CO-ORDINATES DISPLAY
2518 LF(I)=INT(LF(I)*1000+.5)/1000
```

```
2520  PRINT CH(I); TAB( 11);CL(I); TAB( 20);LF(I); TAB( 31);PH(I)
2530  NEXT I
2540 SF =  INT (SF * 1000 + .5) / 1000
2550 XF = INT(XF * 1000 + .5)/1000:YF = INT(YF * 1000 + .5)/1000
2560  IF R$ = "P" THEN 2580
2570  PRINT F2; TAB( 11);SF; TAB( 20);XF; TAB( 31);YF
2575  REM  FIND POLAR ANGLE
2580 LF =  INT (( SQR (XF ^ 2 + YF ^ 2)) * 1000 + .5) / 1000
2590 PH = (R9 * 10 * (SF / K) ^ 2 - 1.213 * (SF / K) ^ 6) / 60
2600  IF L$ = "L" THEN 2640
2610  IF D9 = 1 THEN 2630
2620 D = PH: GOSUB 8050:PT = D: GOTO 2680
2630 PT =  INT (PH * 1000 + .5) / 1000: GOTO 2680
2640 PT = 360 - PH
2650  IF D9 = 1 THEN 2670
2660 D = PT: GOSUB 8050:PT = D: GOTO 2680
2670 PT =  INT (PT * 1000 + .5) / 1000
2680  IF R$ = "R" THEN 2700
2690  PRINT F2; TAB( 11);SF; TAB( 20);LF; TAB( 31);PT
2700  PRINT F3; TAB( 11);"START OF 2ND SPIRAL": PRINT F4; TAB( 11);"F
INAL TANGENT POINT"
2710  PRINT "CHAINAGE OF I.P."; TAB( 20);CI: PRINT "LENGTH OF TANGENT
 "; TAB( 20); INT (TL * 1000 + .5) / 1000
2720  PRINT "========================================"
2730  PRINT : PRINT P$;: INPUT Q$: IF Q$ = "Y" THEN 6200
2735  IF Q$ <  > "N" THEN 2730
2740  GOSUB 9000: PRINT H$
2741 IF DISP% <> 0 THEN 2744
2742  PRINT:PRINT " Do you wish plot of curve (y/n)";:INPUT Q$:IF Q$=
"n" THEN 2750
2744  PRINT:PRINT:PRINT " Select Display ";TAB(18);"<1> Graphics Scre
en":PRINT:PRINT TAB(18);"<2> Plotter":PRINT:PRINT TAB(18);"<3> Contin
ue ";:INPUT DISP%
2746  IF DISP% <1 OR DISP% >3 THEN 2740
2748  ON DISP% GOTO 4000,5000,2750
2750  GOSUB 9000: PRINT H$
2760  PRINT "****************************************"
2770  PRINT : PRINT "DO YOU WISH DATA FOR 2ND SPIRAL (Y/N)";: INPUT Q
$
2780  IF Q$ = "N" THEN 100
2790  IF Q$ <  > "Y" THEN 2770
2795  REM  DATA REQUIRED FOR 2ND SPIRAL
2800 A = 0:P7 = 1:SPIR=2
2810  GOSUB 2210
2820  IF R$ = "P" THEN 2840
2830  PRINT F3; TAB( 11);SF; TAB( 20);XF; TAB( 31);YF
2840  IF L$ = "R" THEN 2890
2850  IF D9 = 1 THEN 2870
2860 D = PH: GOSUB 8050:PT = D: GOTO 2930
2870 PT =  INT (PH * 1000 + .5) / 1000
2880  GOTO 2930
2890 A = 360:PT = A - PH
2900  IF D9 = 1 THEN 2920
2910 D = PT: GOSUB 8050:PT = D: GOTO 2930
2920 PT =  INT (PT * 1000 + .5) / 1000
2930  IF R$ = "R" THEN 2950
2940  PRINT F3; TAB( 11);SF; TAB( 20);LF; TAB( 31);PT
2945  REM  FIND CHAINAGES OF POINTS ON CURVE
2950 C8 = ( INT (F3 / CH) + 1) * CH
2960 C7 = ( INT (F4 / CH)) * CH
2965  REM  FIND NUMBER OF CHORDS
2970 N1 =  INT ((C7 - C8) / CH + .5)
2990 :
3000  FOR IN = 0 TO N1
3010 CH(IN)=INT((C8+IN*CH)*1000+.5)/1000
3020 CL(IN)=INT((F4-CH(IN))*1000+.5)/1000
```

```
3030 SM(IN) = CL(IN)/K:CL(IN) = INT((CL(IN)) * 1000 + .5)/1000
3040 XM(IN) = SM(IN) - ((SM(IN)) ^ 5 / 40):XF(IN) = K * XM(IN)
3050 YM(IN) = (SM(IN))^3/6 - (SM(IN))^7/336:YF(IN) = K * YM(IN)
3060 LF(IN) = INT((SQR((XF(IN)) ^ 2 + (YF(IN)) ^ 2)) * 1000 + .5)/1000
3070 XF(IN) = INT((XF(IN)) * 1000 + .5)/1000:YF(IN) = INT((YF(IN)) * 1000 + .5)/1000
3080 PH(IN) = (R9 * 10 * (SM(IN))^2 - 1.213 * (SM(IN))^6)/60
3090  IF L$ = "R" THEN 3150
3100  IF D9 = 1 THEN 3130
3110 D = PH(IN): GOSUB 8050:PH(IN) = D:
3120  GOTO 3190
3130 PH(IN) =  INT ((PH(IN)) * 1000 + .5) / 1000
3140  GOTO 3190
3150 PH(IN) = 360 - PH(IN)
3160  IF D9 = 1 THEN 3180
3170 D = PH(IN): GOSUB 8050:PH(IN) = D: GOTO 3190
3180 PH(IN) =  INT ((PH(IN)) * 1000 + .5) / 1000
3190  IF R$ = "P" THEN 3220
3195  REM  RECTANGULAR CO-ORDINATE DISPLAY
3200  PRINT CH(IN); TAB( 11);CL(IN); TAB( 20);XF(IN); TAB( 31);YF(IN)
3210  GOTO 3230
3215  REM  POLAR CO-ORDINATE DISPLAY
3220  PRINT CH(IN); TAB( 11);CL(IN); TAB( 20);LF(IN); TAB( 31);PH(IN)
3230  NEXT IN
3240  IF R$ = "P" THEN 3270
3250 PRINT F4;TAB(11);" 0";TAB(20);" 0";TAB(31);" 0"
3260  GOTO 3280
3270 PRINT F4;TAB(11);" 0";TAB(20);" 0";TAB(31);A
3280  PRINT "****************************************"
3290  PRINT
3300  PRINT P$;: INPUT Q$
3310  IF Q$ = "Y" THEN 6500
3320  IF Q$ <  > "N" THEN 3300
3330  GOSUB 9000: PRINT H$
3335 IF DISP% <> 0 THEN 3350
3340  PRINT:PRINT " Do you wish plot of curve (y/n)";:INPUT Q$:IF Q$="n" THEN 100
3350  PRINT:PRINT:PRINT " Select Display ";TAB(18);"<1> Graphics Screen":PRINT:PRINT TAB(18);"<2> Plotter":PRINT:PRINT TAB(18);"<3> Continue ";:INPUT DISP%
3360  IF DISP% <1 OR DISP% >3 THEN 3330
3370  ON DISP% GOTO 4000,5000,100
3995 REM ***** Graphics Screen Display *****
4000  CLS:GOSUB 9000
4010  IF SR%=1 THEN RGB%=2 ELSE RGB%=3
4020 WINDOW (0,0)-(320*SR%,200)
4030 LINE (8,18)-(310*SR%,180),RGB%,B
4040 IF L$="L" THEN MG=1 ELSE MG=-1
4100 IF L$="L" THEN TPLY=50 ELSE TPLY=150
4110 IF L$="L" THEN TPRY=50 ELSE TPRY=150
4120 SCRM=(250*SR%)/(XF*1.6)
4150 TPLX=160*SR%-(SCRM*1.6*XF/2)
4160 TPRX=160*SR%+(SCRM*1.6*XF/2)
4170 IF L$="L" AND SPIR=1 THEN TOLY=0
4172 IF L$="L" AND SPIR=2 THEN TOLY=YF*2*SCRM
4174 IF L$="R" AND SPIR=1 THEN TOLY=0
4176 IF L$="R" AND SPIR=2 THEN TOLY=YF*2*SCRM
4180 IF L$="L" AND SPIR=1 THEN TORY=YF*2*SCRM
4182 IF L$="L" AND SPIR=2 THEN TORY=0
4184 IF L$="R" AND SPIR=1 THEN TORY=YF*2*SCRM
4186 IF L$="R" AND SPIR=2 THEN TORY=0
4200 LINE (TPLX,TPLY)-(TPRX,TPRY),3
4210 LINE (TPLX,TPLY)-(TPLX,(TPLY+TOLY*MG)),3
4220 LINE (TPRX,TPRY)-(TPRX,(TPRY+TORY*MG)),3
4230 IF SPIR=2 THEN 4252
```

```
4232 CCOL=(TPLX*SR%-32)/8
4234 IF CCOL<3 THEN CCOL=3
4240 LOCATE 21,CCOL:PRINT F1
4242 CCOL=(TPRX*SR%-32)/8
4244 IF CCOL>29 THEN CCOL=29
4246 LOCATE 21,CCOL:PRINT F2
4250 GOTO 4270
4252 CCOL=(TPLX*SR%-32)/8
4254 IF CCOL<3 THEN CCOL=3
4260 LOCATE 21,CCOL:PRINT F3
4262 CCOL=(TPRX*SR%-32)/8
4264 IF CCOL>29 THEN CCOL=29
4266 LOCATE 21,CCOL:PRINT F4
4270 LOCATE 4,3*SR%:PRINT "R =";INT(R*1000+.5)/1000
4280 LOCATE 1,2:PRINT CC$
4290 LOCATE 2,2:PRINT "Defln Angle = ";DF
4295 IF SPIR=2 THEN 4350
4300 PSET (TPLX,(TPLY+TOLY)),1
4310 FOR J=0 TO N
4320 XP(J)=TPLX+XF(J)*SCRM*1.6*SR%:YP(J)=TPLY+YF(J)*2*SCRM*MG
4330 LINE -(XP(J),YP(J)),1
4340 NEXT J:LINE -(TPRX,(TPRY+TORY*MG)),1
4345 GOTO 4400
4350 PSET (TPLX,(TPLY+TOLY*MG)),1
4360 FOR J=0 TO N
4370 XP(J)=TPRX-XF(J)*SCRM*1.6*SR%:YP(J)=TPRY+YF(J)*2*SCRM*MG
4380 LINE -(XP(J),YP(J)),1
4390 NEXT J:LINE -(TPRX,(TPRY+TORY)),1
4400 LOCATE 24,2:PRINT "Screen Dump (y/n) ";
4410 Q$=INKEY$:IF Q$ = "" THEN 4410
4420 IF INSTR("yn",Q$)=0 THEN 4400
4430 IF Q$ = "n" THEN 4450
4440 LCOPY
4450 DISP%=3
4460 IF SPIR=1 THEN 2740
4470 IF SPIR=2 THEN 3330
4995 REM ***** Plotter Output *****
5000 GX=0:GY=0:SX=180:SY=120:MX=2670/SX:PX=GX+SX:PY=GY+SY
5010 SCRM=150/XF:XCEN=GX+SX/2
5020 IF L$="L" THEN MG=1 ELSE MG=-1
5050 TPLX=XCEN-(SCRM*XF/2)
5060 TPRX=XCEN+(SCRM*XF/2)
5070 IF L$="L" AND SPIR=1 THEN TOLY=0:TPLY=GY+60
5072 IF L$="L" AND SPIR=2 THEN TOLY=YF*2*SCRM:TPLY=GY+10
5074 IF L$="R" AND SPIR=1 THEN TOLY=0:TPLY=GY+60
5076 IF L$="R" AND SPIR=2 THEN TOLY=YF*2*SCRM:TPLY=GY+30
5080 IF L$="L" AND SPIR=1 THEN TORY=YF*2*SCRM:TPRY=GY+60
5082 IF L$="L" AND SPIR=2 THEN TORY=0:TPRY=GY+10
5084 IF L$="R" AND SPIR=1 THEN TORY=YF*2*SCRM:TPRY=GY+60
5086 IF L$="R" AND SPIR=2 THEN TORY=0:TPRY=GY+30
5100 GOSUB 9000
5110 PRINT:PRINT " Select Plotter -";TAB(18);"<1> Apple 410":PRINT:PR
INT TAB(18);"<2> Epson HI-80"
5120 PRINT:PRINT TAB(18);"<3> Hewlett-Packard":PRINT:PRINT TAB(18);"<
4> QUIT          ";:INPUT PLOT%
5130 IF PLOT% <1 OR PLOT% >4 THEN 5100
5140 IF PLOT%=4 THEN 900
5145 IF PLOT%=1 THEN HP%=3
5150 IF PLOT%=2 THEN HP%=4
5155 IF PLOT%=3 THEN HP%=1
5160 PRINT:PRINT " Place paper on Plotter and switch on"
5170 PRINT:PRINT " To Continue Press <Enter> ":WHILE R$="":R$=INKEY$:
WEND
5180 GOSUB 50000
5190 ON HP% GOTO 5200,5200,8200,10200
5195 REM ***** HP-GL Plotter (HP-7475A) *****
```

```
5200 GOSUB 51000
5210 PRINT#2,"SP4;"
5220 PRINT#2,"PU;PA";TPLX;",";TPLY;";"
5230 PRINT#2,"PD;PA";TPRX;",";TPRY;";"
5240 PRINT#2,"PA";TPRX;",";(TPRY+TORY*MG);";"
5250 PRINT#2,"PU;PA";TPLX;",";TPLY;";"
5260 PRINT#2,"PD;PA";TPLX;",";(TPLY+TOLY*MG);";"
5290 PRINT#2,"SP3;"
5295 IF SPIR=2 THEN 5350
5300 PRINT#2,"PU;PA";TPLX;",";TPLY;";"
5310 FOR J=0 TO N
5315 XP(J)=TPLX+XF(J)*SCRM:YP(J)=TPLY+YF(J)*2*SCRM*MG
5320 PRINT#2,"PD;PA";XP(J);",";YP(J);";"
5330 NEXT J
5340 PRINT#2,"PA";TPRX;",";(TPRY+TORY*MG);";"
5345 GOTO 5395
5350 PRINT#2,"PU;PA";TPLX;",";(TPLY+TOLY*MG)";"
5360 FOR J=0 TO N
5365 XP(J)=TPRX-XF(J)*SCRM:YP(J)=TPRY+YF(J)*2*SCRM*MG
5370 PRINT#2,"PD;PA";XP(J);",";YP(J);";"
5380 NEXT J
5390 PRINT#2,"PA";TPRX;",";(TPRY+TORY);";"
5395 PRINT#2,"SP2;":PRINT#2,"SI0.2,0.3;"
5400 PRINT#2,"PU;PA";(TPLX-8);",";(TPLY-3*MG);";"
5405 IF SPIR=1 THEN TPLL=F1 ELSE TPLL=F3
5410 PRINT#2,"LB";TPLL
5420 PRINT#2,CHR$(3)
5430 PRINT#2,"PU;PA";(TPRX-8);",";(TPRY-3*MG);";"
5435 IF SPIR=1 THEN TPRL=F2 ELSE TPRL=F4
5440 PRINT#2,"LB";TPRL
5450 PRINT#2,CHR$(3)
5460 IF PTXT=1 THEN 5930
5660 PRINT#2,"PU;PA";(GX+5);",";(GY+SY-8);";"
5670 PRINT#2,"LBR =";INT(R*1000+.5)/1000
5680 PRINT#2,CHR$(3)
5700 PRINT#2,"PU;PA";(GX+5);",";(GY+SY-12);";"
5710 PRINT#2,"LBDF=";DF
5720 PRINT#2,CHR$(3)
5730 PRINT#2,"PU;PA";(GX+5);",";(GY+SY-16);";"
5740 PRINT#2,"LBTH=";TH
5750 PRINT#2,CHR$(3)
5760 PRINT#2,"PU;PA";(GX+5);",";(GY+SY-20);";"
5770 PRINT#2,"LBDC=";DC
5780 PRINT#2,CHR$(3)
5800 PRINT#2,"PU;PA";(GX+5);",";(GY+SY-24);";"
5810 PRINT#2,"LBSF=";SF
5820 PRINT#2,CHR$(3)
5830 PRINT#2,"PU;PA";(GX+5);",";(GY+SY-28);";"
5840 PRINT#2,"LBCL=";INT(CL*1000+.5)/1000
5850 PRINT#2,CHR$(3)
5860 PRINT#2,"PU;PA";(GX+5);",";(GY+SY-32);";"
5870 PRINT#2,"LBTC=";INT(TC*1000+.5)/1000
5880 PRINT#2,CHR$(3)
5890 IF TL=0 THEN 5894
5891 PRINT#2,"PU;PA";(GX+5);",";(GY+SY-36);";"
5892 PRINT#2,"LBTL=";INT(TL*1000+.5)/1000
5893 PRINT#2,CHR$(3)
5894 PRINT#2,"PU;PA";(GX+5);",";(GY+SY-40);";"
5895 PRINT#2,"LBCI=";INT(CI*1000+.5)/1000
5896 PRINT#2,CHR$(3)
5900 PRINT#2,"PU,PA5,-3;"
5910 PRINT#2,"LBPC-Transition Curves - ";A$;" : Date : ";DATE$
5920 PRINT#2,CHR$(3)
5930 CLOSE#2:PTXT=1
5940 IF SPIR=1 THEN 2740
5950 IF SPIR=2 THEN 3330
```

```
5985  REM  ***********************************************
5990  REM  PRINTOUT OF LOCATION, ETC
6000 LPRINT "Location of TCurve :- ";A$
6010 LPRINT "***********************************************"
6020 LPRINT "OPERATOR'S NAME :- ";B$
6030 LPRINT "DATE OF CALCS.  :- ";C$
6040 LPRINT "INSTRUMENT USED :- ";K$
6060 LPRINT "***********************************************"
6070 LPRINT
6080  GOTO 1000
6093  REM  ***********************************************
6095  REM  PRINTOUT OF DESIGN CRITERIA ETC.
6100 LPRINT H$
6120 LPRINT "****************************************"
6122 LPRINT "DESIGN CRITERIA FOR CURVE :-"
6123 IF VØ=Ø THEN 6125
6124 LPRINT "VELOCITY (KPH)              = ";VØ
6125 LPRINT "VELOCITY (MPS)              = ";INT(V*1ØØØ+.5)/1ØØØ
6126 LPRINT "RADIAL ACCELERATION (MPS3) = ";C
6127 LPRINT "CENTRIPETAL RATIO           = ";INT(F*1ØØØ+.5)/1ØØØ
6128 LPRINT "MINIMUM RADIUS (M)          = ";INT(R*1ØØØ+.5)/1ØØØ
6129 LPRINT
6130 LPRINT "MULTIPLYING CONSTANT        = ";INT(K*1ØØØ+.5)/1ØØØ
6131 LPRINT "DEFLECTION ANGLE AT I.P.    = ";DF
6132 LPRINT "TRANSITION CURVE ANGLE      = ";TH
6133 LPRINT "CIRCULAR CURVE ANGLE        = ";DC
6134 LPRINT "SPIRAL TRANSITION LENGTH    = ";INT(SF*1ØØØ+.5)/1ØØØ
6135 LPRINT "CIRCULAR CURVE LENGTH       = ";INT(CL*1ØØØ+.5)/1ØØØ
6136 LPRINT "TOTAL CURVE LENGTH          = ";INT(TC*1ØØØ+.5)/1ØØØ
6137 IF TL=Ø THEN 6139
6138 LPRINT "MEASURED TANGENT LENGTH     = ";TL
6139 LPRINT "****************************************"
6140 LPRINT
6150  GOTO 1900
6193  REM  ***********************************************
6195  REM  PRINTOUT FOR FIRST TRANSITION
6200 LPRINT H$
6210 A=Ø:P7=Ø
6220 LPRINT "*******************************************************
***************"
6230 LPRINT "CHAINAGE OF I.P.  = ";TAB(24);CI
6240 LPRINT "LENGTH OF TANGENT = ";TAB(24);INT(TL*1ØØØ+.5)/1ØØØ
6250 LPRINT "*******************************************************
***************"
6260 LPRINT "CHAINAGE";TAB(12);"LENGTH";TAB(24);"TANG.DIST";TAB(36);"
TANG.OFF";TAB(48);"POLAR DIST";TAB(6Ø);"ANGLE"
6270  IF P7 = 1 THEN  RETURN
6280 LPRINT "=======================================================
===============":IF L$="L" THEN A=36Ø
6290 LPRINT F1;TAB(12);" Ø";TAB(24);" Ø";TAB(36);" Ø";TAB(48);" Ø";TA
B(6Ø);A
6300  FOR J = Ø TO N
6310 LPRINT CH(J);TAB(12);CL(J);TAB(24);XF(J);TAB(36);YF(J);TAB(48);L
F(J);TAB(6Ø);PH(J)
6320 NEXT J
6330 LPRINT F2;TAB(12);SF;TAB(24);XF;TAB(36);YF;TAB(48);LF;TAB(6Ø);PT
6340 LPRINT F3;TAB(12);"START OF 2ND SPIRAL"
6350 LPRINT F4;TAB(12);"FINAL TANGENT POINT"
6360 LPRINT "=======================================================
==============="
6370 LPRINT
6390  GOTO 274Ø
6493  REM  ***********************************************
6495  REM  PRINTOUT FOR 2ND SPIRAL
6500 LPRINT H$
6510 A=Ø:P7=1
```

```
652Ø  GOSUB 625Ø
653Ø LPRINT "===============================================================
==============="
654Ø  IF L$ = "R" THEN A = 36Ø
655Ø LPRINT F3;TAB(12);SF;TAB(24);XF;TAB(36);YF;TAB(48);LF;TAB(6Ø);PT
656Ø FOR JN=Ø TO N1
657Ø LPRINT CH(JN);TAB(12);CL(JN);TAB(24);XF(JN);TAB(36);YF(JN);TAB(4
8);LF(JN);TAB(6Ø);PH(JN)
658Ø  NEXT JN
659Ø LPRINT F4;TAB(12);" Ø";TAB(24);" Ø";TAB(36);" Ø";TAB(48);" Ø";TA
B(6Ø);A
66ØØ LPRINT "***************************************************************
***************"
661Ø LPRINT
663Ø  GOTO 333Ø
7993  REM  ****************************************************
7995  REM  ROUTINE DD.MMSS TO DECIMALS
8ØØØ DØ =  INT (D5)
8ØØ5  REM  FIND NO. OF MINUTES,MØ
8Ø1Ø M1 = (D5 - DØ) * 1ØØ:MØ = INT(M1 + .1)
8Ø15  REM  FIND NO. OF SECONDS,SØ
8Ø2Ø SØ =  INT (M1 * 1ØØ - MØ * 1ØØ + .5)
8Ø22  IF MØ < Ø OR MØ > 59 THEN 8Ø3Ø
8Ø24  IF SØ < Ø OR SØ > 59.99 THEN 8Ø3Ø
8Ø25  REM  COLLECT DEGS,MINS,SECS
8Ø26 D5 = DØ + MØ / 6Ø + SØ / 36ØØ
8Ø28 CK = Ø: RETURN
8Ø3Ø  GOSUB 91ØØ: REM  DATA ERROR
8Ø4Ø CK = 1: RETURN
8Ø43  REM  ****************************************************
8Ø45  REM  ROUTINE DECIMAL TO DD.MMSS
8Ø5Ø D1 =  INT (D)
8Ø55  REM  FIND TOTAL NO. OF SECONDS
8Ø6Ø D2 = (D - D1) * 36ØØ
8Ø65  REM  FIND NO. OF MINUTES
8Ø7Ø M =  INT (D2 / 6Ø)
8Ø75  REM  FIND NO. OF SECONDS
8Ø8Ø S = D2 - 6Ø * M + .5
8Ø9Ø  IF S < 6Ø THEN 813Ø
81ØØ M = M + 1:S = Ø
811Ø  IF M < 6Ø THEN 813Ø
812Ø D1 = D1 + 1:M = Ø
813Ø D = D1 * 1ØØØØ + M * 1ØØ + S
814Ø D =  INT (D) / 1ØØØØ: RETURN
8195 REM ***** Apple 41Ø Plotter *****
82ØØ GOSUB 53ØØØ
821Ø PRINT#2,"PS4;"
822Ø PRINT#2,"MA";TPLX;",";TPLY;";"
823Ø PRINT#2,"DA";TPRX;",";TPRY;";"
824Ø PRINT#2,"DA";TPRX;",";(TPRY+TORY*MG);";"
825Ø PRINT#2,"MA";TPLX;",";TPLY;";"
826Ø PRINT#2,"DA";TPLX;",";(TPLY+TOLY*MG);";"
829Ø PRINT#2,"PS3;"
8295 IF SPIR=2 THEN 835Ø
83ØØ PRINT#2,"MA";TPLX;",";TPLY;";"
831Ø FOR J=Ø TO N
8315 XP(J)=TPLX+XF(J)*SCRM:YP(J)=TPLY+YF(J)*2*SCRM*MG
832Ø PRINT#2,"DA";XP(J);",";YP(J);";"
833Ø NEXT J
834Ø PRINT#2,"DA";TPRX;",";(TPRY+TORY*MG);";"
8345 GOTO 8395
835Ø PRINT#2,"MA";TPLX;",";(TPLY+TOLY*MG)";"
836Ø FOR J=Ø TO N
8365 XP(J)=TPRX-XF(J)*SCRM:YP(J)=TPRY+YF(J)*2*SCRM*MG
837Ø PRINT#2,"DA";XP(J);",";YP(J);";"
838Ø NEXT J
```

```
8390 PRINT#2,"DA";TPRX;",";(TPRY+TORY);";"
8395 PRINT#2,"PS2;":PRINT#2,"LS2;"
8400 PRINT#2,"MA";(TPLX-8);",";(TPLY-3*MG);";"
8405 IF SPIR=1 THEN TPLL=F1 ELSE TPLL=F3
8410 PRINT#2,"PL";TPLL
8420 PRINT#2,CHR$(13) CHR$(10) CHR$(3)
8430 PRINT#2,"MA";(TPRX-8);",";(TPLY-3*MG);";"
8435 IF SPIR=1 THEN TPRL=F2 ELSE TPRL=F4
8440 PRINT#2,"PL";TPRL
8450 PRINT#2,CHR$(13) CHR$(10) CHR$(3)
8460 IF PTXT=1 THEN 8930
8660 PRINT#2,"MA";(GX+5);",";(GY+SY-8);";"
8670 PRINT#2,"PLR =";INT(R*1000+.5)/1000
8680 PRINT#2,CHR$(13) CHR$(10) CHR$(3)
8700 PRINT#2,"MA";(GX+5);",";(GY+SY-12);";"
8710 PRINT#2,"PLDF=";DF
8720 PRINT#2,CHR$(13) CHR$(10) CHR$(3)
8730 PRINT#2,"MA";(GX+5);",";(GY+SY-16);";"
8740 PRINT#2,"PLTH=";TH
8750 PRINT#2,CHR$(13) CHR$(10) CHR$(3)
8760 PRINT#2,"MA";(GX+5);",";(GY+SY-20);";"
8770 PRINT#2,"PLDC=";DC
8780 PRINT#2,CHR$(13) CHR$(10) CHR$(3)
8800 PRINT#2,"MA";(GX+5);",";(GY+SY-24);";"
8810 PRINT#2,"PLSF=";SF
8820 PRINT#2,CHR$(13) CHR$(10) CHR$(3)
8830 PRINT#2,"MA";(GX+5);",";(GY+SY-28);";"
8840 PRINT#2,"PLCL=";INT(CL*1000+.5)/1000
8850 PRINT#2,CHR$(13) CHR$(10) CHR$(3)
8860 PRINT#2,"MA";(GX+5);",";(GY+SY-32);";"
8870 PRINT#2,"PLTC=";INT(TC*1000+.5)/1000
8880 PRINT#2,CHR$(13) CHR$(10) CHR$(3)
8890 IF TL=0 THEN 8894
8891 PRINT#2,"MA";(GX+5);",";(GY+SY-36);";"
8892 PRINT#2,"PLTL=";INT(TL*1000+.5)/1000
8893 PRINT#2,CHR$(13) CHR$(10) CHR$(3)
8894 PRINT#2,"MA";(GX+5);",";(GY+SY-40);";"
8895 PRINT#2,"PLCI=";INT(CI*1000+.5)/1000
8896 PRINT#2,CHR$(13) CHR$(10) CHR$(3)
8900 PRINT#2,"MA5,-4;"
8910 PRINT#2,"PLPC-Transition Curves - ";A$;" : Date : ";DATE$
8920 PRINT#2,CHR$(13) CHR$(10) CHR$(3)
8930 CLOSE#2:PTXT=1
8940 IF SPIR=1 THEN 2740
8950 IF SPIR=2 THEN 3330
8985  REM  ****************************************
8990  REM  GOSUB ROUTINE TO CLEAR SCREEN
9000  CLS:T$="PC-Curves":CC$="PC-TCurve"
9010 LOCATE 1,2:PRINT T$
9020 LOCATE 1,30:PRINT DATE$
9030 LOCATE 2,2:PRINT CC$
9040 LOCATE 2,30:PRINT TIME$
9045 IF DISP%=1 THEN 9070
9050 PRINT "====================================="
9060 PRINT
9070  RETURN
9093  REM  **************************************************
9095  REM  GOSUB ROUTINE FOR ERRORS
9100 BEEP
9110  PRINT : PRINT E0$
9120  PRINT
9130  RETURN
9493  REM  **************************************************
9495  REM  END OF PROGRAM
9500  GOSUB 9000
9510  PRINT
```

```
9520  PRINT "      END TRANSITION CURVE PROGRAM"
9530  PRINT
9540  PRINT "***************************************"
9542  FOR I = 1 TO 500
9544  NEXT I
9550  REM  PROGRAM PREPARED BY
9560  REM  DR. P.H. MILNE
9610  REM  **************************
9620  END
10195 REM ***** Epson HI-80 Plotter *****
10200 GOSUB 54000
10210 PRINT#2,"SP4"
10220 PRINT#2,"MA";TPLX*MX;",";TPLY*MX
10230 PRINT#2,"DA";TPRX*MX;",";TPRY*MX
10240 PRINT#2,"DA";TPRX*MX;",";(TPRY+TORY*MG)*MX
10250 PRINT#2,"MA";TPLX*MX;",";TPLY*MX
10260 PRINT#2,"DA";TPLX*MX;",";(TPLY+TOLY*MG)*MX
10290 PRINT#2,"SP3"
10295 IF SPIR=2 THEN 10350
10300 PRINT#2,"MA";TPLX*MX;",";TPLY*MX
10310 FOR J=0 TO N
10315 XP(J)=(TPLX+XF(J)*SCRM)*MX:YP(J)=(TPLY+YF(J)*2*SCRM*MG)*MX
10320 PRINT#2,"DA";XP(J);",";YP(J)
10330 NEXT J
10340 PRINT#2,"DA";TPRX*MX;",";(TPRY+TORY*MG)*MX
10345 GOTO 10395
10350 PRINT#2,"MA";TPLX*MX;",";(TPLY+TOLY*MG)*MX
10360 FOR J=0 TO N
10365 XP(J)=(TPRX-XF(J)*SCRM)*MX:YP(J)=(TPRY+YF(J)*2*SCRM*MG)*MX
10370 PRINT#2,"DA";XP(J);",";YP(J)
10380 NEXT J
10390 PRINT#2,"DA";TPRX*MX;",";(TPRY+TORY)*MX
10395 PRINT#2,"SP2":PRINT#2,"SI25,20"
10400 PRINT#2,"MA";(TPLX-8)*MX;",";(TPLY-3*MG)*MX
10405 IF SPIR=1 THEN TPLL=F1 ELSE TPLL=F3
10410 PRINT#2,"LA";TPLL
10430 PRINT#2,"MA";(TPRX-8)*MX;",";(TPLY-3*MG)*MX
10435 IF SPIR=1 THEN TPRL=F2 ELSE TPRL=F4
10440 PRINT#2,"LA";TPRL
10460 IF PTXT=1 THEN 10930
10660 PRINT#2,"MA";(GX+5)*MX;",";(GY+SY-8)*MX
10670 PRINT#2,"LAR =";INT(R*1000+.5)/1000
10700 PRINT#2,"MA";(GX+5)*MX;",";(GY+SY-12)*MX
10710 PRINT#2,"LADF=";DF
10730 PRINT#2,"MA";(GX+5)*MX;",";(GY+SY-16)*MX
10740 PRINT#2,"LATH=";TH
10760 PRINT#2,"MA";(GX+5)*MX;",";(GY+SY-20)*MX
10770 PRINT#2,"LADC=";DC
10800 PRINT#2,"MA";(GX+5)*MX;",";(GY+SY-24)*MX
10810 PRINT#2,"LASF=";SF
10830 PRINT#2,"MA";(GX+5)*MX;",";(GY+SY-28)*MX
10840 PRINT#2,"LACL=";INT(CL*1000+.5)/1000
10860 PRINT#2,"MA";(GX+5)*MX;",";(GY+SY-32)*MX
10870 PRINT#2,"LATC=";INT(TC*1000+.5)/1000
10890 IF TL=0 THEN 10894
10891 PRINT#2,"MA";(GX+5)*MX;",";(GY+SY-36)*MX
10892 PRINT#2,"LATL=";INT(TL*1000+.5)/1000
10894 PRINT#2,"MA";(GX+5)*MX;",";(GY+SY-40)*MX
10895 PRINT#2,"LACI=";INT(CI*1000+.5)/1000
10900 PRINT#2,"MA";5*MX;",";+2*MX
10905 PRINT#2,"SI30,25"
10910 PRINT#2,"LAPC-Curves - ";A$;" : Date : ";DATE$
10930 CLOSE#2:PTXT=1
10940 IF SPIR=1 THEN 2740
10950 IF SPIR=2 THEN 3330
```

4.3.5 PC-TCurve – computer printouts

(a) Transition curve details for Figs 4.6, 4.8 and 4.10.

```
Location of TCurve :- Strathclyde
**************************************************
OPERATOR'S NAME :- John Smith
DATE OF CALCS.  :- 01/01/87
INSTRUMENT USED :- Sokkisha DT20E
**************************************************

 PC-Transition Curve Design
****************************************
DESIGN CRITERIA FOR CURVE :-
VELOCITY (KPH)                =  120
VELOCITY (MPS)                =  33.333
RADIAL ACCELERATION (MPS3) =  .46
CENTRIPETAL RATIO             =  .25
MINIMUM RADIUS (M)            =  453.052

MULTIPLYING CONSTANT          =  283.752
DEFLECTION ANGLE AT I.P.      =  120
TRANSITION CURVE ANGLE        =  11.1415
CIRCULAR CURVE ANGLE          =  97.3129
SPIRAL TRANSITION LENGTH      =  177.717
CIRCULAR CURVE LENGTH         =  771.153
TOTAL CURVE LENGTH            =  1126.588
****************************************

 PC-Transition Curve Design
***********************************************************************
CHAINAGE OF I.P.  =      1000
LENGTH OF TANGENT =      878.479
***********************************************************************
```

CHAINAGE	LENGTH	TANG.DIST	TANG.OFF	POLAR DIST	ANGLE
121.521	0	0	0	0	360
150	28.479	28.479	.048	28.479	359.5414
180	58.479	58.476	.414	58.478	359.354
210	88.479	88.458	1.434	88.47	359.0418
240	118.479	118.389	3.441	118.439	358.2007
270	148.479	148.201	6.767	148.355	357.2309
299.239	177.717	177.034	11.587	177.413	356.1519
1070.392	START OF 2ND SPIRAL				
1248.11	FINAL TANGENT POINT				

```
 PC-Transition Curve Design
***********************************************************************
```

CHAINAGE	LENGTH	TANG.DIST	TANG.OFF	POLAR DIST	ANGLE
1070.392	177.717	177.034	11.587	177.413	3.4441
1080	168.11	167.592	9.813	167.879	3.2103
1110	138.11	137.916	5.448	138.024	2.1543
1140	108.11	108.053	2.615	108.085	1.231
1170	78.11	78.099	.986	78.10501	.4325
1200	48.11	48.109	.23	48.11	.1628
1230	18.11	18.11	.012	18.11	.022
1248.11	0	0	0	0	0

(b) Transition curve details for Figs 4.7, 4.9 and 4.11.

```
Location of TCurve :- Townhead
*************************************************
OPERATOR'S NAME :- A. N. Other
DATE OF CALCS.  :- 01/01/87
INSTRUMENT USED :- Zeiss Jena T20A
*************************************************

 PC-Transition Curve Design
****************************************
DESIGN CRITERIA FOR CURVE :-
VELOCITY (KPH)              =  160
VELOCITY (MPS)              =  44.444
RADIAL ACCELERATION (MPS3)  =  .46
CENTRIPETAL RATIO           =  .25
MINIMUM RADIUS (M)          =  805.427

MULTIPLYING CONSTANT        =  436.865
DEFLECTION ANGLE AT I.P.    =  75
TRANSITION CURVE ANGLE      =  8.2542
CIRCULAR CURVE ANGLE        =  58.0837
SPIRAL TRANSITION LENGTH    =  236.957
CIRCULAR CURVE LENGTH       =  817.344
TOTAL CURVE LENGTH          =  1291.257
****************************************

 PC-Transition Curve Design
**********************************************************************
CHAINAGE OF I.P.  =      1862.3
LENGTH OF TANGENT =      738.643
**********************************************************************
```

CHAINAGE	LENGTH	TANG.DIST	TANG.OFF	POLAR DIST	ANGLE
1123.657	0	0	0	0	0
1140	16.343	16.343	.004	16.343	.0048
1170	46.343	46.343	.087	46.343	.0627
1200	76.343	76.341	.389	76.342	.173
1230	106.343	106.334	1.05	106.339	.3357
1260	136.343	136.311	2.213	136.329	.5548
1290	166.343	166.256	4.018	166.304	1.2304
1320	196.343	196.143	6.605	196.254	1.5543
1350	226.343	225.935	10.113	226.162	2.3347
1360.613	236.957	236.444	11.601	236.728	2.4832
2177.958	START OF 2ND SPIRAL				
2414.914	FINAL TANGENT POINT				

```
 PC-Transition Curve Design
**********************************************************************
```

CHAINAGE	LENGTH	TANG.DIST	TANG.OFF	POLAR DIST	ANGLE
2177.958	236.957	236.444	11.601	236.728	357.1128
2190	224.914	224.519	9.923	224.738	357.2809
2220	194.914	194.721	6.462	194.828	358.0557
2250	164.914	164.83	3.915	164.877	358.3821
2280	134.914	134.883	2.144	134.9	359.0521
2310	104.914	104.905	1.008	104.91	359.2657
2340	74.914	74.912	.367	74.913	359.4309
2370	44.914	44.914	.079	44.914	359.5357
2400	14.914	14.914	.003	14.914	359.592
2414.914	0	0	0	0	360

4.4 VERTICAL CURVES

The two previous sections in this chapter have dealt with linking straights using curves in the horizontal plane. However, due to level changes in the topography there are also changes in gradient along the straight. These changes are dictated by studying the mass-haul diagram, which was discussed in Section 2.3 under Earthworks. As with horizontal curves, vertical curves are designed for a particular design speed, and a parabolic arc is fitted at the change of gradient. In the United Kingdom the Department of the Environment specifies the maximum permissible gradients allowable on roads, which are normally 3% for motorways and 4% for rural roads. The percentage method is to be preferred for expressing a gradient, that is 4% rather than 1 in 25. Accordingly, a road which rises m metres in 100 m in the direction of chainage would have a gradient of $m\%$. If the road rises in the direction of chainage it has a positive gradient (i.e. $+m\%$), and if the road falls in the direction of chainage it has a negative gradient (e.g. $-n\%$).

Where there is a change of sign from plus to minus between two gradients there will be a summit or *crest* curve, and if from minus to plus there will be a valley or *sag* curve. In addition to these two specific curves there are also four other combinations of gradient, making a total of six different types of vertical curve. In calculating the difference in gradients it should be noted that it is the *algebraic* difference (A) of the gradients which is required.

In designing a vertical curve, there are several criteria. The first is the provision of adequate visibility for vehicles at the design speed to be able to stop or overtake safely. This requirement was met in the past by the use of *sight distances*, but these have now been mostly replaced by tables of *K-values*, published by the Department of the Environment. The second and third criteria to be met are that of passenger comfort and vehicle safety. A sharp crest such as a humpback bridge can cause a vehicle to either leave the road or to straddle the crest. Similarly a sharp sag curve could cause problems with a long low-loader vehicle. Hence in curve design, it is essential to restrict the gradients and introduce the vertical curve gradually.

4.4.1 PC-VCurve – subroutine index

	Line numbers	*Function*
(a)	10–70	Initialization and control
(b)	100–230	Screen header display
(c)	500–680	Entry of curve location, etc.
(d)	1000–1190	Keyboard entry of design criteria
(e)	1200–1520	Vertical curve setting out details
(f)	4000–4450	Graphics display and screen dump

	Line numbers	*Function*
(g)	5000–5190	Plotter initialization and selection
(h)	5200–5940	Plotter routine for HPGL
(i)	6000–6080	Routine to print curve location, etc.
(j)	6100–6270	Routine to print curve design criteria
(k)	6300–6470	Routine to print curve setting-out details
(l)	8200–8940	Plotter routine for Apple 410
(m)	9000–9070	Routine to clear screen
(n)	9100–9130	Routine to display error message
(o)	9500–9620	Termination of program
(p)	10200–10940	Plotter routine for Epson HI-80

(a) Initialization and control Line numbers 10–70

All numeric variables and required string variables are initialized, and four arrays dimensioned. If more than 100 points are required on each curve, the dimensions should be changed accordingly.

(b) Screen header display Line numbers 100–230

At the start of each program selection the screen is cleared using **GOSUB 9000** to identify on screen the program suite and title.

(c) Entry of curve location, etc. Line numbers 500–680

This routine is provided to keep a record of site location, operator, date, etc., with an opportunity to obtain a printout if required.

(d) Keyboard entry of design criteria Line numbers 1000–1190

All the necessary data required to compute a vertical curve design solution are requested, namely, gradients of both straights, K-value, acceptable minimum length of curve, chainage and level of the intersection point. The chainage of the two tangent points is also calculated. A printout option is provided for data record purposes.

(e) Vertical curve setting-out details Line numbers 1200–1520

The only keyboard entry requested is that of the chord interval to be used to set out the curve. The chainage of the first point on the curve, the first chord length and the number of chords are then calculated in a similar manner to that used in Sections 4.2. and 4.3.

(f) Graphics display and screen dump Line numbers 4000–4450

The display screen chosen for the presentation of the longitudinal profile of the curve is shown in Fig. 1.5 with text at the top and bottom of the graphics window.

The curve chainage is considered from left to right across the screen, with the chainage of the tangent point on the first straight appearing in the lower left-hand corner of the screen. The vertical curve is plotted centrally on the screen tangentially to the two straights whose gradients are also noted. The intersection point of the two straights is plotted in the centre of the screen for convenience and a vertical exaggeration (1:10) given to show the curve more clearly. This can easily be changed by altering the value of **VEX** in line 4000. After plotting, a screen dump can be obtained on a dot-matrix printer for record purposes. Examples of both sag and crest curves are given in Figs 4.12 and 4.13.

(g) Plotter initialization and selection Line numbers 5000–5190

After initializing the plotter variables to give a maximum plot width, a choice of three plotters is given, Apple 410, Epson HI-80 and Hewlett-Packard. All details are presented on the plot for a vertical curve. In addition to the chainages of the tangent points and the intersection point, their levels are all also given together with the gradients of the straights. Examples of sag and crest curves are shown in Figs 4.14 and 4.15.

(h) Plotter routine for HPGL Line numbers 5200–5940

This routine is suitable for Hewlett-Packard plotters and their equivalents using HPGL as discussed in routine (k) of Section 2.2. After scaling the plotter into user units, a frame is drawn round the plotting area by pen 1 (black). Pen 4 (blue) is chosen to draw the longitudinal profile of the two straights and then the curve plotted by pen 3 (green). Finally the chainages, levels, gradients and labels are printed by pen 2 (red). The profiles of both sag and crest curves are shown in Figs 4.14 and 4.15.

(i) Routine to print curve locations, etc. Line numbers 6000–6080

This routine is provided for record purposes to annotate the printout, assuming a standard IBM PC type printer. It may require alteration for different microcomputers and printers.

(j) Routine to print curve design criteria Line numbers 6100–6270

This printout provides a record of data entered at the keyboard.

(k) Routine to print curve setting-out details Line numbers 6300–6470

The complete setting-out details for each point, chainage, distance, height difference and reduced level are printed for record purposes.

(l) Plotter routine for Apple 410 Line numbers 8200–8950

This routine is similar to (h) where the plotter commands are altered as shown in Table 2.1.

Fig. 4.12 PC-VCurve – screen dump of sag curve showing both gradients on a vertical curve.

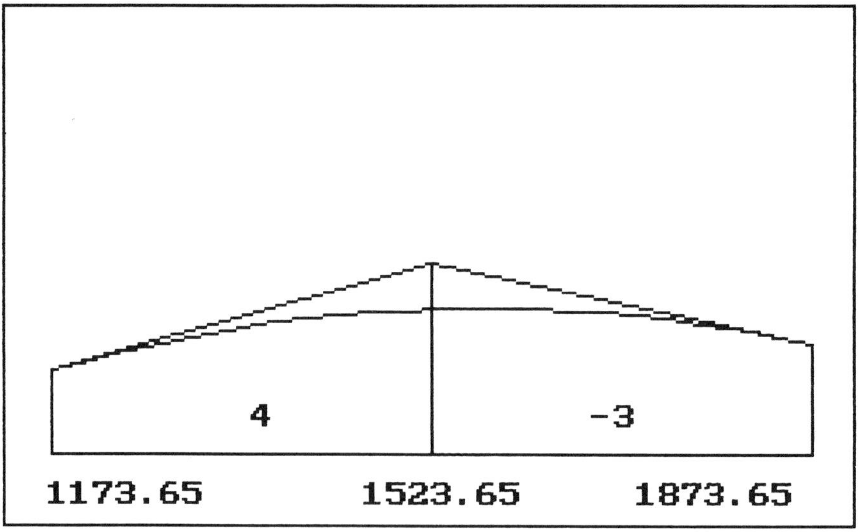

Fig. 4.13 PC-VCurve – screen dump of crest curve showing both gradients on a vertical curve.

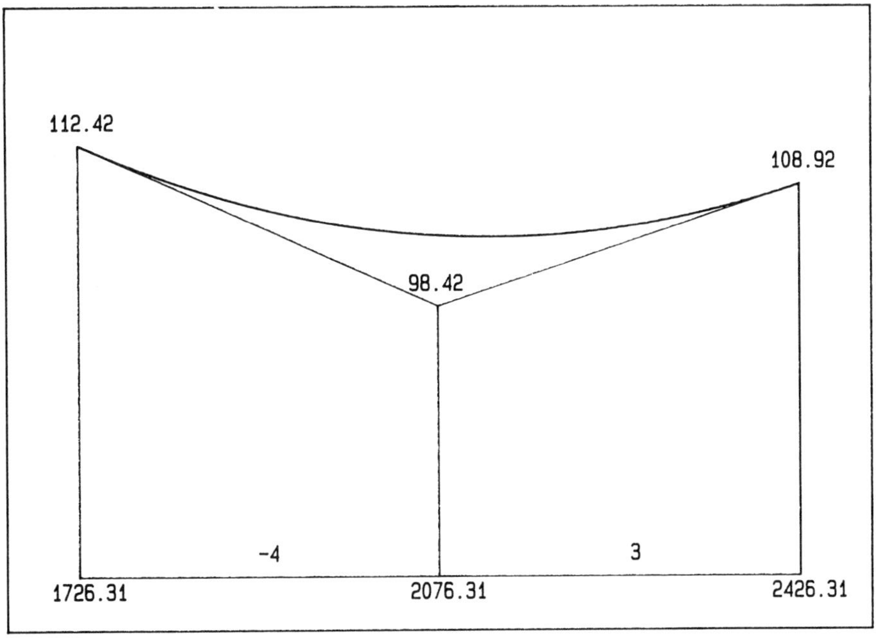

Fig. 4.14 PC-VCurve – plot of sag curve showing gradients, chainages and levels.

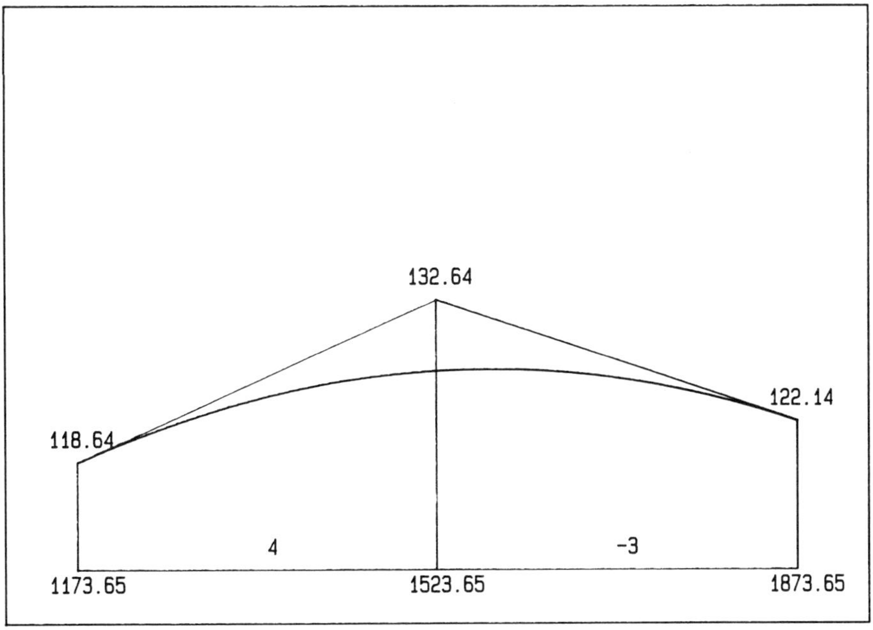

Fig. 4.15 PC-VCurve – plot of crest curve showing gradients, chainages and levels.

(m) Routine to clear screen Line numbers 9000–9070

This routine is used to clear the screen and provide two lines of text at the head of each screen giving the title of the selected software suite and the selected program, together with the date and time of calculation.

(n) Routine to display error message Line numbers 9100–9130

If a wrong keyboard entry has been made in entering data, the error is trapped and this error message displayed and re-entry requested.

(o) Termination of program Line numbers 9500–9620

When the **QUIT** selection is made in the program, an end-of-program message is displayed to advise the user that computations are complete.

(p) Plotter routine for Epson HI-80 Line numbers 10200–10940

This routine is similar to (h) where the plotter commands are altered as shown in Table 2.1.

4.4.2 PC-VCurve – numeric variables

A	=	Algebraic difference in gradients
C1	=	First chord in vertical circle
C9	=	Chainage of last point on curve
CB	=	Chainage to *B*, highest/lowest point
CCOL	=	Column number for text display on screen
CH	=	Multiple chord interval
CH(1ØØ)	=	Chainage array along curve
CI	=	Chainage of IP
DISP%	=	Choice of display, screen or plotter
F1–F3	=	Chainages of points on vertical curve
GX	=	*X*-coordinate at lower left-hand corner of plot
GY	=	*Y*-coordinate at lower left-hand corner of plot
HB	=	Vertical height difference at *B*
HD(1ØØ)	=	Vertical height difference array along curve
HP%	=	Integer variable to identify plotter
IPX	=	*X*-coordinate of IP
IPY	=	*Y*-coordinate of IP
K	=	*K*-value
L,LØ	=	Length of vertical curve
LM	=	Minimum length of vertical curve
M	=	Gradient of first straight
MX	=	Multiplication factor for Epson HI-80 plotter
N	=	Gradient of second straight
NC	=	Number of setting out chords
P5,P7	=	Counters in printout routines
PLOT%	=	Integer variable for plotter selection

PX	=	*X*-coordinate at upper right-hand corner of plot
PY	=	*Y*-coordinate at upper right-hand corner of plot
RA	=	Reduced level of *A*
RAY	=	*Y*-coordinate of *A*
RB	=	Reduced level of *B*
RC	=	Reduced level of *C*
RCY	=	*Y*-coordinate of *C*
RGB%	=	Integer variable to select screen colour
RI	=	Reduced level of IP
RL(1ØØ)	=	Reduced level array
SCD%	=	Monitor display, monochrome/colour
SCR%	=	Screen resolution (installation file)
SCRM	=	Scale magnification to fit screen/plotter
SR%	=	Screen resolution
SX	=	Horizontal display width in user units
SY	=	Vertical display height in user units
TPLX	=	*X*-coordinate of tangent point on left
TPLY	=	*Y*-coordinate of tangent point on left
TPRX	=	*X*-coordinate of tangent point on right
TPRY	=	*Y*-coordinate of tangent point on right
U	=	Menu selection at start
VEX	=	Vertical exaggeration for display
Y(1ØØ)	=	Distance along curve display

4.4.3 PC-VCurve – string variables

A$		Storage for location of curve
B$		Storage for operator's name
C$		Storage for date of calculation
CC$	=	**"PC-VCurve"**
DATE$		Variable used to return the date of the computer's calendar/clock in US format MM–DD–YYYY
EO$	=	**"SORRY, DATA ERROR . . . PLEASE RE-ENTER"**
H$	=	**"PC-Vertical Curve Design"**
K$		Storage for instrument used
P$	=	**"DO YOU WISH PRINTOUT OF ABOVE DATA (Y/N)"**
Q$	=	Answer response to question (Y/N)
T$	=	**"PC-Curves"**
TIME$		Variable used to return the time of the built-in clock in HH:MM:SS format

4.4.4 PC-VCurve – BASIC program

```
10  REM  <PC-VCURV> Vertical Curves Program for IBM PC or Compatible
12  REM  BASIC Programs for Land Surveying, Chapter 5
13  REM  + Computer Graphics for Surveying, Chapter 4
14  REM  (C)  COPYRIGHT  1984-87  P.H.Milne
16  REM  ALL RIGHTS RESERVED
18  REM  VERSION PC-1.0, 01-01-87
20 CLEAR
25 DEFDBL C,F,H,L,R,Y
30 A = 0:C = 0:C1 = 0:CH = 0:K = 0:L = 0:L0 = 0:LM = 0:M = 0:N = 0:NC
 = 0:P5 = 0:P7 = 0:U = 0:SR%=1
35  DIM CH(100),HD(100),RL(100),Y(100)
40 E0$ = "SORRY, DATA ERROR ... PLEASE RE-ENTER."
42 H$ = " PC-Vertical Curve Design"
45 P$ = "DO YOU WISH PRINTOUT OF ABOVE DATA (Y/N)"
50 PCF$="PCSCREEN"
52 OPEN "I",#3,PCF$
54 INPUT#3,SCR%
56 INPUT#3,SCD%
58 CLOSE#3
60 ON (SCR%+1) GOTO 62,64,66
62 SCREEN 0:WIDTH 40:GOTO 80
64 SCREEN 1:GOTO 68
66 SCREEN 1:GOTO 80
68 IF SCD%=1 THEN 80
70 COLOR 8,0
80 KEY OFF
95  REM  **************************************************
100  GOSUB 9000
190  PRINT
200  PRINT " DO YOU WISH TO ENTER <1> LOCATION,      <2> NEW DATA OR
<3> QUIT ";
210  INPUT U
220  IF U < 1 OR U > 3 THEN 200
230  ON U GOTO 500,1000,9500
500  GOSUB 9000
540  PRINT
550  PRINT "ENTER LOCATION OF WORK, ETC. WHEN       REQUESTED, AND PR
ESS <Enter>"
560  PRINT:PRINT "ENTER LOCATION OF CURVE :-"
570  INPUT A$
580  PRINT "ENTER OPERATOR'S NAME :-"
590  INPUT B$
600  PRINT "ENTER DATE OF CALCULATION AS DD/MM/YY -"
610  INPUT C$
620  PRINT "ENTER INSTRUMENT USED :-"
630  INPUT K$
640  PRINT
650  PRINT "DO YOU WISH PRINTOUT OF LOCATION ETC.,  (Y/N) ";
660  INPUT Q$
670  IF Q$ = "Y" THEN 6000
680  IF Q$ <  > "N" THEN 650
993  REM  ************************************************
995  REM  KEYBOARD ENTRY OF DESIGN CRITERIA
1000  GOSUB 9000: PRINT H$
1010  PRINT "***************************************"
1020  PRINT : PRINT "ENTER GRADIENT OF 1ST STRAIGHT ";: INPUT M
1030  PRINT : PRINT "ENTER GRADIENT OF 2ND STRAIGHT ";: INPUT N
1035  REM  ALGEBRAIC DIFF.= A
1040 A = M - N
1050  PRINT : PRINT "ENTER K-VALUE ";: INPUT K
1052  IF K <  = 0 THEN 1050
1055  REM  MINIMUM CURVE LENGTH = L
```

```
1060 LM = K *  ABS (A)
1070  PRINT : PRINT "MINIMUM LENGTH OF CURVE IS = "; INT (LM * 1000 +
 .5) / 1000
1080  PRINT : PRINT "IS THIS ACCEPTABLE (Y/N) ";: INPUT Q$: IF Q$ = "
Y" THEN 1120
1090  IF Q$ <  > "N" THEN 1080
1100  PRINT : PRINT "ENTER ACCEPTABLE LENGTH ";: INPUT L0: IF L0 < LM
 THEN 1100
1110 L = L0
1120  GOSUB 9000: PRINT H$: IF L0 < LM THEN L = LM
1130  PRINT : PRINT "ENTER CHAINAGE OF I.P. ";: INPUT CI
1132  IF CI < 0 THEN 1130
1135  REM  F1 IS CHAINAGE OF 1ST TANGENT PT.
1140 F1 =  INT ((CI - L / 2) * 1000 + .5) / 1000
1145  REM F3 IS CHAINAGE OF 2ND TANGENT PT.
1150 F3 =  INT ((CI + L / 2) * 1000 + .5) / 1000
1160  PRINT : PRINT "ENTER LEVEL OF I.P. ";: INPUT RI
1170  PRINT : PRINT P$;: INPUT Q$
1180  IF Q$ = "Y" THEN 6100
1190  IF Q$ <  > "N" THEN 1170
1195  REM  START CALCULATIONS
1200 RA =  INT ((RI - (M * L / 200)) * 1000 + .5) / 1000
1205  REM  REDUCED LEVELS OF A AND C ARE RA AND RC
1210 RC =  INT ((RI + (N * L / 200)) * 1000 + .5) / 1000
1220  PRINT : PRINT "ENTER MULTIPLE CHORD INTERVALS ";: INPUT CH
1222  IF CH <  = 0 THEN 1220
1224  IF L / CH < 100 THEN 1230
1226  GOSUB 9100: PRINT "TOO SMALL A CHORD INTERVAL !"
1228  GOTO 1220
1229  REM  FIND CHAINAGE OF 1ST POINT ON CURVE
1230 CH(0) = ( INT (F1 / CH) + 1) * CH
1235  REM  FIND LENGTH 1ST CHORD
1240 C1 =  INT ((CH(0) - F1) * 1000 + .5) / 1000:Y(0) = C1
1245  REM  FIND LAST POINT ON CURVE
1250 C9 = ( INT (F3 / CH)) * CH
1255  REM  FIND NUMBER OF CHORDS
1260 NC =  INT ((L - C1) / CH)
1265  REM  CALCULATE HIGHEST/LOWEST POINT ON CURVE AT CB, REDUCED LEV
EL RB
1270 CB =  INT ((L *  ABS (M) /  ABS (A)) * 1000 + .5) / 1000:F2 = F1
 + CB
1280 HB = L * M *  ABS (M) / (200 *  ABS (A)):RB = RA + HB
1290 HB =  INT (HB * 1000 + .5) / 1000:RB =  INT (RB * 1000 + .5) / 1
000
1295  REM  SCREEN DISPLAY
1300  GOSUB 9000: PRINT H$:P5 = 0
1310  PRINT "***************************************"
1320  PRINT "CHAINAGE  LENGTH   HT.DIFF.   RED.LEVEL"
1330  PRINT "======================================="
1340 PRINT F1;TAB(12);" 0";TAB(21);" 0";TAB(31);RA
1345  REM  COLLECT DATA IN ARRAY
1350  FOR I = 0 TO NC
1360 Y(I) = Y(0) + I * CH
1370 CH(I) = CH(0) + I * CH
1380 HD(I) = (M * Y(I) / 100) - (A * (Y(I)) ^ 2 / (200 * L))
1390 RL(I) =  INT ((RA + HD(I)) * 1000 + .5) / 1000
1400 HD(I) =  INT ((HD(I)) * 1000 + .5) / 1000
1410  IF P5 = 1 THEN 1450
1420  IF F2 > CH(I) THEN 1450
1430  PRINT F2; TAB( 12);CB; TAB( 21);HB; TAB( 31);RB
1440 P5 = 1
1450  PRINT CH(I); TAB( 12);Y(I); TAB( 21);HD(I); TAB( 31);RL(I)
1460  NEXT I
1470  PRINT F3;TAB(12);L;TAB(21);INT((RC-RA)*1000+.5)/1000;TAB(31);RC
1480  PRINT "***************************************"
```

```
1490  PRINT : PRINT P$;: INPUT Q$
1500  IF Q$ = "Y" THEN 6300
1510  IF Q$ <  > "N" THEN 1490
1520  GOSUB 9000:PRINT H$
1525 IF DISP% <> 0 THEN 1550
1530 PRINT:PRINT " Do you wish plot of curve (y/n)";:INPUT Q$
1540 IF Q$="n" THEN 100
1550 PRINT:PRINT:PRINT " Select Display";TAB(18);"<1> Graphics Screen
":PRINT:PRINT TAB(18);"<2> Plotter":PRINT:PRINT TAB(18);"<3> Continue
";:INPUT DISP%
1560 IF DISP% <1 OR DISP% >3 THEN 1550
1570 ON DISP% GOTO 4000,5000,100
3995 REM ***** Graphics Screen Display *****
4000  CLS:GOSUB 9000:VEX=10
4010  IF SR%=1 THEN RGB%=2 ELSE RGB%=3
4020 WINDOW (0,0)-(320*SR%,200)
4030 LINE (8,18)-(310*SR%,180),RGB%,B
4100 TPLY=40:TPRY=40
4110 SCRM=(270*SR%)/(L*1.6)
4130 IPX=160*SR%
4140 IPY=100
4150 TPLX=IPX-(SCRM*1.6*L/2)
4160 TPRX=IPX+(SCRM*1.6*L/2)
4180 RAY=(IPY-(RI-RA)*SCRM*VEX)
4190 RCY=(IPY-(RI-RC)*SCRM*VEX)
4200 LINE (TPLX,TPLY)-(TPLX,RAY),3
4210 LINE -(IPX,IPY),3
4215 LINE -(TPRX,RCY),3
4220 LINE -(TPRX,TPRY),3
4225 LINE -(TPLX,TPLY),3
4230 LINE (IPX,TPLY)-(IPX,IPY),3
4232 CCOL=(TPLX*SR%-32)/8
4234 IF CCOL<3 THEN CCOL=3
4240 LOCATE 22,CCOL:PRINT F1
4242 CCOL=(TPRX*SR%-32)/8
4244 IF CCOL>29 THEN CCOL=29
4250 LOCATE 22,CCOL:PRINT F3
4260 LOCATE 22,17*SR%:PRINT CI
4270 LOCATE 19,12*SR%:PRINT M
4275 LOCATE 19,28*SR%:PRINT N
4300 PSET (TPLX,RAY),1
4310 FOR J=0 TO NC
4320 LINE -((TPLX+Y(J)*SCRM*1.6*SR%),(RAY+HD(J)*SCRM*VEX)),1
4330 NEXT J
4340 LINE -(TPRX,RCY),1
4390 P1=0
4400 LOCATE 24,2:PRINT "Screen Dump (y/n) ";
4410 Q$=INKEY$:IF Q$ = "" THEN 4410
4420 IF INSTR("yn",Q$)=0 THEN 4400
4430 IF Q$ = "n" THEN 4450
4440 LCOPY
4450 DISP%=3
4460 GOTO 1520
4995 REM ***** Plotter Output *****
5000 GX=0:GY=0:SX=180:SY=120:MX=2670/SX:PX=GX+SX:PY=GY+SY
5010 SCRM=150/L
5020 VEX=10
5030 IPX=GX+SX/2
5040 IPY=GY+SY/2
5050 TPLX=IPX-(SCRM*L/2)
5060 TPRX=IPX+(SCRM*L/2)
5070 TPLY=GY+10:TPRY=GY+10
5080 RAY=(IPY-(RI-RA)*SCRM*VEX)
5090 RCY=(IPY-(RI-RC)*SCRM*VEX)
5100 GOSUB 9000
```

```
511Ø PRINT:PRINT " Select Plotter -";TAB(18);"<1> Apple 41Ø":PRINT:PR
INT TAB(18);"<2> Epson HI-8Ø"
512Ø PRINT:PRINT TAB(18);"<3> Hewlett-Packard":PRINT:PRINT TAB(18);"<
4> QUIT          ";:INPUT PLOT%
513Ø IF PLOT% <1 OR PLOT% >4 THEN 51ØØ
514Ø IF PLOT%=4 THEN 9ØØ
5145 IF PLOT%=1 THEN HP%=3
515Ø IF PLOT%=2 THEN HP%=4
5155 IF PLOT%=3 THEN HP%=1
516Ø PRINT:PRINT " Place paper on Plotter and switch on"
517Ø PRINT:PRINT " To Continue Press <Enter> ":WHILE R$="":R$=INKEY$:
WEND
518Ø GOSUB 5ØØØØ
519Ø ON HP% GOTO 52ØØ,52ØØ,82ØØ,1Ø2ØØ
5195 REM ***** HP-GL Plotter (HP-7475A) *****
52ØØ GOSUB 51ØØØ
521Ø PRINT#2,"SP4;"
522Ø PRINT#2,"PU;PA";TPLX;",";TPLY;";"
5225 PRINT#2,"PD;PA";TPLX;",";RAY;";"
523Ø PRINT#2,"PA";IPX;",";IPY;";"
524Ø PRINT#2,"PA";TPRX;",";RCY;";"
525Ø PRINT#2,"PA";TPRX;",";TPRY;";"
526Ø PRINT#2,"PA";TPLX;",";TPLY;";"
527Ø PRINT#2,"PU;PA";IPX;",";TPLY;";"
528Ø PRINT#2,"PD;PA";IPX;",";IPY;";"
529Ø PRINT#2,"SP3;"
53ØØ PRINT#2,"PU;PA";TPLX;",";RAY";"
531Ø FOR J=Ø TO NC
532Ø PRINT#2,"PD;PA";(TPLX+Y(J)*SCRM);",";(RAY+HD(J)*SCRM*VEX);";"
533Ø NEXT J
534Ø PRINT#2,"PA";TPRX;",";RCY;";"
539Ø PRINT#2,"SP2;":PRINT#2,"SIØ.2,Ø.3;"
54ØØ PRINT#2,"PU;PA";(TPLX-8);",";(TPLY-3);";"
541Ø PRINT#2,"LB";F1
542Ø PRINT#2,CHR$(3)
543Ø PRINT#2,"PU;PA";(TPRX-8);",";(TPLY-3);";"
544Ø PRINT#2,"LB";F3
545Ø PRINT#2,CHR$(3)
546Ø PRINT#2,"PU;PA";(IPX-8);",";(TPLY-3);";"
547Ø PRINT#2,"LB";CI
548Ø PRINT#2,CHR$(3)
55ØØ PRINT#2,"PU;PA";(IPX-SCRM*L/4);",";(TPLY+3);";"
551Ø PRINT#2,"LB";M
552Ø PRINT#2,CHR$(3)
553Ø PRINT#2,"PU;PA";(IPX+SCRM*L/4);",";(TPLY+3);";"
554Ø PRINT#2,"LB";N
555Ø PRINT#2,CHR$(3)
556Ø PRINT#2,"PU;PA";(TPLX-8);",";(RAY+3);";"
557Ø PRINT#2,"LB";RA
558Ø PRINT#2,CHR$(3)
56ØØ PRINT#2,"PU;PA";(IPX-8);",";(IPY+3);";"
561Ø PRINT#2,"LB";RI
562Ø PRINT#2,CHR$(3)
563Ø PRINT#2,"PU;PA";(TPRX-8);",";(RCY+3);";"
564Ø PRINT#2,"LB";RC
565Ø PRINT#2,CHR$(3)
59ØØ PRINT#2,"PU,PA5,-3;"
591Ø PRINT#2,"LBPC-Vertical Curves - ";A$;" : Date : ";DATE$
592Ø PRINT#2,CHR$(3)
593Ø CLOSE#2
594Ø GOTO 152Ø
5985  REM  ************************************************
599Ø  REM  PRINTOUT OF LOCATION, ETC
6ØØØ LPRINT "Location of VCurve :- ";A$
601Ø LPRINT "************************************************"
```

```
6020 LPRINT "OPERATOR'S NAME :- ";B$
6030 LPRINT "DATE OF CALCS.  :- ";C$
6040 LPRINT "INSTRUMENT USED :- ";K$
6060 LPRINT "***************************************************"
6070 LPRINT
6080  GOTO 1000
6093  REM  **************************************************
6095  REM  PRINTOUT OF DESIGN CRITERIA ETC.
6100 LPRINT H$
6120 LPRINT "*****************************************"
6130 LPRINT "GRADIENT OF 1ST STRAIGHT = ";M
6140 LPRINT "GRADIENT OF 2ND STRAIGHT = ";N
6150 LPRINT "ALGEBRAIC DIFFERENCE M-N = ";A
6160 LPRINT "K-VALUE                  = ";K
6170 LPRINT "MINIMUM LENGTH OF CURVE  = ";LM
6180 LPRINT "CURVE LENGTH CHOSEN      = ";L
6190 LPRINT
6200 LPRINT "CHAINAGE OF INTERSECTION = ";CI
6210 LPRINT "REDUCED LEVEL OF I.P.    = ";RI
6220 LPRINT "CHAINAGE OF 1ST TANG.PT. = ";F1
6230 LPRINT "CHAINAGE OF 2ND TANG.PT. = ";F3
6240 LPRINT "*****************************************"
6250 P7 = 1
6260 LPRINT
6270  GOTO 1200
6293  REM  **************************************************
6295  REM  PRINTOUT OF RESULTS
6300 P5=0:IF P7=1 THEN 6340
6320 LPRINT H$
6330 LPRINT "*****************************************"
6340 LPRINT "CHAINAGE  LENGTH   HT.DIFF.   RED.LEVEL"
6350 LPRINT "========================================="
6360 LPRINT F1;TAB(11);" 0";TAB(20);" 0";TAB(30);RA
6370  FOR J = 0 TO NC
6380  IF P5 = 1 THEN 6420
6390  IF F2 > CH(J) THEN 6420
6400 LPRINT F2;TAB(11);CB;TAB(20);HB;TAB(30);RB
6410 P5 = 1
6420 LPRINT CH(J);TAB(11);Y(J);TAB(20);HD(J);TAB(30);RL(J)
6430  NEXT J
6440 LPRINT F3;TAB(11);L;TAB(20);INT((RC-RA)*1000+.5)/1000;TAB(30);RC
6450 LPRINT "*****************************************"
6460 LPRINT
6470  GOTO 100
8195 REM ***** Apple 410 Plotter *****
8200 GOSUB 53000
8210 PRINT#2,"PS4;"
8220 PRINT#2,"MA";TPLX;",";TPLY;";"
8225 PRINT#2,"DA";TPLX;",";RAY;";"
8230 PRINT#2,"DA";IPX;",";IPY;";"
8240 PRINT#2,"DA";TPRX;",";RCY;";"
8250 PRINT#2,"DA";TPRX;",";TPRY;";"
8260 PRINT#2,"DA";TPLX;",";TPLY;";"
8270 PRINT#2,"MA";IPX;",";TPLY;";"
8280 PRINT#2,"DA";IPX;",";IPY;";"
8290 PRINT#2,"PS3;"
8300 PRINT#2,"MA";TPLX;",";RAY";"
8310 FOR J=0 TO NC
8320 PRINT#2,"DA";(TPLX+Y(J)*SCRM);",";(RAY+HD(J)*SCRM*VEX);";"
8330 NEXT J
8340 PRINT#2,"DA";TPRX;",";RCY;";"
8390 PRINT#2,"PS2;":PRINT#2,"LS2;"
8400 PRINT#2,"MA";(TPLX-8);",";(TPLY-3);";"
8410 PRINT#2,"PL";F1
8420 PRINT#2,CHR$(13) CHR$(10) CHR$(3)
```

```
8430 PRINT#2,"MA";(TPRX-8);",";(TPLY-3);";"
8440 PRINT#2,"PL";F3
8450 PRINT#2,CHR$(13) CHR$(10) CHR$(3)
8460 PRINT#2,"MA";(IPX-8);",";(TPLY-3);";"
8470 PRINT#2,"PL";CI
8480 PRINT#2,CHR$(13) CHR$(10) CHR$(3)
8500 PRINT#2,"MA";(IPX-SCRM*L/4);",";(TPLY+3);";"
8510 PRINT#2,"PL";M
8520 PRINT#2,CHR$(13) CHR$(10) CHR$(3)
8530 PRINT#2,"MA";(IPX+SCRM*L/4);",";(TPLY+3);";"
8540 PRINT#2,"PL";N
8550 PRINT#2,CHR$(13) CHR$(10) CHR$(3)
8560 PRINT#2,"MA";(TPLX-8);",";(RAY+3);";"
8570 PRINT#2,"PL";RA
8580 PRINT#2,CHR$(13) CHR$(10) CHR$(3)
8600 PRINT#2,"MA";(IPX-8);",";(IPY+3);";"
8610 PRINT#2,"PL";RI
8620 PRINT#2,CHR$(13) CHR$(10) CHR$(3)
8630 PRINT#2,"MA";(TPRX-8);",";(RCY+3);";"
8640 PRINT#2,"PL";RC
8650 PRINT#2,CHR$(13) CHR$(10) CHR$(3)
8900 PRINT#2,"MA5,-4;"
8910 PRINT#2,"PLPC-Vertical Curves - ";A$;" : Date : ";DATE$
8920 PRINT#2,CHR$(13) CHR$(10) CHR$(3)
8930 CLOSE#2
8940 GOTO 1520
8985  REM  *****************************************
8990  REM  GOSUB ROUTINE TO CLEAR SCREEN
9000  CLS:T$="PC-Curves":CC$="PC-VCurve"
9010 LOCATE 1,2:PRINT T$
9020 LOCATE 1,30:PRINT DATE$
9030 LOCATE 2,2:PRINT CC$
9040 LOCATE 2,30:PRINT TIME$
9045 IF DISP%=1 THEN 9070
9050 PRINT "======================================="
9060 PRINT
9070  RETURN
9093  REM  ***************************************************
9095  REM  GOSUB ROUTINE FOR ERRORS
9100 BEEP
9110  PRINT : PRINT E0$
9120  PRINT
9130  RETURN
9493  REM  ***************************************************
9495  REM  END OF PROGRAM
9500  GOSUB 9000
9510  PRINT
9520  PRINT "      END VERTICAL CURVE PROGRAM"
9530  PRINT
9540  PRINT "**************************************"
9542  FOR I = 1 TO 500
9544  NEXT I
9550  REM  PROGRAM PREPARED BY
9560  REM  DR. P.H. MILNE
9610  REM  ***************************
9620  END
10195 REM ***** Epson HI-80 Plotter *****
10200 GOSUB 54000
10210 PRINT#2,"SP4"
10220 PRINT#2,"MA";TPLX*MX;",";TPLY*MX
10225 PRINT#2,"DA";TPLX*MX;",";RAY*MX
10230 PRINT#2,"DA";IPX*MX;",";IPY*MX
10240 PRINT#2,"DA";TPRX*MX;",";RCY*MX
10250 PRINT#2,"DA";TPRX*MX;",";TPRY*MX
10260 PRINT#2,"DA";TPLX*MX;",";TPLY*MX
```

```
10270 PRINT#2,"MA";IPX*MX;",";TPLY*MX
10280 PRINT#2,"DA";IPX*MX;",";IPY*MX
10290 PRINT#2,"SP3"
10300 PRINT#2,"MA";TPLX*MX;",";RAY*MX
10310 FOR J=0 TO NC
10320 PRINT#2,"DA";(TPLX+Y(J)*SCRM)*MX;",";(RAY+HD(J)*SCRM*VEX)*MX
10330 NEXT J
10340 PRINT#2,"DA";TPRX*MX;",";RCY*MX
10390 PRINT#2,"SP2":PRINT#2,"SI25,20"
10400 PRINT#2,"MA";(TPLX-8)*MX;",";(TPLY-3)*MX
10410 PRINT#2,"LA";F1
10430 PRINT#2,"MA";(TPRX-8)*MX;",";(TPLY-3)*MX
10440 PRINT#2,"LA";F3
10460 PRINT#2,"MA";(IPX-8)*MX;",";(TPLY-3)*MX
10470 PRINT#2,"LA";CI
10500 PRINT#2,"MA";(IPX-SCRM*L/4)*MX;",";(TPLY+3)*MX
10510 PRINT#2,"LA";M
10530 PRINT#2,"MA";(IPX+SCRM*L/4)*MX;",";(TPLY+3)*MX
10540 PRINT#2,"LA";N
10560 PRINT#2,"MA";(TPLX-8)*MX;",";(RAY+3)*MX
10570 PRINT#2,"LA";RA
10600 PRINT#2,"MA";(IPX-8)*MX;",";(IPY+3)*MX
10610 PRINT#2,"LA";RI
10630 PRINT#2,"MA";(TPRX-8)*MX;",";(RCY+3)*MX
10640 PRINT#2,"LA";RC
10890 PRINT#2,"SI30,25"
10900 PRINT#2,"MA";5*MX;",";+2*MX
10910 PRINT#2,"LAPC-Vertical Curves - ";A$;" : Date : ";DATE$
10930 CLOSE#2
10940 GOTO 1520
```

4.4.5 PC-VCurve – computer printouts

(a) Vertical (sag) curve details for Figs 4.12 and 4.14

```
Location of VCurve :- Renfrew
**************************************************
OPERATOR'S NAME :- A.N. Other
DATE OF CALCS.  :- 01/01/87
INSTRUMENT USED :- Epson-PC
**************************************************

 PC-Vertical Curve Design
****************************************
GRADIENT OF 1ST STRAIGHT = -4
GRADIENT OF 2ND STRAIGHT =  3
ALGEBRAIC DIFFERENCE M-N = -7
K-VALUE                  =  100
MINIMUM LENGTH OF CURVE  =  700
CURVE LENGTH CHOSEN      =  700

CHAINAGE OF INTERSECTION =  2076.31
REDUCED LEVEL OF I.P.    =  98.42
CHAINAGE OF 1ST TANG.PT. =  1726.31
CHAINAGE OF 2ND TANG.PT. =  2426.31
****************************************

CHAINAGE   LENGTH   HT.DIFF.   RED.LEVEL
========================================
 1726.31   0        0          112.42
 1740      13.69    -.538      111.882
 1770      43.69    -1.652     110.768
```

```
1800       73.69   -2.676      109.744
1830      103.69   -3.61       108.81
1860      133.69   -4.454      107.966
1890      163.69   -5.208      107.212
1920      193.69   -5.872      106.548
1950      223.69   -6.446      105.974
1980      253.69   -6.93       105.49
2010      283.69   -7.324      105.096
2040      313.69   -7.628      104.792
2070      343.69   -7.841      104.579
2100      373.69   -7.965      104.455
2126.31   400      -8          104.42
2130      403.69   -7.999      104.421
2160      433.69   -7.943      104.477
2190      463.69   -7.797      104.623
2220      493.69   -7.561      104.859
2250      523.69   -7.235      105.185
2280      553.69   -6.819      105.601
2310      583.69   -6.313      106.107
2340      613.69   -5.717      106.703
2370      643.69   -5.031      107.389
2400      673.69   -4.255      108.165
2426.31   700      -3.5        108.92
****************************************
```

(b) Vertical (crest) curve details for Figs 4.13 and 4.15

```
Location of VCurve :- Strathclyde
***************************************************
OPERATOR'S NAME :- John Smith
DATE OF CALCS.  :- 01/01/87
INSTRUMENT USED :- Epson-PC
***************************************************

 PC-Vertical Curve Design
****************************************
GRADIENT OF 1ST STRAIGHT =  4
GRADIENT OF 2ND STRAIGHT = -3
ALGEBRAIC DIFFERENCE M-N =  7
K-VALUE                  =  100
MINIMUM LENGTH OF CURVE  =  700
CURVE LENGTH CHOSEN      =  700

CHAINAGE OF INTERSECTION =  1523.65
REDUCED LEVEL OF I.P.    =  132.64
CHAINAGE OF 1ST TANG.PT. =  1173.65
CHAINAGE OF 2ND TANG.PT. =  1873.65
****************************************

CHAINAGE  LENGTH   HT.DIFF.   RED.LEVEL
========================================
 1173.65   0        0          118.64
 1200      26.35    1.019      119.659
 1230      56.35    2.095      120.735
 1260      86.35    3.081      121.721
 1290      116.35   3.977      122.617
 1320      146.35   4.783      123.423
 1350      176.35   5.499      124.139
 1380      206.35   6.125      124.765
 1410      236.35   6.661      125.301
 1440      266.35   7.107      125.747
 1470      296.35   7.463      126.103
 1500      326.35   7.729      126.369
```

```
1530       356.35   7.905      126.545
1560       386.35   7.991      126.631
1573.65    400      8          126.64
1590       416.35   7.987      126.627
1620       446.35   7.893      126.533
1650       476.35   7.709      126.349
1680       506.35   7.434      126.074
1710       536.35   7.07       125.71
1740       566.35   6.616      125.256
1770       596.35   6.072      124.712
1800       626.35   5.438      124.078
1830       656.35   4.714      123.354
1860       686.35   3.9        122.54
1873.65    700      3.5        122.14
*************************************
```

4.5 HIGHWAY DESIGN

The previous three sections of this chapter have discussed either horizontal or vertical alignment in isolation. However, in the real world, horizontal and vertical curves are often found to overlap and thus require to be phased together to ensure compatibility. Where the horizontal curves are designed first, for example the spirals of a transition curve followed by the central circular portion, this is normally quite a complex operation. In these cases the vertical alignment is considered in conjunction with the mass-haul diagrams (Chapter 2) and the ground topography (Chapter 3).

To cater for such complexity, an expert program called RODEOS (*RO*ad and *D*esign *E*xpert *O*n *S*etting-out) was designed round five curve modules (Fig. 4.16):

(i) Module-1, curve design
(ii) Module-2, transition curve
(iii) Module-3, circular curve
(iv) Module-4, vertical curve
(v) Module-5, setting-out

For the first-time user, with no prior knowledge of the curve parameters, the program will start with Module-1 and steadily build up enough knowledge to set out the whole curve.

The flowchart in Fig. 4.16 shows the inter-relationship of each of the five modules. If, for example, only a horizontal curve is required, the program is directed through modules 1, 3 and 5. If only vertical curve calculations are required it is possible to proceed directly to modules 4 and 5. The five modules will only be accessed if a transition curve has a central circular portion with a change of gradient. The choice of disc storage of the designed curve parameters can be especially useful if the initial calculations are carried out in head office, before the full site setting-out details are known in the field. An additional advantage of disc storage is that if plotting facilities are not available during the initial calculations, the curve parameters can be

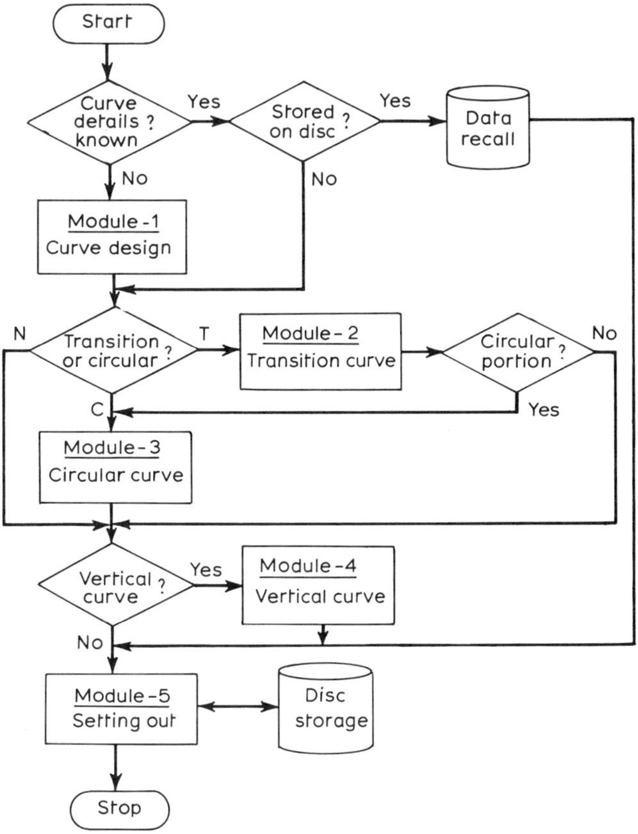

Fig. 4.16 Flow chart for RODEOS expert program for highway route alignment (reproduced by kind permission of the Computational Mechanics Institute from Proceedings of the 1st International Conference on Applications of Artificial Intelligence to Engineering Problems, April 1986, University of Southampton).

quickly recalled without the need for reprocessing of the initial curve design parameters.

4.5.1 PC-RODEOS – subroutine index

	Line numbers	*Function*
(a)	10–195	Initialization and control
(b)	200–330	Initial consultation
(c)	335–996	Module-1, curve design
(d)	1000–1568	Module-2, transition curve
(e)	2000–2544	Module-3, circular curve
(f)	3000–3246	Module-4, vertical curve

	Line numbers	*Function*
(g)	4000–4500	Module-5, graphics display and screen dump
(h)	5000–5190	Module-5, plotter initialization and selection
(i)	5200–5940	Plotter routine for HPGL
(j)	6000–6080	Routine to print curve location, etc.
(k)	6100–8746	Routine to print curve design details
(l)	9000–9070	Routine to clear screen
(m)	9100–9330	Routine to display error message, etc.
(n)	9500–9620	Termination of program
(o)	10200–10950	Plotter routine for Epson HI-80
(p)	11200–11950	Plotter routine for Apple 410
(q)	20000–20770	Disc write and recall routines

(a) Initialization and control Line numbers 10–195

The program is initialized and the installation file checked for the availability of graphics screens and colour for subsequent display of the total curve outline.

(b) Initial consultation Line numbers 200–330

The user enters information on the location, type of curve and how much detail is known about the curve in this routine. Depending upon this information the RODEOS expert system will branch to the appropriate module (Fig. 4.16).

(c) Module-1, curve design Line numbers 335–996

This module is only used if no design parameters are known for the curve, that is, if the radius and/or deviation angle are unknown. The user is given a choice of four design methods:

⟨1⟩ IP inaccessible, or linear measurement only
⟨2⟩ tangent to 3 straights
⟨3⟩ curve through 3 points
⟨4⟩ curve through 1 point

In each case, all the design parameters are computed, stored and listed for reference as discussed in Chapter 5 of *BASIC Programs for Land Surveying*.

(d) Module-2, transition curve Line numbers 1000–1568

This is a condensed version of the earlier program PC-TCurve (Section 4.3) which uses the curve design parameters of module 1.

(e) Module-3, circular curve Line numbers 2000–2544

This is a condensed version of the earlier program PC-HCurve (Section 4.2) which uses the curve design parameters of modules 1 and 2. If the circular curve is the central portion of a transition curve, then the design parameters are calculated automatically.

(f) Module-4, vertical curve Line numbers 3000–3246

This is a condensed version of the earlier program PC-VCurve (Section 4.4) and offers the user an opportunity to phase in a vertical curve with the previously designed horizontal curve. The same chord length, as used for the transition and circular curve, will be chosen for setting out the vertical curve, to ensure that levels are calculated at each of the previous horizontal setting-out point chainages.

(g) Module-5, graphics display and screen dump
Line numbers 4000–4500

The display screen chosen for the presentation of the layout of the curve is shown in Fig. 1.5 with text at the top and bottom of the graphics window. The intersection point *I* is centred on the screen and the combined horizontal curves, i.e., transition spirals and central circular portion plotted symmetrically on either side of the centre line. There is no need to specify a scale for the plot, as the display will automatically present the maximum plot, either for the long chord or the central line. The chainages of the first and last tangent points together with the chainage of the intersection point are displayed. The deviation angle between the straights and the minimum radius are also noted in the top left-hand corner of the screen for reference. The user can obtain a screen dump to a dot-matrix graphics printer before returning to the main menu and the plotter selection, Fig. 4.17.

(h) Module-5, plotter initialization and selection
Line numbers 5000–5190

After initializing the plotter variables to give a maximum plot width or height, a choice of three plotters is given, Apple 410, Epson HI-80 and Hewlett-Packard. As discussed earlier in routine (g), the curve is placed centrally on the plotter and the setting-out design details listed in the upper left-hand corner. The chainages of the first and last tangent points, together with the chainage of the intersection point are displayed on the plot, Fig. 4.18.

(i) Plotter routine for HPGL Line numbers 5200–5940

This routine is suitable for Hewlett-Packard plotters and their equivalents using HPGL as discussed in routine (k) of Section 2.2. After scaling the plotter into user units, a frame is drawn round the plotting area by pen 1 (black). Pen 4 (blue) is chosen to draw the centre line, tangents, total chord and long chord (if a centre circular portion) and then the curve is plotted by

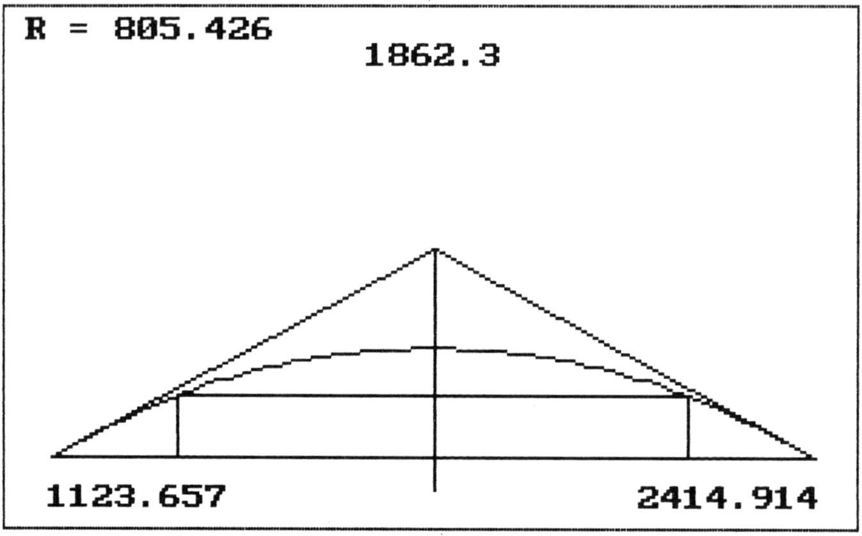

Fig. 4.17 PC-RODEOS – screen dump of a right-handed horizontal curve showing the merging of the circular curve with the transition spirals.

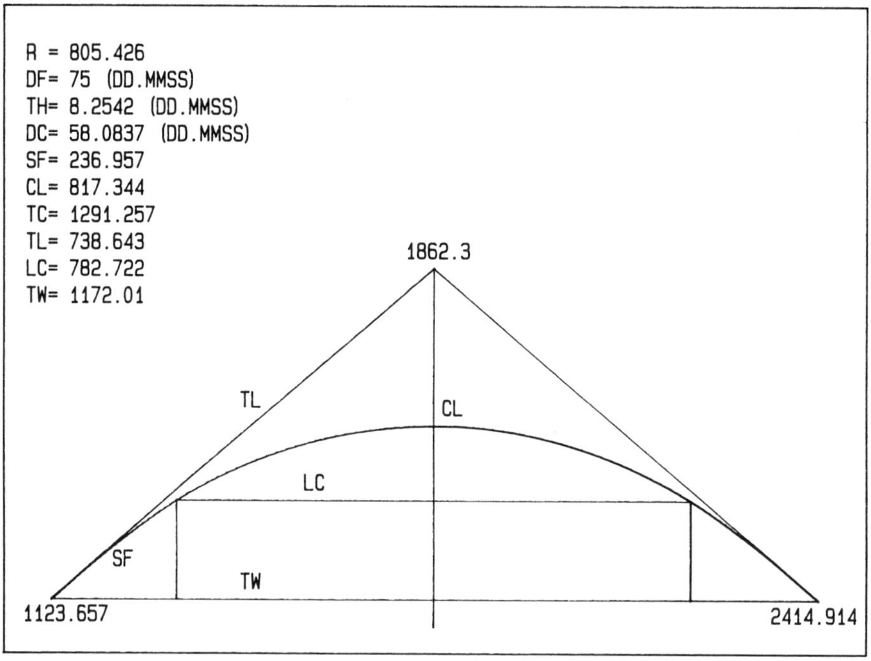

Fig. 4.18 PC-RODEOS – plot of a right-handed horizontal curve showing the merging of the circular curve with transition spirals.

pen 3 (green). Finally the chainages, labels and design details are printed by pen 2 (red).

(j) Routine to print curve location, etc. Line numbers 6000–6080

This routine is provided for record purposes to annotate the printout, assuming a standard IBM PC type printer. It may require alteration for different microcomputers and printers.

(k) Routines to print curve design details Line numbers 6100–8746

These routines to provide a printout of the curve design and setting-out details have been discussed previously for PC-HCurve, PC-TCurve and PC-VCurve in Sections 4.2, 4.3 and 4.4 respectively.

(l) Routine to clear screen Line numbers 9000–9070

This routine is used to clear the screen and provide two lines of text at the head of each screen giving the title of the selected software suite and the selected program, together with the date and time of calculation.

(m) Routine to display error message, etc. Line numbers 9100–9330

If a wrong keyboard entry has been made, the error is trapped and an error message displayed and re-entry requested.

(n) Termination of program Line numbers 9500–9620

When the **QUIT** selection is made in the program, an end-of-program message is displayed to advise the user that computations are complete.

(o) Plotter routine for Epson HI-80 Line numbers 10200–10950

This routine is similar to (i) where the commands are altered as shown in Table 2.1.

(p) Plotter routine for Apple 410 Line numbers 11200–11950

This routine is similar to (i) where the commands are altered as shown in Table 2.1.

(q) Disc write and recall routines Line numbers 20000–20770

These routines allow the user to store the curve design data for subsequent computation and plotting as shown in Fig. 4.16.

4.5.2 PC-RODEOS – numeric variables

A	= Angle or algebraic difference in gradients
A∅	= Chord setting out angle
A1–A3	= Angles of triangle
A9	= Angle solution for arcsine
A(5∅)	= Array containing setting-out angle
B	= Polar angle for given point
BT	= Distance *BT* in Fig. 4.4
C	= Rate of change of radial acceleration
C1	= First chord on curve
C2	= Last chord on curve
C7, **C8**	= Variables to calculate numbers of chords
C9	= Chainage of last point on curve
CA	= Storage for chainage **CI**
CB	= Chainage to *B*, highest/lowest point on curve
CCOL	= Column number for text display on screen
CF	= Chainage of **R1**
CH	= Chord length
CH(1∅∅)	= Array of chord chainages
CI	= Chainage of IP
CIT	= Chainage of IP for transition
CK	= Counter in error check routine
CL	= Curve length
CL(1∅∅)	= Array for distance along curve
CPLX	= *X*-coordinate of circular tangent point (left)
CPLY	= *Y*-coordinate of circular tangent point (left)
CPRX	= *X*-coordinate of circular tangent point (right)
CPRY	= *Y*-coordinate of circular tangent point (right)
CT	= Choice of transition or circular curve
CT1	= Storage for chainage of first tangent (**F1**)
CT2	= Storage for chainage of second tangent (**F2**)
D	= Degrees of angle
D5	= Degrees of angle
D9	= Angles in decimals ⟨1⟩ or DD.MMSS ⟨2⟩
DA	= Deflection angle D_F
DA(5∅)	= Array of angles used in offsets from long chord solution
DC	= Angular arc for circular curve
DF	= Deflection angle at IP
DF(5∅)	= Array of deflection angles at centre of curve
DFT	= Deflection angle (total for transition)
DI	= Internal angle at IP
DISP%	= Choice of display – screen or plotter
DT	= Angle in triangle *HIT* (Fig. 4.4)
EM	= External distance at mid-ordinate
F	= Centripetal force ratio
F1–F4	= Chainages of tangent points

FT1	= Chainage at end of first transition
FT2	= Chainage at start of second transition
G1	= Gradient of first straight
G2	= Gradient of second straight
GVX	= Acceleration due to gravity
GX	= *X*-coordinate at lower left-hand corner of plot
GY	= *Y*-coordinate at lower left-hand corner of plot
H1	= Distance R_1 to R_3
H2	= Distance R_1 to IP
H3	= Distance R_3 to IP
HB	= Distance *HB* in Fig. 4.4
HD(1ØØ)	= Vertical height difference array along curve
HI	= Distance *HI* in Fig. 4.4.
HP	= Distance *HP* in Fig. 4.4
HP%	= Integer variable to identify plotter
HT	= Distance *HT* in Fig. 4.4
HTP(5Ø)	= Array for horizontal distance from tangent point
HV	= Vertical height difference along curve
I	= Loop variable
IN	= Loop variable
IPX	= *X*-coordinate of intersection point
IPY	= *Y*-coordinate of intersection point
J	= Loop variable
K	= Multiplying constant
L,LØ	= Length of vertical curve
L1,L2	= Sub-tangent lengths
LC	= Length of long chord
LF	= Polar distance to end point
LF(5Ø)	= Polar distance array
LFF(5Ø)	= Storage of above **LF** array
LM	= Minimum length of vertical curve
MO	= Mid-ordinate
MX	= Multiplication factor for Epson HI-80 plotter
N,N1,NC,NLC,NT	= Variables to hold number of chords or offsets
P	= Half perimeter of triangle
PØ–P9	= Flags and counters in routines
PC	= Type of known data
PH	= Total angle ϕ at end point
PH(5Ø)	= Polar angle array for setting-out
PHD(5Ø)	= Storage for first **PH** array
PHN(5Ø)	= Storage for second **PH** array
PT	= Total angle ϕ set out
PX	= *X*-coordinate at upper right-hand corner of plot
PY	= *Y*-coordinate at upper right-hand corner of plot
R	= Minimum radius
R(2)	= Array to hold point identification numbers
R1	= Point identification number

R3	= Point identification number
R9	= 57.2957795
RA	= Reduced level of point *A*
RB	= Reduced level of point *B*
RC	= Reduced level of point *C*
RGB%	= Integer variable to select screen colour
RL(1ØØ)	= Array to hold reduced level of points
S,SØ	= Seconds of angle
S1–S3	= Sides of triangle
S9	= Angle solution for arcsine
SCD%	= Monitor display, monochrone/colour
SCR%	= Screen resolution (installation file)
SCRM	= Scale magnification to fit screen/plotter
SF	= Full scale transition curve length
SM	= Model scale transition curve length
SM(5Ø)	= Array of model scale measurements
SR%	= Screen resolution
SX	= Horizontal display width in user units
SY	= Vertical display height in user units
T1–T3	= Temporary angle storage
TANG	= Half angle D_F
TC	= Total transition curve length
TD	= Tangent distance in model
TH	= Angle θ
THD	= Storage of angle θ in DD.MMSS
TL	= Tangent distance at full scale
TPF1	= Chainage of first tangent point
TPF2	= Chainage of last tangent point
TPLX	= *X*-coordinate of transition tangent point (left)
TPLY	= *Y*-coordinate of transition tangent point (left)
TPRX	= *X*-coordinate of transition tangent point (right)
TPRY	= *Y*-coordinate of transition tangent point (right)
TTL	= Storage for transition tangent length *TL*
TW	= Horizontal distance between transition tangent points
U(5Ø)	= Array of chord or tangent distances
U1–U3	= Angles of triangle
U8,U9	= Angles used in arcsine solutions
UM	= Maximum tangent distance at mid-ordinate
V	= Design velocity (ms^{-1})
V(5Ø)	= Array of chord or tangent offset distances
VØ	= Design velocity (km hr^{-1})
V1,V3	= Computed decimal angles
VH	= Vertical height of centre line in display
VM	= Tangent offset at mid-ordinate
VTP(5Ø)	= Array for vertical distance from tangent point
W	= Offset interval
X(2)	= Array to hold *X*-coordinate of points
X1	= *X*-coordinate of point R_1

X3	= X-coordinate of point R_3
XF	= Full scale value of X at end of transition
XF(5Ø)	= Array of X-measurements
XM	= Model scale value of X at end of transition
XM(5Ø)	= Array of model scale X-measurements
XP	= X-distance to point P
Y(1ØØ)	= Distance along curve array
Y1	= Y- coordinate at point R_1
Y3	= Y-coordinate of point R_3
YF	= Full scale value of Y at end of transition
YF(5Ø)	= Array of Y-measurements
YM	= Model scale value of Y at end of transition
YM(5Ø)	= Array of model scale Y-measurements
YP	= Y-distance to point P
Z1,Z3	= Entered horizontal angles
ZP	= Z-distance to point P

4.5.3 PC-RODEOS – string variables

A$	Storage for location of curve
B$	Storage for operator's name
C$	Storage for date of calculation
CC$	= **"RODEOS Expert"**
DATE$	Variable used to return the date of the computer's calendar/clock in US format MM–DD–YYYY
DD$	= **"(DD.DEC)"** or **"(DD.MMSS)"** depending on angle entry
EØ$	= **"SORRY, DATA ERROR . . . PLEASE RE-ENTER"**
H$	= **"RODEOS CURVE DESIGN PROGRAM"**
HC$	= **"CIRCULAR CURVE SETTING OUT PROGRAM"**
K$	Storage for instrument used
L$	= **"L"** for left and **"R"** for right-handed curve
P$	= **"DO YOU WISH PRINTOUT OF ABOVE DATA (Y/N)"**
P1$	= **"I.P. INACCESSIBLE"**
P2$	= **"TANGENT TO 3 STRAIGHTS"**
P3$	= **"CURVE THROUGH 3 POINTS"**
P4$	= **"CURVE THROUGH A GIVEN POINT"**
PCF$	= **"PCSCREEN"**
PR$	= **"P"** for polar and **"R"** for rectangular coordinates
Q$	Answer response to question (Y/N)
R$	Variable for transition data
S$	Variable for deflection angle
T$	= **"PC-Curves"**
TIME$	Variable used to return the time of the built-in clock in HH:MM:SS format
X$	= **"ENTER X(E) COORD.OF STN."**
XX$	= Answer response to phasing of curves (Y/N)
Y$	= **"ENTER Y(N) COORD.OF STN."**

4.5.4 PC-RODEOS – BASIC program

```
1Ø  REM  <PC-RODEO> Expert Program for IBM PC or Compatible
11  REM  VERSION PC-1.Ø, Ø1-Ø1-87
12  REM  (C) COPYRIGHT  1984-87  P.H.Milne
16  REM  ALL RIGHTS RESERVED
2Ø  CLEAR:KEY OFF
25 CH = Ø:CK = Ø:CT=Ø:P7 = Ø:P8=Ø:P9 = Ø:DISP%=Ø:SR%=1
28 CØ = Ø:P1 = Ø:P2 = Ø:VØ = Ø:CL = Ø:TL = Ø:GVX = 9.81ØØØ1
3Ø A = Ø:B = Ø:E = Ø:G = Ø:J = Ø:K = Ø:N = Ø:S = Ø:U = Ø:V = Ø:W = Ø
35  DIM A(5Ø),CH(1ØØ),CL(1ØØ),DA(5Ø),DF(5Ø),HD(1ØØ),RL(1ØØ),U(5Ø),V(5
Ø),Y(1ØØ)
38  DIM LF(5Ø),PH(5Ø),SM(5Ø),XF(5Ø),XM(5Ø),YF(5Ø),YM(5Ø),PHD(5Ø),PHN(
5Ø),HTP(5Ø),VTP(5Ø),LFF(5Ø)
4Ø EØ$ = "SORRY, DATA ERROR ... PLEASE RE-ENTER."
42 H$ = " RODEOS CURVE DESIGN PROGRAM"
45 P$ = "DO YOU WISH PRINTOUT OF ABOVE DATA(Y/N)"
5Ø PCF$="PCSCREEN"
52 OPEN "I",#3,PCF$
54 INPUT#3,SCR%
56 INPUT#3,SCD%
58 CLOSE#3
6Ø ON (SCR%+1) GOTO 62,64,66
62 SCREEN Ø:WIDTH 4Ø:GOTO 8Ø
64 SCREEN 1:GOTO 68
66 SCREEN 1:GOTO 8Ø
68 IF SCD%=1 THEN 8Ø
7Ø COLOR 8,Ø
8Ø  DEF FN C9(C9) = (1.57Ø7964# - ATN(C9/SQR(1 - C9 * C9)))*R9
9Ø  DEF FN S9(S9) = ATN(S9/SQR(1 - S9*S9))*R9
1ØØ  GOSUB 9ØØØ:PRINT H$
11Ø  PRINT:PRINT " Are curve details known (y/n)";:INPUT Q$:IF Q$="n"
 THEN 16Ø
12Ø  IF Q$ <> "y" THEN 1ØØ
13Ø  PRINT:PRINT " Are details stored on disc (y/n)";:INPUT Q$:IF Q$=
"n" THEN 16Ø
14Ø  IF Q$ <> "y" THEN 13Ø
15Ø  GOTO 2Ø5ØØ
16Ø  PRINT:PRINT "ARE ANGLES IN DECIMALS <1>, DD.MMSS <2> OR DO YOU W
ISH TO QUIT <3>."
17Ø  INPUT D9:IF D9 < 1 OR D9 > 3 THEN 16Ø
18Ø  ON D9 GOTO 185,185,95ØØ
185 R9 = 57.2957795#:P1 = Ø
19Ø IF D9=1 THEN DD$="(DD.DEC)"
195 IF D9=2 THEN DD$="(DD.MMSS)"
2ØØ  GOSUB 9ØØØ
21Ø  PRINT H$
215  PRINT "****************************************"
22Ø  PRINT
225  PRINT "ENTER LOCATION OF WORK, ETC. WHEN":PRINT "REQUESTED, AND
PRESS <RETURN>"
23Ø  PRINT "ENTER LOCATION OF CURVE :-"
235  INPUT A$
24Ø  PRINT "ENTER OPERATOR'S NAME :-"
245  INPUT B$
25Ø  PRINT "ENTER DATE OF CALCULATION AS DD/MM/YY :-"
255  INPUT C$
26Ø  PRINT "ENTER INSTRUMENT USED :-"
265  INPUT K$
27Ø  PRINT
275  PRINT "DO YOU WISH PRINTOUT OF LOCATION ETC.,  (Y/N) ";
28Ø  INPUT Q$
285  IF Q$ = "Y" THEN 6ØØØ
29Ø  IF Q$ <  > "N" THEN 275
```

```
300  GOSUB 9000:PRINT H$
301 PRINT:PRINT "SELECT TYPE OF CURVE  :-":PRINT "HORIZONTAL COMBINAT
ION (H) OR VERTICAL":PRINT "ONLY (V) ";:INPUT Q$
302 IF Q$<>"H" AND Q$<>"V" THEN 301
303 IF Q$="V" GOTO 3100
305  PRINT:PRINT "IS DEFLECTION ANGLE KNOWN (Y/N)";: INPUT S$
310  IF S$ <  > "Y" AND S$ <  > "N" THEN 305
315  PRINT : PRINT "ARE EITHER (1) RADIUS, (2) DEGREE(100M)":PRINT "O
R (3) TANGENT LENGTH KNOWN, OR OTHER"
316  PRINT "TRANSITION CURVE DATA KNOWN (Y/N) ";: INPUT R$
320  IF R$ <  > "Y" AND R$ <  > "N" THEN 315
330  IF S$ = "Y" AND R$ = "Y" THEN 1000
335  GOSUB 9000
340  PRINT : PRINT "SELECT CURVE PROGRAM FROM MENU :-"
345  PRINT
350  PRINT "    <1> I.P. INACESSIBLE, OR":PRINT "         LINEAR MEASU
REMENT ONLY"
355  PRINT "    <2> TANGENT TO 3 STRAIGHTS"
360  PRINT "    <3> CURVE THROUGH 3 POINTS"
365  PRINT "    <4> CURVE THROUGH 1 POINT"
370  PRINT "    <5> QUIT"
375  INPUT P1: IF P1 < 1 OR P1 > 5 THEN 335
380  ON P1 GOTO 500,650,800,900,9500
500  GOSUB 9000: PRINT H$: PRINT "***********************************
*****"
502  PRINT : PRINT "SELECT ";TAB(10);"<1> I.P. INACESSIBLE,": PRINT "
OR";TAB(10);"<2> LINEAR MEASUREMENT ONLY";: INPUT P0: IF P0 < 1 OR P0
 > 2 THEN 502
504  ON P0 GOTO 506,576
506 P1$ = "<1> I.P. INACESSIBLE": PRINT : PRINT P1$: PRINT "=========
============================="
508  PRINT : PRINT "ENTER REF. NO. OF 1ST POINT ";: INPUT R1
510  PRINT "ENTER CHAINAGE OF ";R1;: INPUT CF
512  IF CF < 0 THEN 510
514  PRINT : PRINT "ENTER REF. NO. OF 2ND POINT ";: INPUT R3
516  IF R3 = R1 THEN 514
518  PRINT : PRINT "ARE COORDS OF ";R1;" AND ";R3;" KNOWN (Y/N) ";: I
NPUT Q$
520  IF Q$ = "N" THEN 534
522  IF Q$ <  > "Y" THEN 518
524  PRINT : PRINT X$;R1;: INPUT X1
526  PRINT Y$;R1;: INPUT Y1
528  PRINT : PRINT X$;R3;: INPUT X3
530  PRINT Y$;R3;: INPUT Y3
532 H1 =  SQR ((X3 - X1) ^ 2 + (Y3 - Y1) ^ 2): GOTO 538
534  PRINT : PRINT "ENTER DISTANCE ";R1;"-";R3;: INPUT H1
536  IF H1 <  = 0 THEN 534
538  IF P1 = 2 THEN 658
540  PRINT : PRINT "ENTER ANGLE 1ST STR.-";R1;"-";R3;: INPUT Z1
542  IF Z1 <  = 0 OR Z1 > 179.999 THEN 540
544 V1 = Z1: IF D9 = 1 THEN 550
546 D5 = Z1: GOSUB 8000: IF CK = 1 THEN 540
548 V1 = D5
550  PRINT "ENTER ANGLE 2ND STR.-";R3;"-";R1;: INPUT Z3
552  IF Z3 <  = 0 OR Z3 > 179.999 THEN 550
554 V3 = Z3: IF D9 = 1 THEN 560
556 D5 = Z3: GOSUB 8000: IF CK = 1 THEN 550
558 V3 = D5
560 U1 = 180 - V1:U3 = 180 - V3
562 DF = U1 + U3
564 A1 = U1:S1 = H1:A3 = U3: GOSUB 7750
566 H2 = S2:H3 = S3:U2 = A2
568 CI = CF + H2
570 CI =  INT (CI * 1000 + .5) / 1000
572  IF P1 = 2 THEN 688
```

```
574  GOTO 596
576 P1$ = "<1> LINEAR MEASUREMENT ONLY": PRINT : PRINT P1$
578  PRINT "========================================"
580  PRINT : PRINT "ENTER CHAINAGE OF I.P.  ";: INPUT CI
582  IF CI < 0 THEN 580
584  PRINT : PRINT "ENTER DIST. ALONG STRAIGHTS ";: INPUT H2
586  IF H2 <  = 0 THEN 584
588  PRINT : PRINT "ENTER DIST. JOINING STRAIGHTS ";: INPUT H1
590  IF H1 <  = 0 THEN 588
592 U9 = H1 / (2 * H2)
594 U1 =  FN C9(U9):DF = 2 * U1
596  IF D9 = 1 THEN 600
598 D = DF: GOSUB 8050:DF = D: GOTO 602
600 DF =  INT (DF * 1000 + .5) / 1000
602  GOSUB 9000
604  PRINT P1$
606  PRINT "****************************************"
608  PRINT "DEFLECTION ANGLE ";TAB(20);" = ";DF
610  PRINT "CHAINAGE OF I.P. ";TAB(20);" = ";CI
612  PRINT:PRINT P$;: INPUT Q$
614  IF Q$ = "Y" THEN 6100
616  IF Q$ <  > "N" THEN 612
618  GOTO 1000
650  GOSUB 9000:P2$ = "<2> TANGENT TO 3 STRAIGHTS"
652  PRINT H$: PRINT P2$
654  PRINT "****************************************"
656  GOTO 508
658  PRINT : PRINT "ENTER <1> FOR DISTANCES TO I.P. <2> FOR ANGLES AT
 ";R1;" AND ";R3
660  INPUT P2: IF P2 < 1 OR P2 > 2 THEN 658
662  IF P2 = 2 THEN 540
664  PRINT "ENTER DISTANCE ";R1;"-I.P.";: INPUT H2
666  IF H2 <  = 0 THEN 664
668  PRINT "ENTER DISTANCE ";R3;"-I.P.";: INPUT H3
670  IF H3 <  = 0 THEN 668
672  IF (H2 + H3) > H1 THEN 678
674  GOSUB 9100
676  GOTO 664
678 S1 = H1:S2 = H2:S3 = H3: GOSUB 7800
680 U1 = A1:U2 = A2:U3 = A3
682 CI = CF + H2
684 CI =  INT (CI * 1000 + .5) / 1000
686 DF = U1 + U3
688 R = H1 / ( TAN (U1 / (2 * R9)) +  TAN (U3 / (2 * R9)))
690 L1 =  INT ((R *  TAN (U1 / (2 * R9))) * 1000 + .5) / 1000
692 L2 =  INT ((R *  TAN (U3 / (2 * R9))) * 1000 + .5) / 1000
694 CL =  INT ((R * DF / R9) * 1000 + .5) / 1000
696 F1 = CF - L1:F2 = F1 + CL
698 TL =  INT ((R *  TAN (DF / (2 * R9))) * 1000 + .5) / 1000
700 LC =  INT ((2 * R *  SIN (DF / (2 * R9))) * 1000 + .5) / 1000
702 MO =  INT ((R * (1 -  COS (DF / (2 * R9)))) * 1000 + .5) / 1000
704 EM =  INT ((TL *  TAN (DF / (4 * R9))) * 1000 + .5) / 1000
706 R =  INT (R * 1000 + .5) / 1000
708  IF D9 = 1 THEN 712
710 D = DF: GOSUB 8050:DF = D: GOTO 714
712 DF =  INT (DF * 1000 + .5) / 1000
714  GOSUB 9000:P7 = 0
716  PRINT H$: PRINT P2$: PRINT "************************************
****"
718  PRINT "RADIUS";TAB(20);" = "; INT (R * 1000 + .5) / 1000
720  PRINT "DEFLECTION ANGLE";TAB(20);" = ";DF
722  PRINT "CHAINAGE I.P.";TAB(20);" = ";CI
724  PRINT "TANGENT LENGTH";TAB(20);" = ";TL
726  PRINT "CHAINAGE TP1";TAB(20);" = "; INT (F1 * 1000 + .5) / 1000
728  PRINT "CURVE LENGTH";TAB(20);" = ";CL
```

```
730  PRINT "CHAINAGE TP2";TAB(2Ø);" = "; INT (F2 * 1ØØØ + .5) / 1ØØØ
732  IF P1 <  > 2 THEN 74Ø
734  PRINT "1ST TANGENT L1";TAB(2Ø);" = ";L1
736  PRINT "2ND TANGENT L2";TAB(2Ø);" = ";L2
738  PRINT "DISTANCE R1-R3";TAB(2Ø);" = ";INT(H1 * 1ØØØ + .5)/1ØØØ
74Ø  PRINT "LONG CHORD";TAB(2Ø);" = ";LC
742  PRINT "MID-ORDINATE";TAB(2Ø);" = ";MO
744  PRINT "EXTERNAL MO-DIST";TAB(2Ø);" = ";EM
745  IF P1 = 4 THEN  RETURN
746  IF P7 = 1 THEN  RETURN
748  PRINT P$;: INPUT Q$
75Ø  IF Q$ = "Y" THEN 6136
752  IF Q$ <  > "N" THEN 748
754  GOTO 1ØØØ
8ØØ  GOSUB 9ØØØ:P3$ = "<3> CURVE THROUGH 3 POINTS"
8Ø2  PRINT H$: PRINT P3$
8Ø4  PRINT "****************************************"
8Ø6  FOR I = Ø TO 2
8Ø8  PRINT "ENTER REF. NO. OF POINT ";I + 1;: INPUT R(I)
81Ø  NEXT I
812  PRINT "ARE COORDS OF 3 POINTS KNOWN (Y/N)";
814  INPUT Q$
816  IF Q$ = "N" THEN 836
818  IF Q$ <  > "Y" THEN 812
82Ø  FOR J = Ø TO 2
822  PRINT X$;R(J);: INPUT X(J)
824  PRINT Y$;R(J);: INPUT Y(J)
826  NEXT J
828 H1 =  SQR ((X(Ø) - X(2)) ^ 2 + (Y(Ø) - Y(2)) ^ 2)
83Ø H2 =  SQR ((X(1) - X(Ø)) ^ 2 + (Y(1) - Y(Ø)) ^ 2)
832 H3 =  SQR ((X(2) - X(1)) ^ 2 + (Y(2) - Y(1)) ^ 2)
834  GOTO 854
836  PRINT "ENTER DISTANCE ";R(Ø);"-";R(2);: INPUT H1
838  IF H1 <  = Ø THEN 836
84Ø  PRINT "ENTER DISTANCE ";R(Ø);"-";R(1);: INPUT H2
842  IF H2 <  = Ø THEN 84Ø
844  PRINT "ENTER DISTANCE ";R(1);"-";R(2);: INPUT H3
846  IF H3 <  = Ø THEN 844
848  IF (H2 + H3) > H1 THEN 854
85Ø  GOSUB 91ØØ
852  GOTO 836
854 S1 = H1:S2 = H2:S3 = H3
856  GOSUB 78ØØ
858 U1 = A1:U2 = A2:U3 = A3
86Ø R = H1 / (2 *  SIN (U2 / R9))
862 R =  INT (R * 1ØØØ + .5) / 1ØØØ
864  PRINT : PRINT P3$:P7 = Ø
866  PRINT "****************************************"
868  PRINT "RADIUS = ";R
87Ø  PRINT "DISTANCE ";R(Ø);"-";R(2);TAB(2Ø);" = "; INT (H1 * 1ØØØ +
.5) / 1ØØØ
872  PRINT "DISTANCE ";R(Ø);"-";R(1);TAB(2Ø);" = "; INT (H2 * 1ØØØ +
.5) / 1ØØØ
874  PRINT "DISTANCE ";R(1);"-";R(2);TAB(2Ø);" = "; INT (H3 * 1ØØØ +
.5) / 1ØØØ: IF P7 = 1 THEN 6162
876  PRINT P$;: INPUT Q$
878  IF Q$ = "Y" THEN 6152
88Ø  IF Q$ <  > "N" THEN 876
882  GOTO 1ØØØ
9ØØ  GOSUB 9ØØØ:P4$ = "<4> CURVE THROUGH A GIVEN POINT"
9Ø2  PRINT H$: PRINT P4$
9Ø4  PRINT "****************************************"
9Ø6  PRINT "ENTER <1> FOR X AND Y DISTANCES OR <2> FOR POLAR COORDS."
;
9Ø8  INPUT P4: IF P4 < 1 OR P4 > 2 THEN 9Ø6
```

```
910  IF P4 = 2 THEN 922
912  PRINT "ENTER DIST. ALONG TANGENT ";: INPUT XP
914  IF XP <  = 0 THEN 912
916  PRINT "ENTER DIST. PERP. TANGENT ";: INPUT YP
918  IF YP <  = 0 THEN 916
920 ZP =  SQR (XP ^ 2 + YP ^ 2):B =  ATN (YP / XP) * R9: GOTO 936
922  PRINT "ENTER DIST. FROM I.P. TO POINT ";: INPUT ZP
924  IF ZP <  = 0 THEN 922
926  PRINT "ENTER ANGLE FROM TANGENT ";: INPUT B
928  IF B <  = 0 OR B > 179.999 THEN 926
930  IF D9 = 1 THEN 936
932 D5 = B: GOSUB 8000: IF CK = 1 THEN 926
934 B = D5
936  PRINT P$;: INPUT Q$: IF Q$ = "Y" THEN 6168
938  IF Q$ <  > "N" THEN 936
940  PRINT "ENTER DEFLECTION ANGLE ";: INPUT DF: IF DF <  = 0 OR DF >
 179.999 THEN 940
942  IF D9 = 1 THEN 948
944 D5 = DF: GOSUB 8000: IF CK = 1 THEN 940
946 DF = D5
948 U1 = 90 - (B + DF / 2): IF U1 < .01 THEN 956
950 U9 =  SIN (U1 / R9) /  SIN ((B + U1) / R9)
952 U8 =  FN S9(U9):U2 = U8 - U1:U3 = 180 - U1 - U2
954 R = ZP *  SIN (U1 / R9) /  SIN (U2 / R9): GOTO 958
956 R = ZP / (1 /  COS (DF / (2 * R9)) - 1)
958 TL =  INT ((R *  TAN (DF / (2 * R9))) * 1000 + .5) / 1000
960 CL =  INT ((R * DF / R9) * 1000 + .5) / 1000
962 LC =  INT ((2 * R *  SIN (DF / (2 * R9))) * 1000 + .5) / 1000
964 MO =  INT ((R * (1 -  COS (DF / (2 * R9)))) * 1000 + .5) / 1000
966 EM =  INT ((TL *  TAN (DF / (4 * R9))) * 1000 + .5) / 1000
968  PRINT "IS CHAINAGE OF I.P. KNOWN (Y/N)";: INPUT Q$
970  IF Q$ = "N" THEN 982
972  IF Q$ <  > "Y" THEN 968
974  PRINT "ENTER CHAINAGE OF I.P. ";: INPUT CI
976  IF CI < 0 THEN 974
978 F1 = CI - TL
980 F2 = F1 + CL: GOTO 984
982 F1 = 0:CI = TL:F2 = CL
984  IF D9 = 1 THEN 988
986 D = DF: GOSUB 8050:DF = D
988  PRINT H$: PRINT P4$:P7 = 0: PRINT "******************************
***********"
990  GOSUB 718
994  PRINT P$;: INPUT Q$: IF Q$ = "Y" THEN 6190
996  IF Q$ <  > "N" THEN 994
1000  GOSUB 9000
1002  PRINT H$
1004  PRINT : PRINT "SELECT TYPE OF CURVE :-"
1006  PRINT "TRANSITION (T), OR CIRCULAR (C) ";
1008  INPUT Q$
1010  IF Q$ <  > "T" AND Q$ <  > "C" THEN 1004
1012  IF Q$ = "T" THEN CT = 1
1014  IF Q$ = "C" THEN CT = 2
1016  ON CT GOTO 1100,2000
1100  GOSUB 9000: PRINT H$: PRINT "**********************************
******"
1102  PRINT : PRINT "SELECT CURVE DESIGN CRITERIA :-"
1104  PRINT "VELOCITY (V),"
1106  PRINT "RATE OF CHANGE OF RADIAL ACC. (C),"
1108  PRINT "CENTRIPETAL RATIO (F), MIN. RADIUS (R)"
1110  PRINT : PRINT "<1> V,C,F : <2> V,C,R : <3> F,C,R ";
1112  INPUT P1: IF P1 < 1 OR P1 > 3 THEN 1100
1114  PRINT : PRINT "ENTER DEFLECTION ANGLE ";: INPUT DF
1116  IF DF <  = 0 OR DF > 179.999 THEN 1114
1117 DFT=DF
```

```
1118  IF D9 = 1 THEN 1124
1120 D5 = DF: GOSUB 8000: IF CK = 1 THEN 1114
1122 DF = D5:TANG=DF/2
1124  GOSUB 9000: PRINT H$
1126  PRINT "****************************************"
1128  IF P1 = 3 THEN 1134
1130  PRINT : PRINT "ENTER DESIGN VELOCITY (KM/HR) ";: INPUT V0
1132  IF V0 <  = 0 THEN 1130
1134  PRINT : PRINT "ENTER RATE OF CHANGE OF RADIAL          ACCELERA
TION (M/S3) ";: INPUT C
1136  IF C <  = 0 THEN 1134
1138  IF P1 = 2 THEN 1146
1140  PRINT : PRINT "ENTER CENTRIPETAL RATIO ";: INPUT F
1142  IF F <  = 0 THEN 1140
1144  IF P1 = 1 THEN 1150
1146  PRINT : PRINT "ENTER MINIMUM RADIUS (M) ";: INPUT R
1148  IF R <  = 0 THEN 1146
1150  ON P1 GOTO 1152,1156,1160
1152 V = V0 * 10 / 36:R = V * V /(F*GVX)
1154  GOTO 1162
1156 V = V0 * 10 / 36:F = V * V /(R*GVX)
1158  GOTO 1162
1160 V =  SQR (F * R * GVX)
1162 K=SQR(V^3/C):SF=F*V*GVX/C:P2=0
1164 TH = (SF * SF / (2 * K * K)) * R9
1166  IF TH > (DF / 2) THEN 1172
1168 DC = DF - 2 * TH:CL = R * DC / R9
1170 TC = CL + 2 * SF: GOTO 1216
1172  GOSUB 9100
1174  PRINT "DO YOU WISH TO ENTER NEW CURVE DESIGN":PRINT "CRITERIA (
Y/N): IF (N) TRANSITIONS WILL":PRINT "MEET AT CENTRE ";: INPUT Q$
1176  IF Q$ = "Y" THEN 1100
1178  IF Q$ <  > "N" THEN 1174
1180 TH = DF / 2:SM =  SQR (DF / R9)
1182 XM = SM - (SM ^ 5 / 40):YM = SM ^ 3 / 6 - (SM ^ 7 / 336)
1184 TD = XM + YM *  TAN (TH / R9)
1186  PRINT : PRINT "IS TANGENT DISTANCE FROM I.P. KNOWN":PRINT "(Y/N
) ";: INPUT Q$
1188  IF Q$ = "N" THEN 1204
1190  IF Q$ <  > "Y" THEN 1186
1192  PRINT : PRINT "ENTER TANGENT DISTANCE (M) ";: INPUT TL
1194  IF TL <  = 0 THEN 1192
1196 P2 = 1:K = TL / TD:TTL = TL
1198 V = (K * K * C) ^ (1 / 3):R = V * V / (F * GVX)
1200 SF = SM * K:XF = K * XM:YF = K * YM
1202 TC = 2 * SF: GOTO 1216
1204  PRINT : PRINT "DO YOU WISH TO SPECIFY MIN. RADIUS (Y/N) IF (N)
PROGRAM WILL REVERT TO START";
1206  INPUT Q$: IF Q$ = "N" THEN 100
1208  IF Q$ <  > "Y" THEN 1204
1210  PRINT : PRINT "ENTER MINIMUM RADIUS (M) ";: INPUT R
1212  IF R <  = 0 THEN 1210
1214 P1 = 3: GOTO 1160
1216  IF D9 = 1 THEN 1226
1218 D = DF: GOSUB 8050:DF = D
1220 D = TH: GOSUB 8050:TH = D:THD=TH
1222 D = DC: GOSUB 8050:DC = D
1224  GOTO 1230
1226 TH =  INT (TH * 1000 + .5) / 1000
1228 DC =  INT (DC * 1000 + .5) / 1000
1230  GOSUB 9000: PRINT H$:P7 = 0
1232  PRINT "****************************************"
1234  PRINT "DESIGN CRITERIA FOR CURVE :-"
1236  IF V0 = 0 THEN 1240
1238  PRINT "VELOCITY (KPH)             = ";V0
```

```
1240  PRINT "VELOCITY (MPS)               = ";INT(V * 1000 + .5)/1000
1242  PRINT "RADIAL ACCELERATION (MPS3)= ";C
1244  PRINT "CENTRIPETAL RATIO            = ";INT(F * 1000 + .5)/1000
1246  PRINT "MINIMUM RADIUS (M)           = ";INT(R * 1000 + .5)/1000
1248  PRINT
1250  PRINT "MULTIPLYING CONSTANT         = ";INT(K * 1000 + .5)/1000
1252  PRINT "DEFLECTION ANGLE AT I.P.     = ";DF
1254  PRINT "TRANSITION CURVE ANGLE       = ";TH
1256  PRINT "CIRCULAR CURVE ANGLE         = ";DC
1258  PRINT "SPIRAL TRANSITION LENGTH     = ";INT(SF * 1000 + .5)/1000
1260  PRINT "CIRCULAR CURVE LENGTH        = ";INT(CL * 1000 + .5)/1000
1262  PRINT "TOTAL CURVE LENGTH           = ";INT(TC * 1000 + .5)/1000
1264  IF TL = 0 THEN 1268
1266  PRINT "MEASURED TANGENT LENGTH    = ";TL
1268  PRINT "*****************************************"
1270  IF P7 = 1 THEN  RETURN
1272  PRINT P$;: INPUT Q$
1274  IF Q$ = "Y" THEN 6900
1276  IF Q$ <  > "N" THEN 1272
1278  GOSUB 9000: PRINT H$
1280  PRINT "*****************************************"
1282  PRINT : PRINT "DO YOU WISH TO SET OUT CURVE (Y/N) ";: INPUT Q$
1284  IF Q$ = "N" THEN 100
1286  IF Q$ <  > "Y" THEN 1278
1288  PRINT : PRINT "SELECT POLAR (P) OR RECTANGULAR (R)":PRINT "CO-O
RDINATES (P/R) ";: INPUT PR$
1290  IF PR$ <  > "P" AND PR$ <  > "R" THEN 1288
1292  PRINT : PRINT "ENTER CHAINAGE OF I.P. ";: INPUT CI
1294  IF CI < 0 THEN 1292
1295 CIT=CI
1296  PRINT : PRINT "ENTER MULTIPLE CHORD INTERVALS (M) ";: INPUT CH
1298  IF CH <  = 0 THEN 1296
1300  IF SF / CH < 50 THEN 1306
1302  GOSUB 9100: PRINT "TOO SMALL A CHORD INTERVAL !"
1304  GOTO 1296
1306  PRINT : PRINT "IS CURVE LEFT OR RIGHT HANDED (L/R) ";: INPUT L$
: IF L$ <  > "L" AND L$ <  > "R" THEN 1306
1308  IF P2 = 1 THEN 1327
1310 XF = SF - SF ^ 5 / (40 * K ^ 4):YF = SF ^ 3 / (6 * K ^ 2) - SF ^
 7 / (336 * K ^ 6)
1312  IF D9 = 1 THEN 1318
1314 D5 = DF: GOSUB 8000:DF = D5
1316 D5 = TH: GOSUB 8000:TH = D5
1318 HP = YF / ( TAN (TH / R9)):HB = YF / ( SIN (TH / R9))
1320 BT = R *  TAN ((DF / 2 - TH) / R9):HT = HB + BT
1322 DI = (180 - DF) / 2:DT = (180 - TH - DI)
1324 HI = (HT *  SIN (DT / R9)) /  SIN (DI / R9)
1326 TL = HI + XF - HP:TTL=TL
1327 TW=2*TL*COS(TANG/R9):VH=TL*SIN(TANG/R9)
1328 F1 = CI - TL
1329 TPF1=F1
1330 FT1 = F1 + SF
1332 FT2 = FT1 + CL
1334 F2 = F1 + TC
1335 TPF2=F2
1336 CH(0) = ( INT (F1 / CH) + 1) * CH
1338 C1 =  INT ((CH(0) - F1) * 1000 + .5) / 1000
1340 C9 = ( INT (FT1 / CH)) * CH
1342 N =  INT ((SF - C1) / CH):NT=N
1344 CL(0) = C1
1346 F1 =  INT (F1 * 1000 + .5) / 1000:F2 =  INT (F2 * 1000 + .5) / 1
000
1348 FT1 =  INT (FT1 * 1000 + .5) / 1000:FT2 =  INT (FT2 * 1000 + .5)
 / 1000
1350 A = 0:P7 = 0
```

```
1352  GOSUB 9000: PRINT H$: PRINT "************************************
******"
1354  IF PR$ = "P" THEN 1358
1356  PRINT "CHAINAGE";TAB(12);"LENGTH";TAB(21);"TANG.DIST";TAB(32);"
TANG.OFF": GOTO 1360
1358  PRINT "CHAINAGE";TAB(12);"LENGTH";TAB(21);"POLAR DIST";TAB(32);
"ANGLE"
1360  PRINT "=======================================": IF P7 = 1 THE
N  RETURN
1362  IF PR$ = "P" THEN 1366
1364  PRINT F1; TAB( 12);" 0"; TAB( 21);" 0"; TAB( 32);" 0": GOTO 137
0
1366  IF L$ = "L" THEN A = 360
1368  PRINT F1; TAB( 12);" 0"; TAB( 21);" 0"; TAB( 32);A
1370  FOR I = 0 TO N
1372 CH(I) = CH(0) + CH * I
1374 CL(I) = CL(0) + CH * I
1376 XF(I) = CL(I) - (CL(I)) ^ 5 / (40 * K ^ 4)
1378 YF(I) = (CL(I)) ^ 3 / (6 * K ^ 2) - (CL(I)) ^ 7 / (336 * K ^ 6)
1380 LF(I) =  INT (( SQR ((XF(I)) ^ 2 + (YF(I)) ^ 2)) * 1000 + .5) /
1000
1381 LFF(I)=LF(I)
1382 PH(I) =  ATN (YF(I) / XF(I)) * R9
1383 PHD(I)=PH(I)
1384  IF L$ = "L" THEN 1394
1386  IF D9 = 1 THEN 1390
1388 D = PH(I): GOSUB 8050:PH(I) = D: GOTO 1402
1390 PH(I) =  INT ((PH(I)) * 1000 + .5) / 1000
1392  GOTO 1402
1394 PH(I) = 360 - PH(I)
1396  IF D9 = 1 THEN 1400
1398 D = PH(I): GOSUB 8050:PH(I) = D: GOTO 1402
1400 PH(I) =  INT ((PH(I)) * 1000 + .5) / 1000
1402 XF(I) =  INT ((XF(I)) * 1000 + .5) / 1000
1404 YF(I) =  INT ((YF(I)) * 1000 + .5) / 1000
1406  IF PR$ = "P" THEN 1412
1408  PRINT CH(I); TAB( 11);CL(I); TAB( 20);XF(I); TAB( 31);YF(I)
1410  GOTO 1414
1412  PRINT CH(I); TAB( 11);CL(I); TAB( 20);LF(I); TAB( 31);PH(I)
1414  NEXT I
1416 SF =  INT (SF * 1000 + .5) / 1000
1418 XF =  INT (XF * 1000 + .5) / 1000:YF = INT(YF * 1000 + .5)/1000
1420  IF PR$ = "P" THEN 1424
1422  PRINT FT1; TAB( 11);SF; TAB( 20);XF; TAB( 31);YF
1424 LF =  INT (( SQR (XF ^ 2 + YF ^ 2)) * 1000 + .5) / 1000
1426 PH = (R9 * 10 * (SF / K) ^ 2 - 1.213 * (SF / K) ^ 6) / 60
1428  IF L$ = "L" THEN 1436
1430  IF D9 = 1 THEN 1434
1432 D = PH: GOSUB 8050:PT = D: GOTO 1444
1434 PT =  INT (PH * 1000 + .5) / 1000: GOTO 1444
1436 PT = 360 - PH
1438  IF D9 = 1 THEN 1442
1440 D = PT: GOSUB 8050:PT = D: GOTO 1444
1442 PT =  INT (PT * 1000 + .5) / 1000
1444  IF PR$ = "R" THEN 1448
1446  PRINT FT1; TAB( 11);SF; TAB( 20);LF; TAB( 31);PT
1448  PRINT FT2; TAB( 11);"START OF 2ND SPIRAL": PRINT F2; TAB( 11);"
FINAL TANGENT POINT"
1450  PRINT "CHAINAGE OF I.P."; TAB( 20);CI: PRINT "LENGTH OF TANGENT
 "; TAB( 20); INT (TL * 1000 + .5) / 1000
1452  PRINT "======================================="
1454  PRINT P$;: INPUT Q$: IF Q$ = "Y" THEN 7000
1456  IF Q$ < > "N" THEN 1454
1458  GOSUB 9000: PRINT H$
1460  PRINT "****************************************"
```

```
1462  PRINT : PRINT "DO YOU WISH DATA FOR 2ND SPIRAL (Y/N)";: INPUT Q
$
1464  IF Q$ = "N" THEN 2000
1466  IF Q$ <  > "Y" THEN 1462
1468 A = 0:P7 = 1
1470  GOSUB 1352
1472  IF PR$ = "P" THEN 1476
1474  PRINT FT2; TAB( 11);SF; TAB( 20);XF; TAB( 31);YF
1476  IF L$ = "R" THEN 1486
1478  IF D9 = 1 THEN 1482
1480 D = PH: GOSUB 8050:PT = D: GOTO 1494
1482 PT =  INT (PH * 1000 + .5) / 1000
1484  GOTO 1494
1486 A = 360:PT = A - PH
1488  IF D9 = 1 THEN 1492
1490 D = PT: GOSUB 8050:PT = D: GOTO 1494
1492 PT =  INT (PT * 1000 + .5) / 1000
1494  IF PR$ = "R" THEN 1498
1496  PRINT FT2; TAB( 11);SF; TAB( 20);LF; TAB( 31);PT
1498 C8 = ( INT (FT2 / CH) + 1) * CH
1500 C7 = ( INT (F2 / CH)) * CH
1502 N1 =  INT ((C7 - C8) / CH + .5)
1504  FOR IN = 0 TO N1
1506 CH(IN) = C8 + IN * CH
1508 CL(IN) = F2 - CH(IN)
1510 SM(IN) = CL(IN) / K:CL(IN) =  INT ((CL(IN)) * 1000 + .5) / 1000
1512 XM(IN) = SM(IN) - ((SM(IN)) ^ 5 / 40):XF(IN) = K * XM(IN)
1514 YM(IN) = (SM(IN)) ^ 3 / 6 - (SM(IN)) ^ 7 / 336:YF(IN) = K * YM(I
N)
1516 LF(IN) =  INT (( SQR ((XF(IN)) ^ 2 + (YF(IN)) ^ 2)) * 1000 + .5)
 / 1000
1518 XF(IN) =  INT ((XF(IN)) * 1000 + .5) / 1000:YF(IN) =  INT ((YF(I
N)) * 1000 + .5) / 1000
1520 PH(IN) = (R9 * 10 * (SM(IN)) ^ 2 - 1.213 * (SM(IN)) ^ 6) / 60
1521 PHN(IN)=PH(IN)
1522  IF L$ = "R" THEN 1534
1524  IF D9 = 1 THEN 1530
1526 D = PH(IN): GOSUB 8050:PH(IN) = D:
1528  GOTO 1542
1530 PH(IN) =  INT ((PH(IN)) * 1000 + .5) / 1000
1532  GOTO 1542
1534 PH(IN) = 360 - PH(IN)
1536  IF D9 = 1 THEN 1540
1538 D = PH(IN): GOSUB 8050:PH(IN) = D: GOTO 1542
1540 PH(IN) =  INT ((PH(IN)) * 1000 + .5) / 1000
1542  IF PR$ = "P" THEN 1548
1544  PRINT CH(IN); TAB( 11);CL(IN); TAB( 20);XF(IN); TAB( 31);YF(IN)
1546  GOTO 1550
1548  PRINT CH(IN); TAB( 11);CL(IN); TAB( 20);LF(IN); TAB( 31);PH(IN)
1550  NEXT IN
1552  IF PR$ = "P" THEN 1558
1554  PRINT F2; TAB( 12);" 0"; TAB( 21);" 0"; TAB( 31);" 0"
1556  GOTO 1560
1558  PRINT F2; TAB( 12);" 0"; TAB( 21);" 0"; TAB( 31);A
1560  PRINT "****************************************"
1562  PRINT
1564  PRINT P$;: INPUT Q$
1566  IF Q$ = "Y" THEN 7100
1568  IF Q$ <  > "N" THEN 1564
2000 IF P2=1 THEN 3000
2002 GOSUB 9000:PC=0:PRINT H$
2004 IF CT=2 THEN 2104
2006 CT1=F1:CT2=F2
2008 F1=FT1:F2=FT2
2010 DA=DF:DF=DC
```

```
2012 IF D9=1 THEN 2020
2014 D5=DF:GOSUB 8000:DF=D5:P8=1
2016 D5=DA:GOSUB 8000:DA=D5
2020 TL=INT(R*TAN(DF/(2*R9))*1000+.5)/1000
2022 CA=CI:CI=F1+TL
2024 GOTO 2166
2100  GOSUB 9000
2102  PRINT H$
2104  IF P1 = 0 THEN 2108
2106  ON P1 GOTO 2118,9200,2188,9200
2108  PRINT : PRINT "ENTER CHAINAGE OF I.P. ";: INPUT CI
2110  IF CI < 0 THEN 2108
2112  PRINT : PRINT "ENTER DEFLECTION ANGLE ";: INPUT DF
2114  IF DF <  = 0 OR DF > 179.999 THEN 2112
2116  GOTO 2120
2118  PRINT P1$: PRINT "*****************************************"
2120  IF D9 = 1 THEN 2126
2122 D5 = DF: GOSUB 8000: IF CK = 1 THEN 2112
2124 DF = D5:P8=1
2126  PRINT : PRINT "ENTER KNOWN DATA <1> RADIUS, <2> DEGREE(100M), <
3> TANGENT LENGTH";
2128  INPUT PC: IF PC < 1 OR PC > 3 THEN 2126
2130  ON PC GOTO 2132,2140,2156
2132  PRINT : PRINT "ENTER RADIUS ";: INPUT R
2134  IF R <  = 0 THEN 2132
2136 TL =  INT (R *  TAN (DF / (2 * R9)) * 1000 + .5) / 1000
2138  GOTO 2162
2140  PRINT : PRINT "ENTER ANGLE SUBTENDED BY 100M CHORD";: INPUT A
2142  IF A <  = 0 OR A > 179.999 THEN 2140
2144  IF D9 = 1 THEN 2150
2146 D5 = A: GOSUB 8000: IF CK = 1 THEN 2140
2148 A = D5
2150 R =  INT ((R9 * 100 / A) * 1000 + .5) / 1000
2152 TL =  INT (R *  TAN (DF / (2 * R9)) * 1000 + .5) / 1000
2154  GOTO 2162
2156  PRINT : PRINT "ENTER TANGENT LENGTH ";: INPUT TL
2158  IF TL <  = 0 THEN 2156
2160 R =  INT ((TL /  TAN (DF / (2 * R9))) * 1000 + .5) / 1000
2162 CL =  INT ((R * DF / R9) * 1000 + .5) / 1000
2164 F1 = CI - TL:F2 = F1 + CL
2166 LC =  INT ((2 * R *  SIN (DF / (2 * R9))) * 1000 + .5) / 1000
2168 MO =  INT ((R * (1 -  COS (DF / (2 * R9)))) * 1000 + .5) / 1000
2170 EM =  INT ((TL *  TAN (DF / (4 * R9))) * 1000 + .5) / 1000
2172  IF D9 = 1 THEN 2176
2174 D = DF: GOSUB 8050:DF = D:P8=2
2176  GOSUB 9000: PRINT H$:P7 = 1: PRINT "****************************
*************"
2178  IF PC = 2 THEN  PRINT "100M CHORD ANGLE = ";A
2180  GOSUB 718: PRINT "*****************************************"
2182  PRINT P$;: INPUT Q$: IF Q$ = "Y" THEN 6202
2184  IF Q$ <  > "N" THEN 2182
2186  GOTO 9200
2188  PRINT P3$
2190  PRINT "*****************************************"
2192  PRINT : PRINT "ARE ";R(0);" AND ";R(2);" TANGENT POINTS (Y/N)";
2194  INPUT Q$: IF Q$ = "N" THEN 9200
2196  IF Q$ <  > "Y" THEN 2192
2198 DF = 2 * (U1 + U3)
2200 TL =  INT (R *  TAN (DF / (2 * R9)) * 1000 + .5) / 1000
2202 CL =  INT ((R * DF / R9) * 1000 + .5) / 1000
2204  PRINT : PRINT "ENTER CHAINAGE OF ";R(0);: INPUT F1
2206  IF F1 < 0 THEN 2204
2208 CI = F1 + TL:F2 = F1 + CL
2210 LC =  INT ((2 * R *  SIN (DF / (2 * R9))) * 1000 + .5) / 1000
2212 MO =  INT ((R * (1 -  COS (DF / (2 * R9)))) * 1000 + .5) / 1000
```

```
2214 EM =  INT ((TL *  TAN (DF / (4 * R9))) * 1000 + .5) / 1000
2216  IF D9 = 1 THEN 2220
2218 D = DF: GOSUB 8050:DF = D:P8=2: GOTO 2222
2220 DF =  INT (DF * 1000 + .5) / 1000
2222  GOSUB 9000
2224  PRINT H$: PRINT P3$
2226  PRINT "****************************************"
2228 P7 = 1: GOSUB 718
2230  PRINT "****************************************"
2232  PRINT P$;: INPUT Q$: IF Q$ = "Y" THEN 6220
2234  IF Q$ <  > "N" THEN 2232
2236  GOTO 9200
2238  GOSUB 9000:HC$ = "CIRCULAR CURVE SETTING OUT PROGRAM":PRINT HC$
2239 IF P8=1 THEN 2242: REM  DF  in  DD.DEC
2240 D5=DF:GOSUB 8000:DF=D5:P8=1:REM Convert DF in DD.MMSS to DD.DEC
2242  PRINT : PRINT "SELECT SETTING OUT METHOD FROM MENU :-"
2244  PRINT
2246  PRINT "    <1> OFFSETS FROM TANGENT"
2248  PRINT "    <2> OFFSETS FROM LONG CHORD"
2250  PRINT "    <3> TANGENTIAL DEFLECTION ANGLES"
2252  PRINT "    <4> CONTINUE     <5> QUIT"
2254  INPUT P1
2256  IF P1 < 1 OR P1 > 4 THEN 2238
2258  ON P1 GOTO 2260,2338,2418,3000,9260
2260  GOSUB 9000:P1$ = "<1> OFFSETS FROM TANGENT"
2262  PRINT HC$: PRINT P1$
2264  PRINT "****************************************"
2266 UM =  INT ((TL * R / (EM + R)) * 1000 + .5) / 1000
2268 VM =  INT ((EM * R / (EM + R)) * 1000 + .5) / 1000
2270  PRINT : PRINT "MAX. OFFSET AT MID-ORDINATE = ";UM
2272  PRINT : PRINT "SELECT OFFSET INTERVAL ";: INPUT W
2274  IF W <  = 0 THEN 2272
2276 N =  INT (UM / W)
2278  IF N < 50 THEN 2286
2280  GOSUB 9100
2282  PRINT "TOO MANY OFFSETS !"
2284  GOTO 2272
2286  IF D9 = 1 THEN 2290
2288 D = DF: GOSUB 8050:DF = D:P8=2
2290  GOSUB 9000
2292  PRINT H$: PRINT HC$:PRINT P1$
2294  PRINT "****************************************"
2296  PRINT "PT.";TAB(6);"CHAINAGE";TAB(20);"TAN.DIST";TAB(30);"TAN.O
FFSET"
2298  PRINT "=============================================="
2300  PRINT "TP1"; TAB( 6);F1; TAB( 21);" 0"; TAB( 31);" 0"
2302  FOR I = 0 TO N - 1
2304 U(I) = (I + 1) * W
2306 V(I) =  INT ((R -  SQR (R * R - U(I) * U(I))) * 1000 + .5) / 100
0
2308 A9 =  SQR (U(I) * U(I) + V(I) * V(I)) / (2 * R)
2310 DF(I) = 2 *  FN S9(A9)
2312 CL(I) =  INT ((R * DF(I) / R9) * 1000 + .5) / 1000
2314  PRINT (I + 1); TAB( 6);(F1 + CL(I)); TAB( 20);U(I); TAB( 30);V(
I)
2316  NEXT I
2318  PRINT "MO"; TAB( 6); INT ((F1 + CL / 2) * 1000 + .5) / 1000; TA
B( 20);UM; TAB( 30);VM
2320  FOR J = N - 1 TO 0 STEP  - 1
2322  PRINT (2 * N - J); TAB( 6);F2 - CL(J); TAB( 20);U(J); TAB( 30);
V(J)
2324  NEXT J
2326  PRINT "TP2"; TAB( 6);F2; TAB( 21);" 0"; TAB( 31);" 0"
2328  PRINT "****************************************"
2330  PRINT P$;: INPUT Q$
```

```
2332  IF Q$ = "Y" THEN 625Ø
2334  IF Q$ <  > "N" THEN 233Ø
2336  GOTO 2238
2338  GOSUB 9ØØØ:P2$ = "<2> OFFSETS FROM LONG CHORD"
234Ø  PRINT HC$: PRINT P2$
2342  PRINT "****************************************"
2344  PRINT : PRINT "MAX. OFFSET AT MID-ORDINATE = ";LC / 2
2346  PRINT : PRINT "SELECT OFFSET INTERVAL ";: INPUT W
2348  IF W <  = Ø THEN 2346
235Ø N =  INT (LC / (2 * W)):NLC=N
2352  IF N < 5Ø THEN 236Ø
2354  GOSUB 91ØØ
2356  PRINT "TOO MANY OFFSETS !"
2358  GOTO 2346
236Ø F4 =  INT ((F1 + CL / 2) * 1ØØØ + .5) / 1ØØØ
2362  FOR I = Ø TO N - 1
2364 U(I) = (I + 1) * W
2366 V(I) =  INT (( SQR (R * R - U(I) * U(I)) - (R - MO)) * 1ØØØ + .5
) / 1ØØØ
2368 A9 = U(I) / R
237Ø DA(I) =  FN S9(A9)
2372 CL(I) =  INT ((R * DA(I) / R9) * 1ØØØ + .5) / 1ØØØ
2374  NEXT I
2376  IF D9 = 1 THEN 2378
2377 D = DF: GOSUB 8Ø5Ø:DF = D:P8=2
2378 IF DISP%=1 THEN 435Ø
2379 ON HP% GOTO 535Ø,535Ø,1135Ø,1Ø35Ø
238Ø  GOSUB 9ØØØ
2382  PRINT H$:PRINT HC$: PRINT P2$
2384  PRINT "****************************************"
2386  PRINT "PT.";TAB(6);"CHAINAGE";TAB(2Ø);"CH.DIST";TAB(3Ø);"CH.OFF
SET"
2388  PRINT "========================================"
239Ø  PRINT "TP1"; TAB( 6);F1; TAB( 2Ø);LC / 2; TAB( 31);"Ø"
2392  FOR J = N - 1 TO Ø STEP  - 1
2394  PRINT (N - J); TAB( 6);(F4 - CL(J)); TAB( 2Ø);U(J); TAB( 3Ø);V(
J)
2396  NEXT J
2398  PRINT "MO"; TAB( 6);F4; TAB( 21);"Ø"; TAB( 3Ø);MO
24ØØ  FOR K = Ø TO N - 1
24Ø2  PRINT (N + 1 + K); TAB( 6);(F4 + CL(K)); TAB( 2Ø);U(K); TAB( 3Ø
);V(K)
24Ø4  NEXT K
24Ø6  PRINT "TP2"; TAB( 6);F2; TAB( 2Ø);LC / 2; TAB( 31);"Ø"
24Ø8  PRINT "****************************************"
241Ø  PRINT P$;: INPUT Q$
2412  IF Q$ = "Y" THEN 64ØØ
2414  IF Q$ <  > "N" THEN 241Ø
2416  GOTO 2238
2418  GOSUB 9ØØØ:P3$ = "<3> TANGENTIAL DEFLECTION ANGLES"
242Ø  PRINT HC$: PRINT P3$
2422  PRINT "****************************************"
2424  PRINT : PRINT "TOTAL CURVE LENGTH = ";CL
2426  PRINT : PRINT "SELECT CHORD LENGTH C FROM :-"
2428  PRINT "  1/25ØØ ";TAB(12);"ACCURACY, C < "; INT (R / 1Ø)
243Ø  PRINT "  1/5ØØØ ";TAB(12);"ACCURACY, C < "; INT (R / 14)
2432  PRINT "  1/1ØØØØ ";TAB(12);"ACCURACY, C < "; INT (R / 2Ø)
2434  INPUT CH
2436  IF CL / CH < 5Ø THEN 2444
2438  GOSUB 91ØØ
244Ø  PRINT "TOO MANY CHORDS !"
2442  GOTO 2426
2444 CH(Ø) = ( INT (F1 / CH + .Ø1) + 1) * CH
2446 C1 =  INT ((CH(Ø) - F1) * 1ØØØ + .5) / 1ØØØ
2448 C9 = ( INT (F2 / CH)) * CH
```

```
2450 C2 =  INT ((F2 - C9) * 1000 + .5) / 1000
2452 N =  INT ((CL - C1 - C2) / CH + .5)
2454 A0 = 1718.8 * CH / (R * 60)
2455 A1 = 1718.8 * C1 / (R * 60):A(0) = A1
2456 A2 = 1718.8 * C2 / (R * 60)
2457 IF P8=1 THEN 2460
2458 IF D9=1 THEN 2460
2459 D5=DF:GOSUB 8000:DF=D5:P8=1
2460  PRINT : PRINT "IS CURVE LEFT OR RIGHT HANDED (L/R)";: INPUT L$
2462  IF L$ = "L" THEN 2506
2464  IF L$ <  > "R" THEN 2460
2466  GOSUB 9000
2468  PRINT H$:PRINT HC$: PRINT P3$
2470  PRINT "****************************************"
2472  PRINT "PT.";TAB(6);"CHAINAGE";TAB(20);"CHORD";TAB(29);"ANGLE(DD
.DEC)"
2474  PRINT "========================================"
2476  PRINT "TP1"; TAB( 6);F1; TAB( 30);" .000"
2478  PRINT " 1"; TAB( 6);CH(0); TAB( 20);C1; TAB( 30); INT (A1 * 100
0 + .5) / 1000
2480  FOR I = 1 TO N
2482 CH(I) = CH(0) + I * CH
2484 A(I) = A(0) + I * A0
2486  PRINT (I + 1); TAB( 6);CH(I); TAB( 20);CH; TAB( 30); INT (A(I)
* 1000 + .5) / 1000
2488  NEXT I
2490  PRINT "TP2"; TAB( 6);F2; TAB( 20);C2; TAB( 30); INT ((A(N) + A2
) * 1000 + .5) / 1000
2492  PRINT "========================================"
2494  PRINT "FINAL ANGLE AT TP2 SHOULD BE "; INT ((DF / 2) * 1000 + .
5) / 1000
2496  PRINT "****************************************"
2498  PRINT P$;: INPUT Q$
2500  IF Q$ = "Y" THEN 6500
2502  IF Q$ <  > "N" THEN 2498
2504  GOTO 2238
2506  GOSUB 9000
2508  PRINT H$:PRINT HC$: PRINT P3$
2510  PRINT "****************************************"
2512  PRINT "PT.";TAB(6);"CHAINAGE";TAB(20);"CHORD";TAB(29);"ANGLE(DD
.DEC)"
2514  PRINT "========================================"
2516  PRINT "TP1"; TAB( 6);F1; TAB( 30);" 360.000"
2518  PRINT "1"; TAB( 6);CH(0); TAB( 20);C1; TAB( 30); INT ((360 - A1
) * 1000 + .5) / 1000
2520  FOR I = 1 TO N
2522 CH(I) = CH(0) + I * CH
2524 A(I) = (360 - A(0)) - (I * A0)
2526  PRINT (I + 1); TAB( 6);CH(I); TAB( 20);CH; TAB( 30); INT (A(I)
* 1000 + .5) / 1000
2528  NEXT I
2530  PRINT "TP2"; TAB( 6);F2; TAB( 20);C2; TAB( 30); INT ((A(N) - A2
) * 1000 + .5) / 1000
2532  PRINT "========================================"
2534  PRINT "FINAL ANGLE AT TP2 SHOULD BE "; INT ((360 - DF / 2) * 10
00 + .5) / 1000
2536  PRINT "****************************************"
2538  PRINT P$;: INPUT Q$
2540  IF Q$ = "Y" THEN 6650
2542  IF Q$ <  > "N" THEN 2538
2544  GOTO 2238
3000 GOSUB 9000
3002 PRINT H$
3004 PRINT:PRINT "VERTICAL CURVE REQUIRED (Y/N) ";
3006 INPUT Q$
```

```
3ØØ8 IF Q$="N" THEN 323Ø
3Ø1Ø IF Q$<>"Y" THEN 3ØØ4
3Ø12 IF CT=2 THEN 31ØØ
3Ø14 CL=TC
3Ø16 F1=CT1
3Ø18 F2=CT2
31ØØ  GOSUB 9ØØØ:HC$ = "VERTICAL CURVE DESIGN": PRINT H$:PRINT HC$
31Ø2  PRINT "****************************************"
31Ø4  PRINT : PRINT "ENTER GRADIENT OF 1ST STRAIGHT ";: INPUT G1
31Ø6  PRINT : PRINT "ENTER GRADIENT OF 2ND STRAIGHT ";: INPUT G2
31Ø8 A = G1 - G2
311Ø  PRINT : PRINT "ENTER K-VALUE ";: INPUT K
3112  IF K <  = Ø THEN 311Ø
3114 LM = K *  ABS (A)
3116  PRINT : PRINT "MINIMUM LENGTH OF CURVE IS = "; INT (LM * 1ØØØ +
 .5) / 1ØØØ
3117 IF CT=Ø THEN 312Ø
3118  PRINT : PRINT "LENGTH OF HORIZ.  CURVE IS = "; INT (CL * 1ØØØ +
 .5) / 1ØØØ
312Ø  PRINT : PRINT "IS MINIMUM LENGTH ACCEPTABLE (Y/N) ";: INPUT Q$:
 IF Q$ = "Y" THEN 3134
3122  IF Q$ <  > "N" THEN 312Ø
3123 IF CT=Ø THEN 313Ø
3124  PRINT : PRINT "DO YOU WISH CURVES PHASED (Y/N) ";: INPUT XX$
3126  IF XX$ = "Y" THEN 3142
3128  IF XX$ <  > "N" THEN 3124
313Ø  PRINT : PRINT "ENTER ACCEPTABLE LENGTH ";: INPUT LØ: IF LØ < LM
 THEN 313Ø
3132 L = LØ
3134  GOSUB 9ØØØ: PRINT H$: IF LØ < LM THEN L = LM
3135 IF CT<>Ø THEN 3138
3136 PRINT:PRINT "ENTER CHAINAGE OF I.P. ";:INPUT CI
3137 GOTO 314Ø
3138 CI = (F1 + F2) / 2
314Ø F1 =  INT ((CI - L / 2) * 1ØØØ + .5) / 1ØØØ
3141 F2 =  INT ((CI + L / 2) * 1ØØØ + .5) / 1ØØØ
3142  IF XX$ = "Y" THEN L = CL
3144  PRINT : PRINT "ENTER LEVEL OF I.P. ";: INPUT RI
3146  PRINT : PRINT P$;: INPUT Q$
3148  IF Q$ = "Y" THEN 68ØØ
315Ø  IF Q$ <  > "N" THEN 3146
3152 RA =  INT ((RI - (G1 * L / 2ØØ)) * 1ØØØ + .5) / 1ØØØ
3154 RC =  INT ((RI + (G2 * L / 2ØØ)) * 1ØØØ + .5) / 1ØØØ
3156  IF CH > Ø THEN 3168
3158  PRINT : PRINT "ENTER MULTIPLE CHORD INTERVALS ";: INPUT CH
316Ø  IF CH <  = Ø THEN 3158
3162  IF L / CH < 1ØØ THEN 3168
3164  GOSUB 91ØØ: PRINT "TOO SMALL A CHORD INTERVAL !"
3166  GOTO 3158
3168 CH(Ø) = ( INT (F1 / CH) + 1) * CH
317Ø C1 =  INT ((CH(Ø) - F1) * 1ØØØ + .5) / 1ØØØ:Y(Ø) = C1
3172 C9 = ( INT (F2 / CH)) * CH
3174 NC =  INT ((L - C1) / CH)
3176 CB =  INT ((L *  ABS (G1) /  ABS (A)) * 1ØØØ + .5) / 1ØØØ:F3 = F
1 + CB
3178 HV = L * G1 *  ABS (G1) / (2ØØ *  ABS (A)):RB = RA + HV
318Ø HV =  INT (HV * 1ØØØ + .5) / 1ØØØ:RB =  INT (RB * 1ØØØ + .5) / 1
ØØØ
3182  GOSUB 9ØØØ: PRINT H$:PRINT HC$:P5 = Ø
3184  PRINT "****************************************"
3186  PRINT "CHAINAGE";TAB(12);"LENGTH";TAB(21);"HT.DIFF.";TAB(31);"R
ED.LEVEL"
3188  PRINT "========================================"
319Ø  PRINT F1; TAB( 13);"Ø"; TAB( 22);"Ø"; TAB( 31);RA
3192  FOR I = Ø TO NC
```

```
3194 Y(I) = Y(Ø) + I * CH
3196 CH(I) = CH(Ø) + I * CH
3198 HD(I) = (G1 * Y(I) / 1ØØ) - (A * (Y(I)) ^ 2 / (2ØØ * L))
3200 RL(I) =  INT ((RA + HD(I)) * 1ØØØ + .5) / 1ØØØ
3202 HD(I) =  INT ((HD(I)) * 1ØØØ + .5) / 1ØØØ
3204  IF P5 = 1 THEN 3212
3206  IF F3 > CH(I) THEN 3212
3208  PRINT F3; TAB( 12);CB; TAB( 21);HV; TAB( 31);RB
3210 P5 = 1
3212  PRINT CH(I); TAB( 12);Y(I); TAB( 21);HD(I); TAB( 31);RL(I)
3214  NEXT I
3216  PRINT F2; TAB( 12);L; TAB( 21); INT ((RC - RA) * 1ØØØ + .5) / 1
ØØØ; TAB( 31);RC
3218  PRINT "****************************************"
3220  PRINT : PRINT P$;: INPUT Q$
3222  IF Q$ = "Y" THEN 6838
3224  IF Q$ <  > "N" THEN 3220
3230  GOSUB 9ØØØ:PRINT H$
3235 IF DISP% <> Ø THEN 3242
3240 PRINT:PRINT " Do you wish plot of curve (y/n)";:INPUT Q$:IF Q$="
n" THEN 93ØØ
3242  PRINT:PRINT:PRINT " Select Display";TAB(18);"<1> Graphics Scree
n":PRINT:PRINT TAB(18);"<2> Plotter":PRINT:PRINT TAB(18);"<3> Continu
e ";:INPUT DISP%
3244  IF DISP% <1 OR DISP%> 3 THEN 323Ø
3246  ON DISP% GOTO 4ØØØ,5ØØØ,93ØØ
3995 REM ***** Graphics Screen Display *****
4ØØØ  CLS:GOSUB 9ØØØ
4Ø1Ø  IF SR%=1 THEN RGB%=2 ELSE RGB%=3
4Ø2Ø WINDOW (Ø,Ø)-(32Ø*SR%,2ØØ)
4Ø3Ø LINE (8,18)-(31Ø*SR%,18Ø),RGB%,B
41ØØ TPLY=4Ø:TPRY=4Ø
41Ø5 IF CT=2 THEN TW=LC:VH=MO+EM:CIT=CI:TPLL=F1:TPRL=F2:CPLY=TPLY
411Ø SCRM=(27Ø*SR%)/(TW*1.6)
412Ø IF SCRM*VH > 11Ø THEN SCRM=11Ø/VH
413Ø IPX=16Ø*SR%
414Ø IPY=4Ø+SCRM*VH
415Ø TPLX=16Ø*SR%-(SCRM*1.6*SR%*TW/2)
416Ø TPRX=16Ø*SR%+(SCRM*1.6*SR%*TW/2)
4165 IF CT=2 THEN 42ØØ
417Ø CPLX=TPLX+(LF*SIN((9Ø-TANG+PH)/R9))*SCRM*1.6*SR%
418Ø CPRX=TPRX-(LF*SIN((9Ø-TANG+PH)/R9))*SCRM*1.6*SR%
419Ø CPLY=TPLY+(LF*COS((9Ø-TANG+PH)/R9))*SCRM:CPRY=CPLY
4192 IF L$="R" THEN TPLL=TPF1:TPRL=TPF2
4194 IF L$="L" THEN TPLL=TPF2:TPRL=TPF1
42ØØ LINE (IPX,TPLY-1Ø)-(IPX,IPY),3
42Ø5 LINE -(TPLX,TPLY),3
421Ø LINE -(TPRX,TPRY),3
4215 LINE -(IPX,IPY),3
4216 IF CT=2 THEN 4232
4218 IF P2=1 THEN 4232
422Ø LINE (CPLX,TPLY)-(CPLX,CPLY),3
4225 LINE -(CPRX,CPRY),3
423Ø LINE -(CPRX,TPRY),3
4232 CCOL=(TPLX*SR%-32)/8
4234 IF CCOL<3 THEN CCOL=3
424Ø LOCATE 22,CCOL:PRINT INT(TPLL*1ØØØ+.5)/1ØØØ
4242 CCOL=(TPRX*SR%-32)/8
4244 IF CCOL>29 THEN CCOL=29
425Ø LOCATE 22,CCOL:PRINT INT(TPRL*1ØØØ+.5)/1ØØØ
426Ø LOCATE 5,17*SR%:PRINT CIT
427Ø LOCATE 4,3*SR%:PRINT "R =";INT(R*1ØØØ+.5)/1ØØØ
428Ø LOCATE 1,2:PRINT CC$
429Ø LOCATE 2,2:PRINT "Defln Angle =";DFT;DD$
43ØØ PSET (TPLX,TPLY),1
```

```
4305 IF CT=2 THEN 4345
4310 FOR J=0 TO NT
4312 HTP(J)=LFF(J)*SIN((90-TANG+PHD(J))/R9)*SCRM*1.6*SR%
4314 VTP(J)=LFF(J)*COS((90-TANG+PHD(J))/R9)*SCRM
4320 LINE -((TPLX+HTP(J)),(TPLY+VTP(J))),3
4330 NEXT J
4335 IF P2=1 THEN 4400
4340 LINE -(CPLX,CPLY),3
4345 W=15:GOTO 2350
4350 FOR K=NLC-1 TO 0 STEP -1
4360 LINE -((IPX-U(K)*SCRM*1.6*SR%),(CPLY+V(K)*SCRM)),3
4365 NEXT K
4370 LINE -(IPX,CPLY+MO*SCRM),3
4375 FOR I=0 TO NLC-1
4380 LINE -((IPX+U(I)*SCRM*1.6*SR%),(CPLY+V(I)*SCRM)),3
4385 NEXT I
4388 IF CT=2 THEN 4430
4390 LINE -(CPRX,CPRY),3
4400 FOR J=0 TO N1
4402 HTP(J)=LF(J)*SIN((90-TANG+PHN(J))/R9)*SCRM*1.6*SR%
4404 VTP(J)=LF(J)*COS((90-TANG+PHN(J))/R9)*SCRM
4410 LINE -((TPRX-HTP(J)),(TPRY+VTP(J))),3
4420 NEXT J
4430 LINE -(TPRX,TPRY),3
4450 LOCATE 24,2:PRINT "Screen Dump (y/n) ";
4460 Q$=INKEY$:IF Q$ = "" THEN 4460
4470 IF INSTR("yn",Q$)=0 THEN 4450
4480 IF Q$ = "n" THEN 4500
4490 LCOPY
4500 DISP%=3:GOTO 3230
4995 REM ***** Plotter Output *****
5000 GX=0:GY=0:SX=180:SY=120:MX=2670/SX:PX=GX+SX:PY=GY+SY
5005 IF CT=2 THEN TW=LC:VH=MO+EM:CIT=CI:TPLL=F1:TPRL=F2:DFT=DF:TTL=TL
5010 SCRM=160/TW
5020 IF SCRM*VH > 100 THEN SCRM =100/VH
5030 IPX=GX+SX/2
5040 IPY=GY+10+SCRM*VH
5050 TPLX=IPX-(SCRM*TW/2):TPLY=GY+10
5060 TPRX=IPX+(SCRM*TW/2):TPRY=GY+10
5065 IF CT=2 THEN 5100
5070 CPLX=TPLX+(LF*SIN((90-TANG+PH)/R9))*SCRM
5080 CPRX=TPRX-(LF*SIN((90-TANG+PH)/R9))*SCRM
5090 CPLY=TPLY+(LF*COS((90-TANG+PH)/R9))*SCRM:CPRY=CPLY
5092 IF L$="R" THEN TPLL=TPF1:TPRL=TPF2
5094 IF L$="L" THEN TPLL=TPF2:TPRL=TPF1
5100 GOSUB 9000
5110 PRINT:PRINT " Select Plotter -";TAB(18);"<1> Apple 410":PRINT:PR
INT TAB(18);"<2> Epson HI-80"
5120 PRINT:PRINT TAB(18);"<3> Hewlett-Packard":PRINT:PRINT TAB(18);"<
4> QUIT           ";:INPUT PLOT%
5130 IF PLOT% <1 OR PLOT% >4 THEN 5100
5140 IF PLOT%=4 THEN 900
5145 IF PLOT%=1 THEN HP%=3
5150 IF PLOT%=2 THEN HP%=4
5155 IF PLOT%=3 THEN HP%=1
5160 PRINT:PRINT " Place paper on Plotter and switch on"
5170 PRINT:PRINT " To Continue Press <Enter> ":WHILE E$=" ":E$=INKEY$
:WEND
5180 GOSUB 50000
5190 ON HP% GOTO 5200,5200,11200,10200
5195 REM ***** HP-GL Plotter (HP-7475A) *****
5200 GOSUB 51000
5210 PRINT#2,"SP4;"
5220 PRINT#2,"PU;PA";IPX;",";(TPLY-5);";"
5225 PRINT#2,"PD;PA";IPX;",";IPY;";"
```

```
523Ø PRINT#2,"PA";TPLX;",";TPLY;";"
5235 PRINT#2,"PA";TPRX;",";TPRY;";"
524Ø PRINT#2,"PA";IPX;",";IPY;";"
5242 IF CT=2 THEN 529Ø
5245 IF P2=1 THEN 529Ø
525Ø PRINT#2,"PU;PA";CPLX;",";TPLY;";"
5255 PRINT#2,"PD;PA";CPLX;",";CPLY;";"
526Ø PRINT#2,"PA";CPRX;",";CPRY;";"
5265 PRINT#2,"PA";CPRX;",";TPRY;";"
529Ø PRINT#2,"SP3;"
5295 IF CT=2 THEN CPLY=TPLY
53ØØ PRINT#2,"PU;PA";TPLX;",";TPLY";"
53Ø5 IF CT=2 THEN 5345
531Ø FOR J=Ø TO NT
5312 HTP(J)=LFF(J)*SIN((9Ø-TANG+PHD(J))/R9)*SCRM
5314 VTP(J)=LFF(J)*COS((9Ø-TANG+PHD(J))/R9)*SCRM
532Ø PRINT#2,"PD;PA";(TPLX+HTP(J));",";(TPLY+VTP(J));";"
533Ø NEXT J
5335 IF P2=1 THEN 54ØØ
534Ø PRINT#2,"PA";CPLX;",";CPLY;";"
5345 W=15:GOTO 235Ø
535Ø FOR K=NLC-1 TO Ø STEP -1
536Ø PRINT#2,"PA";(IPX-U(K)*SCRM);",";(CPLY+V(K)*SCRM);";"
5365 NEXT K
537Ø PRINT#2,"PA";IPX;",";(CPLY+MO*SCRM);";"
5375 FOR I=Ø TO NLC-1
538Ø PRINT#2,"PA";(IPX+U(I)*SCRM);",";(CPLY+V(I)*SCRM);";"
5385 NEXT I
5388 IF CT=2 THEN 542Ø
539Ø PRINT#2,"PA";CPRX;",";CPRY;";"
54ØØ FOR J=Ø TO N1
54Ø2 HTP(J)=LF(J)*SIN((9Ø-TANG+PHN(J))/R9)*SCRM
54Ø4 VTP(J)=LF(J)*COS((9Ø-TANG+PHN(J))/R9)*SCRM
541Ø PRINT#2,"PA";(TPRX-HTP(J));",";(TPRY+VTP(J));";"
5415 NEXT J
542Ø PRINT#2,"PA";TPRX;",";TPRY;";"
543Ø PRINT#2,"SP2;":PRINT#2,"SIØ.2,Ø.3;"
544Ø PRINT#2,"PU;PA";(TPLX-8);",";(TPLY-3);";"
545Ø PRINT#2,"LB";INT(TPLL*1ØØØ+.5)/1ØØØ
546Ø PRINT#2,CHR$(3)
547Ø PRINT#2,"PU;PA";(TPRX-8);",";(TPLY-3);";"
548Ø PRINT#2,"LB";INT(TPRL*1ØØØ+.5)/1ØØØ
549Ø PRINT#2,CHR$(3)
55ØØ PRINT#2,"PU;PA";(IPX-8);",";(IPY+2);";"
551Ø PRINT#2,"LB";CIT
552Ø PRINT#2,CHR$(3)
553Ø PRINT#2,"PU;PA";(IPX-SCRM*TW/4);",";((TPLY+IPY)/2+5);";"
554Ø PRINT#2,"LBTL"
555Ø PRINT#2,CHR$(3)
5555 IF CT=2 THEN 56ØØ
556Ø PRINT#2,"PU;PA";(IPX-SCRM*TW/4);",";(TPLY+2);";"
557Ø PRINT#2,"LBTW"
558Ø PRINT#2,CHR$(3)
559Ø IF P2=1 THEN 563Ø
56ØØ PRINT#2,"PU;PA";(IPX-SCRM*LC/4);",";(CPLY+2);";"
561Ø PRINT#2,"LBLC"
562Ø PRINT#2,CHR$(3)
563Ø PRINT#2,"PU;PA";IPX+2;",";(CPLY+MO*SCRM+2);";"
5635 PRINT#2,"LBCL"
564Ø PRINT#2,CHR$(3)
5642 IF CT=2 THEN 566Ø
5645 PRINT#2,"PU;PA";IPX-(SCRM*(TW+LC)/4);",";((TPLY+CPLY)/2-3);";"
565Ø PRINT#2,"LBSF"
5655 PRINT#2,CHR$(3)
566Ø PRINT#2,"PU;PA";(GX+5);",";(GY+SY-8);";"
```

```
5670 PRINT#2,"LBR =";INT(R*1000+.5)/1000
5680 PRINT#2,CHR$(3)
5700 PRINT#2,"PU;PA";(GX+5);",";(GY+SY-12);";"
5710 PRINT#2,"LBDF=";DFT;DD$
5720 PRINT#2,CHR$(3)
5728 IF CT=2 THEN 5830
5730 PRINT#2,"PU;PA";(GX+5);",";(GY+SY-16);";"
5740 PRINT#2,"LBTH=";THD;DD$
5750 PRINT#2,CHR$(3)
5760 PRINT#2,"PU;PA";(GX+5);",";(GY+SY-20);";"
5770 PRINT#2,"LBDC=";DC;DD$
5780 PRINT#2,CHR$(3)
5800 PRINT#2,"PU;PA";(GX+5);",";(GY+SY-24);";"
5810 PRINT#2,"LBSF=";SF
5820 PRINT#2,CHR$(3)
5830 PRINT#2,"PU;PA";(GX+5);",";(GY+SY-28);";"
5840 PRINT#2,"LBCL=";INT(CL*1000+.5)/1000
5850 PRINT#2,CHR$(3)
5858 IF CT=2 THEN 5870
5860 PRINT#2,"PU;PA";(GX+5);",";(GY+SY-32);";"
5862 PRINT#2,"LBTC=";INT(TC*1000+.5)/1000
5864 PRINT#2, CHR$(3)
5870 PRINT#2,"PU;PA";(GX+5);",";(GY+SY-36);";"
5872 PRINT#2,"LBTL=";INT(TTL*1000+.5)/1000
5874 PRINT#2, CHR$(3)
5880 PRINT#2,"PU;PA";(GX+5);",";(GY+SY-40);";"
5882 PRINT#2,"LBLC=";INT(LC*1000+.5)/1000
5884 PRINT#2, CHR$(3)
5888 IF CT=2 THEN 5900
5890 PRINT#2,"PU;PA";(GX+5);",";(GY+SY-44);";"
5892 PRINT#2,"LBTW=";INT(TW*1000+.5)/1000
5894 PRINT#2, CHR$(3)
5900 PRINT#2,"PU,PA5,-3;"
5910 PRINT#2,"LBPC-RODEOS - ";A$;" : Date : ";DATE$
5920 PRINT#2,CHR$(3)
5930 CLOSE#2
5940 DISP%=3
5950 GOTO 3230
6000  LPRINT  H$
6005  LPRINT "LOCATION OF CURVE :- ";TAB(25);A$
6010  LPRINT "********************************************************
*****"
6020  LPRINT "OPERATOR'S NAME :- ";TAB(25);B$
6030  LPRINT "DATE OF CALCS.  :- ";TAB(25);C$
6040  LPRINT "INSTRUMENT USED :- ";TAB(25);K$
6060  LPRINT "********************************************************
*****"
6070  LPRINT
6080  GOTO 300
6100  LPRINT H$: LPRINT P1$
6106  LPRINT "*****************************************": IF P0 = 2 TH
EN 6122
6108  LPRINT "ANGLE 1ST STR-";R1;"-";R3;TAB(30);" = ";Z1
6110  LPRINT "ANGLE 2ND STR-";R3;"-";R1;TAB(30);" = ";Z3
6112  LPRINT "DEFLECTION ANGLE";TAB(30);" = ";DF
6114  LPRINT "DISTANCE ";R1;"-";R3;TAB(30);" = ";INT(H1*1000+.5)/1000
6116  LPRINT "DISTANCE ";R1;"-IP";TAB(30);" = ";INT(H2*1000+.5)/1000
6118  LPRINT "DISTANCE ";R3;"-IP";TAB(30);" = ";INT(H3*1000+.5)/1000
6120  LPRINT "CHAINAGE OF ";R1;TAB(30);" = ";CF
6122  LPRINT "CHAINAGE OF IP";TAB(30);" = ";CI: IF P0 = 1 THEN 6130
6124  LPRINT "DISTANCE ALONG STRAIGHTS";TAB(30);" = ";H2
6126  LPRINT "DISTANCE JOINING STRAIGHTS";TAB(30);" = ";H1
6128  LPRINT "DEFLECTION ANGLE";TAB(30);" = ";DF
6130  LPRINT "=========================================="
6132  LPRINT
```

```
6134  GOTO 1ØØØ
6136  LPRINT H$: LPRINT P2$
6142  LPRINT "*******************************************"
6144 P7 = 1:P9 = 1: GOSUB 8718
6146  LPRINT "*******************************************"
6148  LPRINT
615Ø  GOTO 1ØØØ
6152  LPRINT H$: LPRINT P3$
6154  LPRINT "*******************************************"
6156  LPRINT "RADIUS ";TAB(2Ø);" = ";R
6157  LPRINT "DISTANCE ";R(Ø);"-";R(2);TAB(2Ø);" = "; INT (H1 * 1ØØØ
+ .5) / 1ØØØ
6158  LPRINT "DISTANCE ";R(Ø);"-";R(1);TAB(2Ø);" = "; INT (H2 * 1ØØØ
+ .5) / 1ØØØ
616Ø  LPRINT "DISTANCE ";R(1);"-";R(2);TAB(2Ø);" = "; INT (H3 * 1ØØØ
+ .5) / 1ØØØ
6162  LPRINT "*******************************************"
6164  LPRINT
6166  GOTO 1ØØØ
6168  LPRINT H$: LPRINT P4$
6174  LPRINT "*******************************************"
6176  IF P4 = 2 THEN 6182
6178  LPRINT "DIST. ALONG TANGENT";TAB(21);" = ";XP
618Ø  LPRINT "DIST. PERP. TANGENT";TAB(21);" = ";YP
6182  LPRINT "DIST. FROM  I.P.";TAB(21);" = ";ZP
6184  LPRINT "ANGLE FROM TANGENT";TAB(21);" = "; INT (B * 1ØØØ + .5)
/ 1ØØØ;" (DD.DEC)"
6186  LPRINT "==========================================="
6188  LPRINT : GOTO 94Ø
619Ø P7 = 1: GOSUB 8718
6196  LPRINT "*******************************************"
6198  LPRINT
62ØØ  GOTO 1ØØØ
62Ø2  LPRINT H$:P7 = 1
62Ø8  LPRINT "*******************************************"
621Ø  IF PC = 2 THEN  LPRINT "1ØØM CHORD ANGLE = ";A
6211 IF CT=1 THEN LPRINT "CENTRE CIRCULAR CURVE"
6212  GOSUB 8718
6214  LPRINT "*******************************************"
6216  LPRINT
6218  GOTO 92ØØ
622Ø  LPRINT H$: LPRINT P3$:P7 = 1
6226  LPRINT "*******************************************"
6228  GOSUB 8718
623Ø  LPRINT "*******************************************"
6232  LPRINT
6234  GOTO 92ØØ
625Ø  LPRINT H$: LPRINT P1$:P7 = 1
6256  LPRINT "*******************************************"
6258  IF P9 = 1 THEN 6264
626Ø  GOSUB 8718: LPRINT
6262  LPRINT "==========================================="
6264  LPRINT "PT.  CHAINAGE       TAN.DIST  TAN.OFFSET"
6266  LPRINT "==========================================="
627Ø  LPRINT "TP1"; TAB( 6);F1; TAB( 21);" Ø"; TAB( 31);" Ø"
6276  FOR I = Ø TO N - 1
628Ø  LPRINT (I + 1); TAB( 6);(F1 + CL(I)); TAB( 2Ø);U(I); TAB( 3Ø);V
(I)
6286  NEXT I
629Ø  LPRINT "MO"; TAB( 6); INT ((F1 + CL / 2) * 1ØØØ + .5) / 1ØØØ; T
AB( 2Ø);UM; TAB( 3Ø);VM
6296  FOR J = N - 1 TO Ø STEP  - 1
63ØØ  LPRINT (2 * N - J); TAB( 6);(F2 - CL(J)); TAB( 2Ø);U(J); TAB( 3
Ø);V(J)
63Ø6  NEXT J
631Ø  LPRINT "TP2"; TAB( 6);F2; TAB( 21);" Ø"; TAB( 31);" Ø"
```

```
6316  LPRINT "*******************************************"
6318  LPRINT
6322  GOTO 2238
6400 LPRINT H$: LPRINT P2$
6402 LPRINT "*******************************************"
6406 P7 = 1: IF P9 = 1 THEN 6412
6408  GOSUB 8718: LPRINT
6410  LPRINT "=========================================="
6412  LPRINT "PT.  CHAINAGE      CH.DIST   CH.OFFSET"
6414  LPRINT "=========================================="
6418  LPRINT "TP1"; TAB( 6);F1; TAB( 20);LC / 2; TAB( 31);"0"
6424  FOR J = N - 1 TO 0 STEP  - 1
6428  LPRINT (N - J); TAB( 6);(F4 - CL(J)); TAB( 20);U(J); TAB( 30);V
(J)
6434  NEXT J
6438  LPRINT "MO"; TAB( 6);F4; TAB( 21);"0"; TAB( 30);MO
6444  FOR K = 0 TO N - 1
6448  LPRINT (N + 1 + K); TAB( 6);(F4 + CL(K)); TAB( 20);U(K); TAB( 3
0);V(K)
6454  NEXT K
6458  LPRINT "TP2"; TAB( 6);F2; TAB( 20);LC / 2; TAB( 31);"0"
6464  LPRINT "*******************************************"
6466  LPRINT
6470  GOTO 2238
6500  LPRINT H$: LPRINT P3$:P7 = 1
6502  LPRINT "*******************************************"
6504  IF D9 = 1 THEN 6510
6506 D = DF / 2: GOSUB 8050:DX = D
6508 D = DF: GOSUB 8050:DF = D:P8=2
6510  IF P9 = 1 THEN 6516
6512  GOSUB 8718: LPRINT
6514  LPRINT "=========================================="
6516  LPRINT "PT.";TAB(6);"CHAINAGE";TAB(21);"CHORD";TAB(30);"ANGLE";
DD$
6518  LPRINT "=========================================="
6522  LPRINT "TP1"; TAB( 6);F1; TAB( 30);" 0.000"
6528  IF D9 = 2 THEN 6566
6532  LPRINT "  1"; TAB( 6);CH(0); TAB( 20);C1; TAB( 30); INT (A1 * 1
000 + .5) / 1000
6538  FOR J = 1 TO N
6542  LPRINT (J + 1); TAB( 6);CH(J); TAB( 20);CH; TAB( 30); INT (A(J)
 * 1000 + .5) / 1000
6548  NEXT J
6552  LPRINT "TP2"; TAB( 6);F2; TAB( 20);C2; TAB( 30); INT ((A(N) + A
2) * 1000 + .5) / 1000
6558  LPRINT "========================================"
6560  LPRINT "FINAL ANGLE AT TP2 SHOULD BE  "; INT ((DF / 2) * 1000 +
 .5) / 1000
6562  LPRINT "*******************************************"
6564  GOTO 6608
6566 D = (A(N) + A2): GOSUB 8050:A(N + 1) = D
6568  FOR J = 0 TO N
6570 D = A(J): GOSUB 8050:A(J) = D
6572  NEXT J
6576  LPRINT "1"; TAB( 6);CH(0); TAB( 20);C1; TAB( 30);A(0)
6582  FOR K = 1 TO N
6586  LPRINT (K + 1); TAB( 6);CH(K); TAB( 20);CH; TAB( 30);A(K)
6592  NEXT K
6596  LPRINT "TP2"; TAB( 6);F2; TAB( 20);C2; TAB( 30);A(N + 1)
6602  LPRINT "=========================================="
6604  LPRINT "FINAL ANGLE AT TP2 SHOULD BE  ";DX
6606  LPRINT "*******************************************"
6608  LPRINT
6612  GOTO 2238
6650  LPRINT H$: LPRINT P3$:P7 = 1: IF D9 = 1 THEN 6660
6656 D = (360 - DF / 2): GOSUB 8050:DX = D
```

```
6658 D = DF: GOSUB 8Ø5Ø:DF = D:P8=2
666Ø  IF P9 = 1 THEN 6666
6662  GOSUB 8718: LPRINT
6664  LPRINT "==========================================="
6666  LPRINT "PT.";TAB(6);"CHAINAGE";TAB(21);"CHORD";TAB(31);"ANGLE";
DD$
6668  LPRINT "==========================================="
6672  LPRINT "TP1"; TAB( 6);F1; TAB( 3Ø);"  36Ø.ØØØØ"
6678  IF D9 = 2 THEN 6716
6682  LPRINT " 1"; TAB( 6);CH(Ø); TAB( 2Ø);C1; TAB( 3Ø); INT ((36Ø -
A1) * 1ØØØ + .5) / 1ØØØ
6688  FOR J = 1 TO N
6692  LPRINT (J + 1); TAB( 6);CH(J); TAB( 2Ø);CH; TAB( 3Ø); INT (A(J)
 * 1ØØØ + .5) / 1ØØØ
6698  NEXT J
67Ø2  LPRINT "TP2"; TAB( 6);F2; TAB( 2Ø);C2; TAB( 3Ø); INT ((A(N) - A
2) * 1ØØØ + .5) / 1ØØØ
67Ø8  LPRINT "==========================================="
671Ø  LPRINT "FINAL ANGLE AT TP2 SHOULD BE  "; INT ((36Ø - DF / 2) *
1ØØØ + .5) / 1ØØØ
6712  LPRINT "*******************************************"
6714  GOTO 676Ø
6716 D = (A(N) - A2): GOSUB 8Ø5Ø:A(N + 1) = D
6718 A(Ø) = 36Ø - A1
672Ø  FOR J = Ø TO N
6722 D = A(J): GOSUB 8Ø5Ø:A(J) = D
6724  NEXT J
6728  LPRINT " 1"; TAB( 6);CH(Ø); TAB( 2Ø);C1; TAB( 3Ø);A(Ø)
6734  FOR K = 1 TO N
6738  LPRINT (K + 1); TAB( 6);CH(K); TAB( 2Ø);CH; TAB( 3Ø);A(K)
6744  NEXT K
6748  LPRINT "TP2"; TAB( 6);F2; TAB( 2Ø);C2; TAB( 3Ø);A(N + 1)
6754  LPRINT "==========================================="
6756  LPRINT "FINAL ANGLE AT TP2 SHOULD BE  ";DX
6758  LPRINT "*******************************************"
676Ø  LPRINT
6764  GOTO 2238
68ØØ  LPRINT H$:LPRINT HC$
68Ø6  LPRINT "*******************************************"
68Ø8  LPRINT "GRADIENT OF 1ST STRAIGHT";TAB(3Ø);" = ";G1
681Ø  LPRINT "GRADIENT OF 2ND STRAIGHT";TAB(3Ø);" = ";G2
6812  LPRINT "ALGEBRAIC DIFFERENCE M-N";TAB(3Ø);" = ";A
6814  LPRINT "K-VALUE ";TAB(3Ø);" = ";K
6816  LPRINT "MINIMUM LENGTH OF CURVE";TAB(3Ø);" = ";LM
6818  LPRINT "CURVE LENGTH CHOSEN";TAB(3Ø);" = ";L
682Ø  LPRINT
6822  LPRINT "CHAINAGE OF INTERSECTION";TAB(3Ø);" = ";CI
6824  LPRINT "REDUCED LEVEL OF I.P.";TAB(3Ø);" = ";RI
6826  LPRINT "CHAINAGE OF 1ST TANG.PT.";TAB(3Ø);" = ";F1
6828  LPRINT "CHAINAGE OF 2ND TANG.PT";TAB(3Ø);" = ";F2
683Ø  LPRINT "*******************************************"
6832 P7 = 1
6834  LPRINT
6836  GOTO 3152
6838  P5=Ø:IF P7 = 1 THEN 6848
6844  LPRINT H$:LPRINT HC$
6846  LPRINT "*******************************************"
6848  LPRINT "CHAINAGE  LENGTH    HT.DIFF.   RED.LEVEL"
685Ø  LPRINT "==========================================="
6854  LPRINT F1; TAB( 11);" Ø"; TAB( 21);" Ø"; TAB( 31);RA
686Ø  FOR J = Ø TO NC
6862  IF P5 = 1 THEN 6878
6864  IF F3 > CH(J) THEN 6878
6868  LPRINT F3; TAB( 11);CB; TAB( 21);HV; TAB( 31);RB
6874  P5 = 1
6878  LPRINT CH(J); TAB( 11);Y(J); TAB( 21);HD(J); TAB( 31);RL(J)
```

```
6884  NEXT J
6888  LPRINT F2; TAB( 11);L; TAB( 21); INT ((RC - RA) * 1000 + .5) /
1000; TAB( 31);RC
6894  LPRINT "*******************************************"
6896  LPRINT
6898  GOTO 3230
6900  LPRINT H$:P7 = 1
6904  LPRINT "*******************************************"
6906  GOSUB 8234
6908  LPRINT
6910  GOTO 1278
7000  LPRINT H$:A = 0:P7 = 0
7006  LPRINT "********************************************************
********************"
7010  LPRINT "CHAINAGE OF I.P.   = "; TAB( 24);CI
7012  LPRINT "LENGTH OF TANGENT = "; TAB( 24); INT (TL * 1000 + .5) /
 1000
7020  LPRINT "********************************************************
********************"
7022  LPRINT "CHAINAGE    LENGTH      TANG.DIST   TANG.OFF    POLAR D
IST  ANGLE";DD$
7024  IF P7 = 1 THEN  RETURN
7026  LPRINT "========================================================
====================": IF L$ = "L" THEN A = 360
7036  LPRINT F1;TAB(12); " 0";TAB(24);" 0";TAB(36); " 0";TAB(48);" 0"
;TAB(60); A
7038  FOR J = 0 TO N
7056  LPRINT CH(J);TAB(12);CL(J);TAB(24); XF(J);TAB(36); YF(J);TAB(48
); LF(J);TAB(60);PH(J)
7058  NEXT J
7080  LPRINT FT1;TAB(12);SF;TAB(24); XF;TAB(36);YF;TAB(48);LF;TAB(60)
; PT
7082  LPRINT FT2;TAB(12); "START OF 2ND SPIRAL"
7084  LPRINT F2;TAB(12); "FINAL TANGENT POINT"
7086  LPRINT "========================================================
===================="
7088  LPRINT
7092  GOTO 1458
7100  LPRINT H$:A = 0:P7 = 1
7106  GOSUB 7020
7108  LPRINT "========================================================
===================="
7110  IF L$ = "R" THEN A = 360
7128  LPRINT FT2;TAB(12);SF;TAB(24); XF;TAB(36);YF;TAB(48); LF;TAB(60
); PT
7130  FOR JN = 0 TO N1
7148  LPRINT CH(JN);TAB(12); CL(JN);TAB(24);XF(JN);TAB(36);YF(JN);TAB
(48); LF(JN);TAB(60);PH(JN)
7150  NEXT JN
7160  LPRINT F2;TAB(12); " 0";TAB(24);" 0";TAB(36);" 0";TAB(48);" 0";
TAB(60); A
7162  LPRINT "********************************************************
********************"
7164  LPRINT
7168  GOTO 2000
7745  REM  <TRIANGLE> SOLUTION (A1,S1,A3)
7750 T2 =  -  COS ((A3 + A1) / R9)
7760 A2 =  FN C9(T2)
7770 S2 = S1 *  SIN (A3 / R9) /  SIN (A2 / R9)
7780 S3 = S1 *  COS (A3 / R9) + S2 *  COS (A2 / R9)
7790  RETURN
7795  REM  <TRIANGLE> SOLUTION (S1,S2,S3)
7800 P = (S1 + S2 + S3) / 2
7810 T3 =  SQR (P * (P - S2) / (S1 * S3))
7820 A3 = 2 *  FN C9(T3)
7830 T2 =  SQR (P * (P - S1) / (S2 * S3))
```

```
7840 A2 = 2 *  FN C9(T2)
7850 T1 =  -  COS ((A3 + A2) / R9)
7860 A1 =  FN C9(T1)
7870  RETURN
7995  REM  ROUTINE DD.MMSS TO DECIMALS
8000 D0 =  INT (D5)
8010 M1 = (D5 - D0) * 100:M0 =  INT (M1 + .1)
8020 S0 =  INT (M1 * 100 - M0 * 100 + .5)
8022  IF M0 < 0 OR M0 > 59 THEN 8030
8024  IF S0 < 0 OR S0 > 59.99 THEN 8030
8026 D5 = D0 + M0 / 60 + S0 / 3600
8028 CK = 0: RETURN
8030  GOSUB 9100
8040 CK = 1: RETURN
8045  REM  ROUTINE DECIMAL TO DD.MMSS
8050 D1 =  INT (D)
8060 D2 = (D - D1) * 3600
8070 M =  INT (D2 / 60)
8080 S = D2 - 60 * M + .5
8090  IF S < 60 THEN 8130
8100 M = M + 1:S = 0
8110  IF M < 60 THEN 8130
8120 D1 = D1 + 1:M = 0
8130 D = D1 * 10000 + M * 100 + S
8140 D =  INT (D) / 10000: RETURN
8234  LPRINT "DESIGN CRITERIA FOR CURVE :-"
8236  IF V0 = 0 THEN 8240
8238  LPRINT "VELOCITY (KPH)";TAB(30);" = ";V0
8240  LPRINT "VELOCITY (MPS)";TAB(30);" = ";INT(V*1000+.5)/1000
8242  LPRINT "RADIAL ACCELERATION (MPS^3)";TAB(30);" = ";C
8244  LPRINT "CENTRIPETAL RATIO";TAB(30);" = ";INT(F*1000+.5)/1000
8246  LPRINT "MINIMUM RADIUS (M)";TAB(30);" = ";INT(R*1000+.5)/1000
8248  LPRINT
8250  LPRINT "MULTIPLYING CONSTANT";TAB(30);" = ";INT(K*1000+.5)/1000
8252  LPRINT "DEFLECTION ANGLE AT I.P.";TAB(30);" = ";DF
8254  LPRINT "TRANSITION CURVE ANGLE";TAB(30);" = ";TH
8256  LPRINT "CIRCULAR CURVE ANGLE";TAB(30);" = ";DC
8258  LPRINT "SPIRAL TRANSITION LENGTH";TAB(30);" = ";INT(SF*1000+.5)
/1000
8260  LPRINT "CIRCULAR CURVE LENGTH";TAB(30);" = ";INT(CL*1000+.5)/10
00
8262  LPRINT "TOTAL CURVE LENGTH";TAB(30);" = ";INT(TC*1000+.5)/1000
8264  IF TL = 0 THEN 8268
8266  LPRINT "MEASURED TANGENT LENGTH";TAB(30);" = ";TL
8268  LPRINT "*****************************************"
8270  RETURN
8718  LPRINT "RADIUS";TAB(20);" = ";INT(R*1000+.5)/1000
8720  LPRINT "DEFLECTION ANGLE";TAB(20);" = ";DF
8722  LPRINT "CHAINAGE I.P.";TAB(20);" = ";CI
8724  LPRINT "TANGENT LENGTH";TAB(20);" = ";TL
8726  LPRINT "CHAINAGE TP1";TAB(20);" = ";INT(F1*1000+.5)/1000
8728  LPRINT "CURVE LENGTH";TAB(20);" = ";CL
8730  LPRINT "CHAINAGE TP2";TAB(20);" = ";INT(F2*1000+.5)/1000
8732  IF P1 <  > 2 THEN 8740
8734  LPRINT "1ST TANGENT L1";TAB(20);" = ";L1
8736  LPRINT "2ND TANGENT L2";TAB(20);" = ";L2
8738  LPRINT "DISTANCE R1-R3";TAB(20);" = ";INT(H1*1000+.5)/1000
8740  LPRINT "LONG CHORD";TAB(20);" = ";LC
8742  LPRINT "MID-ORDINATE";TAB(20);" = ";MO
8744  LPRINT "EXTERNAL MO-DIST";TAB(20);" = ";EM
8746  P9=1:RETURN
9000  CLS:T$="PC-Graphics":CC$="RODEOS Expert"
9010 LOCATE 1,2:PRINT T$
9020 LOCATE 1,30:PRINT DATE$
9030 LOCATE 2,2:PRINT CC$
9040 LOCATE 2,30:PRINT TIME$
```

```
9045 IF DISP%=1 THEN 9070
9050 PRINT "========================================":PRINT
9070  RETURN
9100  BEEP
9110  PRINT : PRINT E0$
9120  PRINT:RETURN
9200  GOSUB 9000:PRINT H$
9220  PRINT "****************************************"
9230  PRINT : PRINT "DO YOU WISH TO SET OUT CURVE (Y/N)";: INPUT Q$
9240  IF Q$ = "Y" THEN 2238
9250  IF Q$ <  > "N" THEN 9200
9260  PRINT:PRINT : PRINT "DO YOU WISH TO RUN PROGRAM AGAIN (Y/N)"
9270  INPUT Q$: IF Q$ = "Y" THEN 20
9280  IF Q$ <  > "N" THEN 9260
9290  GOTO 9500
9300  PRINT:PRINT:PRINT " Do you wish to save DATA to disc (y/n)"
9310  INPUT Q$:IF Q$="n" THEN 9260
9320  IF Q$ <> "y" THEN 9300
9330  GOTO 20000
9500  GOSUB 9000
9510  PRINT
9520  PRINT "END RODEOS EXPERT CURVE DESIGN PROGRAM"
9530  PRINT
9540  PRINT "**************************************"
9542  FOR I = 1 TO 500
9544  NEXT I
9620  END
10195 REM ***** Epson HI-80 Plotter *****
10200 GOSUB 54000
10210 PRINT#2,"SP4"
10220 PRINT#2,"MA";IPX*MX;",";(TPLY-5)*MX
10225 PRINT#2,"DA";IPX*MX;",";IPY*MX
10230 PRINT#2,"DA";TPLX*MX;",";TPLY*MX
10235 PRINT#2,"DA";TPRX*MX;",";TPRY*MX
10240 PRINT#2,"DA";IPX*MX;",";IPY*MX
10242 IF CT=2 THEN 10290
10245 IF P2=1 THEN 10290
10250 PRINT#2,"MA";CPLX*MX;",";TPLY*MX
10255 PRINT#2,"DA";CPLX*MX;",";CPLY*MX
10260 PRINT#2,"DA";CPRX*MX;",";CPRY*MX
10265 PRINT#2,"DA";CPRX*MX;",";TPRY*MX
10290 PRINT#2,"SP3"
10300 PRINT#2,"MA";TPLX*MX;",";TPLY*MX
10305 IF CT=2 THEN 10345
10310 FOR J=0 TO NT
10312 HTP(J)=LFF(J)*SIN((90-TANG+PHD(J))/R9)*SCRM
10314 VTP(J)=LFF(J)*COS((90-TANG+PHD(J))/R9)*SCRM
10320 PRINT#2,"DA";(TPLX+HTP(J))*MX;",";(TPLY+VTP(J))*MX
10330 NEXT J
10335 IF P2=1 THEN 10400
10340 PRINT#2,"DA";CPLX*MX;",";CPLY*MX
10345 W=15:GOTO 2350
10350 FOR K=NLC-1 TO 0 STEP -1
10360 PRINT#2,"DA";(IPX-U(K)*SCRM)*MX;",";(CPLY+V(K)*SCRM)*MX
10365 NEXT K
10370 PRINT#2,"DA";IPX*MX;",";(CPLY+MO*SCRM)*MX
10375 FOR I=0 TO NLC-1
10380 PRINT#2,"DA";(IPX+U(I)*SCRM)*MX;",";(CPLY+V(I)*SCRM)*MX
10385 NEXT I
10388 IF CT=2 THEN 10420
10390 PRINT#2,"DA";CPRX*MX;",";CPRY*MX
10400 FOR J=0 TO N1
10402 HTP(J)=LF(J)*SIN((90-TANG+PHN(J))/R9)*SCRM
10404 VTP(J)=LF(J)*COS((90-TANG+PHN(J))/R9)*SCRM
10410 PRINT#2,"DA";(TPRX-HTP(J))*MX;",";(TPRY+VTP(J))*MX
10415 NEXT J
```

```
10420 PRINT#2,"DA";TPRX*MX;",";TPRY*MX
10430 PRINT#2,"SP2":PRINT#2,"SI25,20"
10440 PRINT#2,"MA";(TPLX-8)*MX;",";(TPLY-3)*MX
10450 PRINT#2,"LA";INT(TPLL*1000+.5)/1000
10470 PRINT#2,"MA";(TPRX-8)*MX;",";(TPLY-3)*MX
10480 PRINT#2,"LA";INT(TPRL*1000+.5)/1000
10500 PRINT#2,"MA";(IPX-8)*MX;",";(IPY+2)*MX
10510 PRINT#2,"LA";CIT
10530 PRINT#2,"MA";(IPX-SCRM*TW/4)*MX;",";((TPLY+IPY)/2+5)*MX
10540 PRINT#2,"LATL"
10560 PRINT#2,"MA";(IPX-SCRM*TW/4)*MX;",";(TPLY+2)*MX
10570 PRINT#2,"LATW"
10590 IF P2=1 THEN 10630
10600 PRINT#2,"MA";(IPX-SCRM*LC/4)*MX;",";(CPLY+2)*MX
10610 PRINT#2,"LALC"
10630 PRINT#2,"MA";(IPX+2)*MX;",";(CPLY+MO*SCRM+2)*MX
10635 PRINT#2,"LACL"
10640 PRINT#2,"MA";(GX+5)*MX;",";(GY+SY-8)*MX
10645 PRINT#2,"MA";(IPX-SCRM*(TW+LC)/4)*MX;",";((TPLY+CPLY)/2-3)*MX
10650 PRINT#2,"LASF"
10660 PRINT#2,"MA";(GX+5)*MX;",";(GY+SY-8)*MX
10670 PRINT#2,"LAR =";INT(R*1000+.5)/1000
10700 PRINT#2,"MA";(GX+5)*MX;",";(GY+SY-12)*MX
10710 PRINT#2,"LADF=";DFT;DD$
10730 PRINT#2,"MA";(GX+5)*MX;",";(GY+SY-16)*MX
10740 PRINT#2,"LATH=";THD;DD$
10760 PRINT#2,"MA";(GX+5)*MX;",";(GY+SY-20)*MX
10770 PRINT#2,"LADC=";DC;DD$
10800 PRINT#2,"MA";(GX+5)*MX;",";(GY+SY-24)*MX
10810 PRINT#2,"LASF=";SF
10830 PRINT#2,"MA";(GX+5)*MX;",";(GY+SY-28)*MX
10840 PRINT#2,"LACL=";INT(CL*1000+.5)/1000
10860 PRINT#2,"MA";(GX+5)*MX;",";(GY+SY-32)*MX
10862 PRINT#2,"LATC=";INT(TC*1000+.5)/1000
10870 PRINT#2,"MA";(GX+5)*MX;",";(GY+SY-36)*MX
10872 PRINT#2,"LATL=";INT(TTL*1000+.5)/1000
10880 PRINT#2,"MA";(GX+5)*MX;",";(GY+SY-40)*MX
10882 PRINT#2,"LALC=";INT(LC*1000+.5)/1000
10890 PRINT#2,"MA";(GX+5)*MX;",";(GY+SY-44)*MX
10892 PRINT#2,"LATW=";INT(TW*1000+.5)/1000
10895 PRINT#2,"SI30,25"
10900 PRINT#2,"MA";5*MX;",";+2*MX
10910 PRINT#2,"LAPC-RODEOS - ";A$;" : Date : ";DATE$
10930 CLOSE#2
10940 DISP%=3
10950 GOTO 3230
11195 REM ***** Apple 410 Plotter *****
11200 GOSUB 53000
11210 PRINT#2,"PS4;"
11220 PRINT#2,"MA";IPX;",";(TPLY-5);";"
11225 PRINT#2,"DA";IPX;",";IPY;";"
11230 PRINT#2,"DA";TPLX;",";TPLY;";"
11235 PRINT#2,"DA";TPRX;",";TPRY;";"
11240 PRINT#2,"DA";IPX;",";IPY;";"
11242 IF CT=2 THEN 11290
11245 IF P2=1 THEN 11290
11250 PRINT#2,"MA";CPLX;",";TPLY;";"
11255 PRINT#2,"DA";CPLX;",";CPLY;";"
11260 PRINT#2,"DA";CPRX;",";CPRY;";"
11265 PRINT#2,"DA";CPRX;",";TPRY;";"
11290 PRINT#2,"PS3;"
11295 IF CT=2 THEN CPLY=TPLY
11300 PRINT#2,"MA";TPLX;",";TPLY";"
11305 IF CT=2 THEN 11345
11310 FOR J=0 TO NT
11312 HTP(J)=LFF(J)*SIN((90-TANG+PHD(J))/R9)*SCRM
```

```
11314 VTP(J)=LFF(J)*COS((90-TANG+PHD(J))/R9)*SCRM
11320 PRINT#2,"DA";(TPLX+HTP(J));",";(TPLY+VTP(J));";"
11330 NEXT J
11335 IF P2=1 THEN 11400
11340 PRINT#2,"DA";CPLX;",";CPLY;";"
11345 W=15:GOTO 2350
11350 FOR K=NLC-1 TO 0 STEP -1
11360 PRINT#2,"DA";(IPX-U(K)*SCRM);",";(CPLY+V(K)*SCRM);";"
11365 NEXT K
11370 PRINT#2,"DA";IPX;",";(CPLY+MO*SCRM);";"
11375 FOR I=0 TO NLC-1
11380 PRINT#2,"DA";(IPX+U(I)*SCRM);",";(CPLY+V(I)*SCRM);";"
11385 NEXT I
11388 IF CT=2 THEN 11420
11390 PRINT#2,"DA";CPRX;",";CPRY;";"
11400 FOR J=0 TO N1
11402 HTP(J)=LF(J)*SIN((90-TANG+PHN(J))/R9)*SCRM
11404 VTP(J)=LF(J)*COS((90-TANG+PHN(J))/R9)*SCRM
11410 PRINT#2,"DA";(TPRX-HTP(J));",";(TPRY+VTP(J));";"
11415 NEXT J
11420 PRINT#2,"DA";TPRX;",";TPRY;";"
11430 PRINT#2,"PS2;":PRINT#2,"LS2;"
11440 PRINT#2,"MA";(TPLX-8);",";(TPLY-3);";"
11450 PRINT#2,"PL";INT(TPLL*1000+.5)/1000
11460 PRINT#2,CHR$(13) CHR$(10) CHR$(3)
11470 PRINT#2,"MA";(TPRX-12);",";(TPLY-3);";"
11480 PRINT#2,"PL";INT(TPRL*1000+.5)/1000
11490 PRINT#2,CHR$(13) CHR$(10) CHR$(3)
11500 PRINT#2,"MA";(IPX-8);",";(IPY+2);";"
11510 PRINT#2,"PL";CIT
11520 PRINT#2,CHR$(13) CHR$(10) CHR$(3)
11530 PRINT#2,"MA";(IPX-SCRM*TW/4);",";((TPLY+IPY)/2+5);";"
11540 PRINT#2,"PLTL"
11550 PRINT#2,CHR$(13) CHR$(10) CHR$(3)
11555 IF CT=2 THEN 11600
11560 PRINT#2,"MA";(IPX-SCRM*TW/4);",";(TPLY+2);";"
11570 PRINT#2,"PLTW"
11580 PRINT#2,CHR$(13) CHR$(10) CHR$(3)
11590 IF P2=1 THEN 11630
11600 PRINT#2,"MA";(IPX-SCRM*LC/4);",";(CPLY+2);";"
11610 PRINT#2,"PLLC"
11620 PRINT#2,CHR$(13) CHR$(10) CHR$(3)
11630 PRINT#2,"MA";IPX+2;",";(CPLY+MO*SCRM+2);";"
11635 PRINT#2,"PLCL"
11640 PRINT#2,CHR$(13) CHR$(10) CHR$(3)
11642 IF CT=2 THEN 11660
11645 PRINT#2,"MA";IPX-(SCRM*(TW+LC)/4);",";((TPLY+CPLY)/2-3);";"
11650 PRINT#2,"PLSF"
11655 PRINT#2,CHR$(13) CHR$(10) CHR$(3)
11660 PRINT#2,"MA";(GX+5);",";(GY+SY-8);";"
11670 PRINT#2,"PLR =";INT(R*1000+.5)/1000
11680 PRINT#2,CHR$(13) CHR$(10) CHR$(3)
11700 PRINT#2,"MA";(GX+5);",";(GY+SY-12);";"
11710 PRINT#2,"PLDF=";DFT;DD$
11720 PRINT#2,CHR$(13) CHR$(10) CHR$(3)
11728 IF CT=2 THEN 11830
11730 PRINT#2,"MA";(GX+5);",";(GY+SY-16);";"
11740 PRINT#2,"PLTH=";THD;DD$
11750 PRINT#2,CHR$(13) CHR$(10) CHR$(3)
11760 PRINT#2,"MA";(GX+5);",";(GY+SY-20);";"
11770 PRINT#2,"PLDC=";DC;DD$
11780 PRINT#2,CHR$(13) CHR$(10) CHR$(3)
11800 PRINT#2,"MA";(GX+5);",";(GY+SY-24);";"
11810 PRINT#2,"PLSF=";SF
11820 PRINT#2,CHR$(13) CHR$(10) CHR$(3)
11830 PRINT#2,"MA";(GX+5);",";(GY+SY-28);";"
```

```
11840 PRINT#2,"PLCL=";INT(CL*1000+.5)/1000
11850 PRINT#2,CHR$(13) CHR$(10) CHR$(3)
11858 IF CT=2 THEN 11870
11860 PRINT#2,"MA";(GX+5);",";(GY+SY-32);";"
11862 PRINT#2,"PLTC=";INT(TC*1000+.5)/1000
11864 PRINT#2,CHR$(13) CHR$(10) CHR$(3)
11870 PRINT#2,"MA";(GX+5);",";(GY+SY-36);";"
11872 PRINT#2,"PLTL=";INT(TTL*1000+.5)/1000
11874 PRINT#2,CHR$(13) CHR$(10) CHR$(3)
11880 PRINT#2,"MA";(GX+5);",";(GY+SY-40);";"
11882 PRINT#2,"PLLC=";INT(LC*1000+.5)/1000
11884 PRINT#2,CHR$(13) CHR$(10) CHR$(3)
11888 IF CT=2 THEN 11900
11890 PRINT#2,"MA";(GX+5);",";(GY+SY-44);";"
11892 PRINT#2,"PLTW=";INT(TW*1000+.5)/1000
11894 PRINT#2,CHR$(13) CHR$(10) CHR$(3)
11900 PRINT#2,"MA5,-3;"
11910 PRINT#2,"PLPC-RODEOS - ";A$;" : Date : ";DATE$
11920 PRINT#2,CHR$(13) CHR$(10) CHR$(3)
11930 CLOSE#2
11940 DISP%=3
11950 GOTO 3230
19995  REM ***** Routine to save DATA *****
20000  GOSUB 9000
20010  INPUT " Enter Curve DATA file name ";F$
20020  IF LEN(F$) < 9 THEN 20050
20030  PRINT:PRINT " Maximum 8 letters in file name"
20040  PRINT:GOTO 20010
20050  D$=F$+".DAT"
20060 OPEN "O",#1,D$
20070 PRINT#1,A$;",";B$;",";C$;",";DD$;",";K$;",";L$
20080 PRINT#1,D9;",";R9;",";SR%;",";CT;",";P2
20090 PRINT#1,R;",";CL;",";LC;",";MO;",";EM
20100 PRINT#1,CI;",";CIT;",";F1;",";F2;",";TL
20110 PRINT#1,DF;",";DFT;",";LF;",";TANG;",";PH
20120 PRINT#1,TPF1;",";TPF2;",";NT;",";N1;",";TW;",";VH
20130  FOR I=0 TO NT
20140 PRINT#1,LFF(I)
20150  NEXT I
20160  FOR J=0 TO N1
20170 PRINT#1,LF(J)
20180  NEXT J
20190  FOR I=0 TO NT
20200 PRINT#1,PHD(I)
20210  NEXT I
20220  FOR J=0 TO N1
20230 PRINT#1,PHN(J)
20240  NEXT J
20250 CLOSE#1
20260  PRINT:PRINT:PRINT " DATA file ";F$;" closed"
20270 GOTO 9260
20495  REM ***** Routine to read DATA *****
20500  GOSUB 9000
20510  INPUT " Enter Curve DATA file name ";F$
20520  IF LEN(F$) < 9 THEN 20550
20530  PRINT:PRINT " Maximum 8 letters in file name"
20540  PRINT:GOTO 20510
20550  D$=F$+".DAT"
20560 OPEN "I",#4,D$
20570 INPUT#4,A$,B$,C$,DD$,K$,L$
20580 INPUT#4,D9,R9,SR%,CT,P2
20590 INPUT#4,R,CL,LC,MO,EM
20600 INPUT#4,CI,CIT,F1,F2,TL
20610 INPUT#4,DF,DFT,LF,TANG,PH
20620 INPUT#4,TPF1,TPF2,NT,N1,TW,VH
```

```
20630  FOR I=0 TO NT
20640 INPUT#4,LFF(I)
20650  NEXT I
20660  FOR J=0 TO N1
20670 INPUT#4,LF(J)
20680  NEXT J
20690  FOR I=0 TO NT
20700 INPUT#4,PHD(I)
20710  NEXT I
20720  FOR J=0 TO N1
20730 INPUT#4,PHN(J)
20740  NEXT J
20750 CLOSE#1
20760  PRINT:PRINT:PRINT " DATA file ";F$;" closed"
20770 GOTO 3230
```

4.5.5 PC-RODEOS – computer printout

RODEOS curve details for Figs 4.17 and 4.18

Transition curve printout was given earlier in Section 4.3.5(b)

```
 RODEOS CURVE DESIGN PROGRAM
LOCATION OF CURVE :-    Strathclyde
*************************************************************
OPERATOR'S NAME :-      Dr. P.H. Milne
DATE OF CALCS.  :-      01/01/87
INSTRUMENT USED :-      Epson-PC
*************************************************************

 RODEOS CURVE DESIGN PROGRAM
******************************************
DESIGN CRITERIA FOR CURVE :-
VELOCITY (KPH)                   =  160
VELOCITY (MPS)                   =  44.444
RADIAL ACCELERATION (MPS^3)      =  .46
CENTRIPETAL RATIO                =  .25
MINIMUM RADIUS (M)               =  805.426

MULTIPLYING CONSTANT             =  436.865
DEFLECTION ANGLE AT I.P.         =  75
TRANSITION CURVE ANGLE           =  8.2542
CIRCULAR CURVE ANGLE             =  58.0837
SPIRAL TRANSITION LENGTH         =  236.957
CIRCULAR CURVE LENGTH            =  817.344
TOTAL CURVE LENGTH               =  1291.257
******************************************

 RODEOS CURVE DESIGN PROGRAM
******************************************
CENTRE CIRCULAR CURVE
RADIUS               =  805.426
DEFLECTION ANGLE     =  58.0837
CHAINAGE I.P.        =  1808.39
TANGENT LENGTH       =  447.776
CHAINAGE TP1         =  1360.614
CURVE LENGTH         =  817.3442
CHAINAGE TP2         =  2177.958
LONG CHORD           =  782.722
MID-ORDINATE         =  101.474
EXTERNAL MO-DIST     =  116.102
******************************************
```

```
 RODEOS CURVE DESIGN PROGRAM
<3> TANGENTIAL DEFLECTION ANGLES
******************************************
PT.  CHAINAGE         CHORD     ANGLE(DD.MMSS)
==========================================
TP1   1360.614                   0.000
1     1380          19.386      .4122
 2    1410          30           1.4523
 3    1440          30           2.4925
 4    1470          30           3.5326
 5    1500          30           4.5727
 6    1530          30           6.0128
 7    1560          30           7.053
 8    1590          30           8.0931
 9    1620          30           9.1332
 10   1650          30           10.1733
 11   1680          30           11.2135
 12   1710          30           12.2536
 13   1740          30           13.2937
 14   1770          30           14.3338
 15   1800          30           15.374
 16   1830          30           16.4141
 17   1860          30           17.4542
 18   1890          30           18.4943
 19   1920          30           19.5345
 20   1950          30           20.5746
 21   1980          30           22.0147
 22   2010          30           23.0548
 23   2040          30           24.095
 24   2070          30           25.1351
 25   2100          30           26.1752
 26   2130          30           27.2153
 27   2160          30           28.2555
TP2   2177.958      17.958       29.0414
==========================================
FINAL ANGLE AT TP2 SHOULD BE    29.0419
******************************************


 RODEOS CURVE DESIGN PROGRAM
VERTICAL CURVE DESIGN
******************************************
GRADIENT OF 1ST STRAIGHT        =  2
GRADIENT OF 2ND STRAIGHT        = -2
ALGEBRAIC DIFFERENCE M-N        =  4
K-VALUE                         =  100
MINIMUM LENGTH OF CURVE         =  400
CURVE LENGTH CHOSEN             =  1291.257

CHAINAGE OF INTERSECTION        =  1808.39
REDUCED LEVEL OF I.P.           =  125
CHAINAGE OF 1ST TANG.PT.        =  1123.657
CHAINAGE OF 2ND TANG.PT         =  2414.914
******************************************
```

CHAINAGE	LENGTH	HT.DIFF.	RED.LEVEL
1123.657	0	0	112.087
1140	16.343	.323	112.41
1170	46.343	.894	112.981
1200	76.343	1.437	113.524
1230	106.343	1.952	114.039
1260	136.343	2.439	114.526
1290	166.343	2.898	114.985
1320	196.343	3.33	115.417
1350	226.343	3.733	115.82
1380	256.343	4.109	116.196
1410	286.343	4.457	116.544
1440	316.343	4.777	116.864
1470	346.343	5.069	117.156
1500	376.343	5.333	117.42
1530	406.343	5.569	117.656
1560	436.343	5.778	117.865
1590	466.343	5.958	118.045
1620	496.343	6.111	118.198
1650	526.343	6.236	118.323
1680	556.343	6.333	118.42
1710	586.343	6.402	118.489
1740	616.343	6.443	118.53
1769.286	645.629	6.456	118.543
1770	646.343	6.456	118.543
1800	676.343	6.442	118.529
1830	706.343	6.399	118.486
1860	736.343	6.329	118.416
1890	766.343	6.231	118.318
1920	796.343	6.104	118.191
1950	826.343	5.95	118.037
1980	856.343	5.769	117.856
2010	886.343	5.559	117.646
2040	916.343	5.321	117.408
2070	946.343	5.056	117.143
2100	976.343	4.762	116.849
2130	1006.343	4.441	116.528
2160	1036.343	4.092	116.179
2190	1066.343	3.715	115.802
2220	1096.343	3.31	115.397
2250	1126.343	2.877	114.964
2280	1156.343	2.416	114.503
2310	1186.343	1.928	114.015
2340	1216.343	1.411	113.498
2370	1246.343	.867	112.954
2400	1276.343	.295	112.382
2414.914	1291.257	0	112.087

5

Automated surveying

5.1 INTRODUCTION

The methods used for the recording and processing of field data obtained in land surveying have changed dramatically in the last decade. The development initially of compact EDM (Electromagnetic Distance Measurement) instruments for use in conjunction with the optical or electronic theodolites has latterly led to the increased availability of 'total stations'. Advances in microprocessor technology combined with miniaturization have now led to the production of compact electro-optical field surveying equipment with the capability not only of presenting the difference in distance and level, but also of calculating the coordinates of the observed point. The use of such automated measurement systems is revolutionising the surveying profession.

In traditional survey methods using tapes, levels and theodolites, all the details are recorded in the field book. In addition to the observations, suitable remarks and possibly sketches are also required to identify each reading and aid in the computation in order to produce the final drawing. Such manual recording methods are susceptible to initial reading errors and subsequent recording errors in writing down the data, with consequent errors at the computation and plotting stage.

The introduction of electronic theodolites and EDMs with digital readouts has lessened the risk of misreading the angle and distance observations. The next stage of development in automated surveying was the use of data loggers or field data terminals to record all the observations. Initial data loggers were very basic, had a very small memory capacity, and were really just electronic field books. After use, the data then had to be transferred to a larger computer for processing and analysis.

Modern field data terminals can be connected directly to the measuring system allowing horizontal angle, vertical angle and slope distance to be recorded with a single keystroke. Another advantage is that they can often be programmed in BASIC to accept alphanumeric inputs for point coding and identification of the data. The importance of point coding and identification is most important in the transfer of data between different organizations using different software. Thus, in the United Kingdom, a National Transfer Format has recently been evolved to assist in the transfer

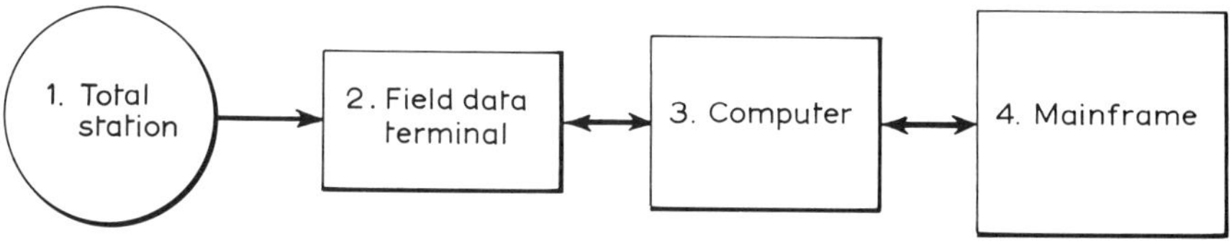

Fig. 5.1 Data transfer from EDM to field data terminal and computer for processing.

of digital map data for use in Geographic Information Systems (GIS).

The choice of a field data terminal for use with a total station will depend on their interfacing capabilities. Equipment manufacturers thus often provide a package including all the equipment, interfacing cables and software. It is therefore most important at purchase time to ensure that the portable field data terminal is capable of connecting to a larger microcomputer for off-line processing of the data for topographic plotting etc. (Fig. 5.1). It is therefore essential that the field data terminal has a standard interface, for example RS-232C as used by the Epson HX-20 (Fig. 5.2). The benefits of

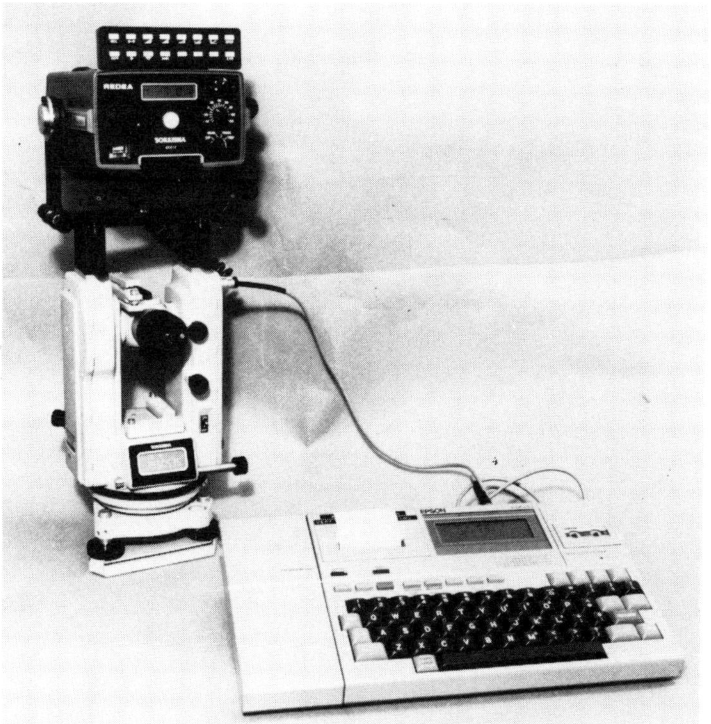

Fig. 5.2 Use of Epson HX-20 as field data terminal to store survey detail from Sokkisha DT20E electronic theodolite and Red 2A EDM.

Fig. 5.3 Epson HX-20 connected to battery operated Epson P-80 printer for field printout of data.

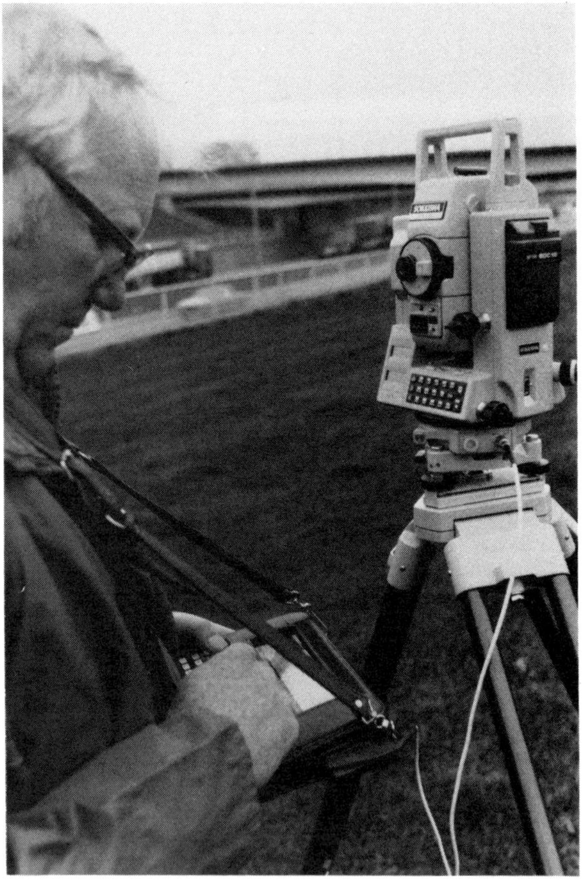

Fig. 5.4. Use of Husky Hunter as field data terminal in conjunction with Sokkisha SET 3 Total Station. (Photograph by courtesy of A. Clarkson & Co. Ltd.)

this are that the field data terminal can be linked in the field to a portable printer for the immediate printout of data for checking prior to transfer (Fig. 5.3). Although the Epson HX-20 is not IBM PC compatible, it is a useful example of the type of field data terminal that can be linked to various total stations. The data is stored in the field on a micro-cassette tape for subsequent transfer to a larger microcomputer. As the Epson HX-20 can be programmed in BASIC, it is also possible to plot the data. However its memory capability is not sufficient to process the data for contour plotting or surface modelling as discussed in Chapter 3. If more memory is required for the storage of data then another versatile microcomputer is the Husky Hunter (Fig. 5.4) which can be used to collect data in the MOSS format for transfer to a mainframe.

If computer graphic facilities for plotting are required in the field then a portable IBM PC or compatible microcomputer should be chosen, like the Data General One (Fig. 5.5), which can run all the programs discussed in

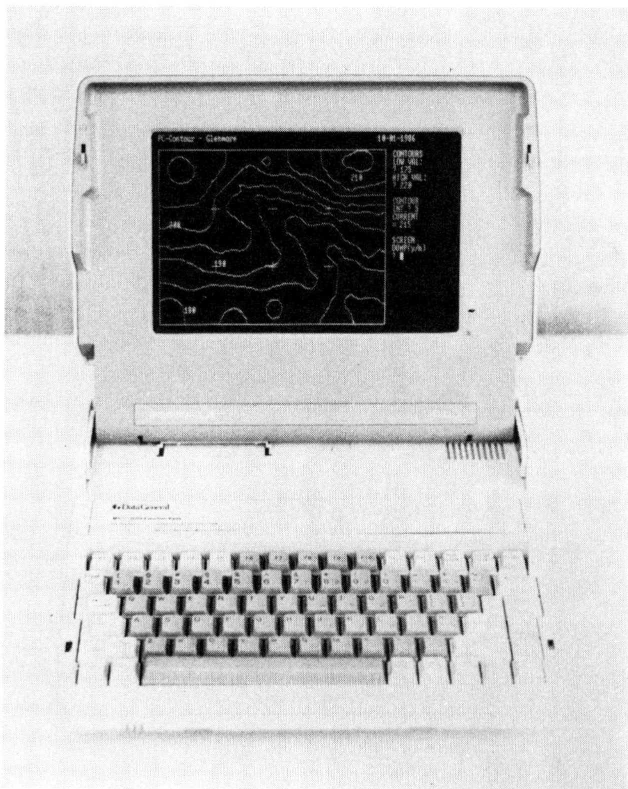

Fig. 5.5 Data General One, Model 2, a portable IBM PC compatible with large graphics display screen for on site plotting. (Equipment kindly supplied for evaluation by Data General Corporation.)

this book. With a microcomputer of this type, it is possible to display the data points as collected for subsequent storage and plotting. Two programs are presented in this chapter, PC-Record for the storing of data on disk during the running of a survey program, and PC-Transfer for the reading of data from disk for use in another program.

5.2 PC-RECORD

If a standard data logger is used for the storage of data, the number of fields in each record will depend upon whether the data has been collected from a tacheometric or EDM survey. However, if the basic data information has already been processed to calculate the *X*, *Y*, *Z* coordinates of each point, then to save double handling of the data, it is more convenient to store the coordinates directly.

The following program provides an opportunity for the storing of data on disc during the running of survey programs like 3-D Tacheometry Plot and 3-D Positions from EDM, two programs which appeared in the previous book, *BASIC Programs for Land Surveying*. The line numbers were chosen to slot conveniently into these existing programs. Users who wish to incorporate the routines into their own programs will need to change the line numbers accordingly.

5.2.1 PC-Record – subroutine index

	Line numbers	*Function*
(a)	2980–2986	Plotter selection
(b)	4695–4870	Data file storage

(a) Plotter selection Line numbers 2980–2986

This routine allows the user to select the plotting area for the data so that it can be scaled accordingly. The input format is the typical Hewlett-Packard entry of (*X*-min, *X*-max, *Y*-min, *Y*-max).

(b) Data file storage Line numbers 4695–4870

The user is asked to enter a name for the data file, which then has **.DAT** added to identify it specifically as data. If the variable **DF** (for data file) equals 1 it means that routine (a) has already been used to enter the coordinates of the plotting area and the routine continues at line 4770. If no plotting has been done previously, the user is returned to routine (a) to collect (*X*-min, *X*-max, *Y*-min, *Y*-max). A new data file is then opened in line 4770 and the relevant data stored, together with the number of data points, their identification number *P* and their *X*, *Y*, *Z* coordinates.

5.2.2 PC-Record – numeric variables

DF = Flag for plotter coordinates
I = Loop variables
N = Number of data points
P(I) = Array to hold point identification number
XN = *X*-coordinate at left-hand side of plot
XM = *X*-coordinate at right-hand side of plot
X(I) = Array to hold *X*-coordinates of points
YN = *Y*-coordinate at foot of plot
YM = *Y*-coordinate at top of plot
Y(I) = Array to hold *Y*-coordinates of points
Z(I) = Array to hold *Z*-coordinates of points

5.2.3 PC-Record – string variables

D$** = **F$ + ."DAT"
F$ = Data file name
Q$ = Question response (Y/N)

5.2.4 PC-Record – BASIC program

```
2965 PRINT:PRINT " Select Plotter";TAB(18);"<1> Apple 410":PRINT TAB(
18);"<2> Epson HI-80":PRINT TAB(18);"<3> Hewlett-Packard"
2966 PRINT TAB(18);:INPUT PX:IF PX<1 OR PX>3 THEN 2965
2967 IF PX=3 THEN HP=1
2968 IF PX=1 THEN HP=3
2969 IF PX=2 THEN HP=4
2980 PRINT:PRINT " Select Plotter Scale -":PRINT " for Natural Plot Y
 should be 75% of X"
2981 PRINT " (X-min, X-max, Y-min, Y-max)":INPUT XN,XX,YN,YX
2982 IF HP=1 THEN YX=YN+(XX-XN)*100/152
2983 IF HP=3 THEN YX=YN+(XX-XN)*2570/3800
2984 IF HP=4 THEN YX=YN+(XX-XN)*1900/2670
2985 IF HP=4 THEN MX=2670/(XX-XN)
2986 IF DF=1 THEN 4770

4695 REM ***** Data File Storage for Random/FData File *****
4700 PRINT:INPUT "Do you wish to store data on disc (Y/N)";Q$
4710 IF Q$="N" THEN 4900
4720 IF Q$<>"Y" THEN 4690
4730 PRINT:PRINT " ENTER DATA FILE NAME (Max. 8 letters)":INPUT F$
4735 IF LEN(F$)>8 THEN 4730
4740 D$=F$+".DAT"
4750 IF DF=1 THEN 4770
4755 REM ***** GO BACK AND COLLECT XN,XX,YN,YX *****
4760 DF=1:GOTO 2965
4770 OPEN "O",#1,D$
4780 PRINT#1,XN
4790 PRINT#1,XX
4800 PRINT#1,YN
4810 PRINT#1,YX
4820 PRINT#1,(N+1)
4830 FOR I=0 TO N
4840 PRINT#1,P(I);X(I);Y(I);Z(I)
4850 NEXT I
4860 CLOSE#1
4870 PRINT:PRINT " DATA FILE : ";F$;" CLOSED"
```

5.2.5 PC-Transfer

If data which has been stored by the previous program PC-Record is to be used for plotting and contouring, it needs to be transferred from disc into a random data generation file like PC-Random for preparation. The following routine is incorporated in Random/FData which was described in Chapter 3. The data collected during a tacheometry survey at Glenmore is shown in Figs 3.1 and 3.2 with the contours plotted on screen in Figs 3.3 and 5.5.

5.2.6 PC-Transfer – BASIC program

```
1010 INPUT "  Enter DATA file name ";F$
1020 D$=F$+".DAT"
1030 OPEN "I",#4,D$
1040 INPUT#4,XL,XR
1050 INPUT#4,YB,YT
1060 INPUT#4,N
1070  DIM P(N),X(N),Y(N),Z(N),R(N),S(N)
1080 PRINT:PRINT
1090 PRINT "  DATA now being read from file"
1100 MAX=-25000
1110 MIN=25000
1120 FOR I=1 TO N
1130 INPUT#4,P(I),X(I),Y(I),Z(I)
1140  IF Z(I) > MAX THEN MAX = Z(I)
1150  IF Z(I) < MIN THEN MIN = Z(I)
1160  NEXT I
1165 CLOSE#4
```

5.3 AUTOMATED SURVEYING – THE FUTURE

At present (1987), although there is a National Transfer Format for digital map data between computer users, as discussed earlier in this chapter, there is no such format for field data terminals. The popularity of some commercial systems has already introduced a *de facto* standardization – one example in this context is MOSS. However, if survey data collected by field data terminals is to be transferrable between computers and companies, it is essential that there is some standardization in the future. This would then allow data collected by any field data terminal, to be transferred to an IBM PC or compatible for subsequent processing and plotting. At the present time, data collected by a field data terminal, e.g. Husky Hunter (Fig. 5.4), often has to be transferred to an IBM PC (or compatible) for post-processing and preparation before being submitted as data input to MOSS files on a mainframe (Fig. 5.1). If the field data terminal is IBM PC compatible, like the Data General One (Fig. 5.5), then one of the links in the chain in Fig. 5.1 is removed, thus reducing the overall processing time.

If they are not to become servants of technology, future surveyors must become more aware of the potential presented by increased microprocessing power to improve their analysis of data capture and processing, and must be familiar with the mathematical algorithms used in the computations.

Bibliography

Angell, I. O. (1981) *A Practical Introduction to Computer Graphics*, Macmillan, London.

Brodlie, K. W. (ed.) (1980) *Mathematical Methods in Computer Graphics and Design*, Academic Press, London.

Chasen, S. H. (1978) *Geometric Principles and Procedures for Computer Graphics Applications*, Prentice-Hall, Inc., New Jersey.

Institution of Civil Engineers (1977) *Surface Modelling by Computer*, Thomas Telford Ltd, London.

MacDougall, E. B. (1976) *Computer Programming for Spatial Problems*, Edward Arnold, London.

Milne, P. H., (1983) Computer graphic applications in surveying, 2nd Educational Computing Conference, London, Technical Session **14**, 1–13.

Milne, P. H. (1983) Demonstration graphics, *Educational Computing*, **4**(1), 27.

Milne, P. H. (1984) Computer graphic applications in engineering surveying, *Civil Engineering Surveyor*, **IX**(2), 14–17.

Milne, P. H. (1984) *BASIC Programs for Land Surveying*, E. & F. N. Spon Ltd, London.

Milne, P. H. (1985) Computer-aided instruction in engineering surveying, 2nd Intl Conf. on Civil and Structural Engineering Computing, Vol. 2, ICE, London, Dec., 483–7.

Milne, P. H. (1986) An expert system for road curve design and setting-out, in *Applications of Artificial Intelligence to Engineering Problems*, Vol. II (eds D. Sriram and R. Adey), a Computational Mechanics Publication, Springer-Verlag, Berlin, 733–743.

Milne, P. H. (1986) Educational use of CAI in engineering surveying, Proc. 18th Intl FIG Congress for Surveyors, Toronto, Commission 2, PS 24.2/1–12.

Ordnance Survey (1987) National Transfer Format, Report by Working Party to produce National Standards for the Transfer of Digital Map Data, Ordnance Survey, Southampton.

Sabin, M. A. (1985) Contouring – the state of the art, in *Fundamental Algorithms for Computer Graphics* (ed. R. A. Earnslaw), Springer-Verlag, Berlin.

Smallshire, V. W. (1985) Use of Portable Microcomputer as a Survey Data Logger, *Civil Engineering Surveyor*, **X**(6), 14–19.

Waite, M. and Morgan, C. L. (1983) *Graphics Primer for the IBM PC*, Osborne/McGraw-Hill, Berkeley.

Walker, B. S., Gurd, J. R. and Drawneek, E. A. (1975) *Interactive Computer Graphics*, Crane Russak, New York.

Index